MW00744117

Table of Contents

Table of Contents (continued)

Table of Contents (continued)

List of Tables in Text

List of Figures in Text

Part I: Introduction and Overview

INTRODUCTION

The *Stanford Achievement Test* Series, with a rich history dating from the early twentieth century, assesses students' school achievement in reading, mathematics, spelling, language, science, social science, and listening. The *Stanford Achievement Test* first appeared in 1923. New editions were published in 1929, 1940, 1953, 1964, 1973, 1982, 1989, and 1996. Each edition was developed in order to:

- update content to align the test with current educational and curriculum trends;

- update the normative information to make score interpretations more valid;

- increase and improve the kinds of information available from testing; and

- revise the look of the test to make it more relevant to students.

Each new edition has built upon the strong standard of reliability, validity, and technical excellence for which the *Stanford Achievement Test* Series is known.

The *Stanford Achievement Test* Series, Tenth Edition (Stanford 10), continues the tradition of testing excellence. While retaining the distinguished features of previous editions, Stanford 10 advances standardized testing with new features that help educators assess student achievement as well as meet today's accountability requirements.

The Stanford 10 multiple-choice assessment offers newly developed standards-based content that yields 2002 norm-referenced information and, as desired, customized, criterion-referenced information. To assess student progress toward high academic standards, the test items included in Stanford 10 reflect an extensive review of national and state instructional standards, content-specific curricula, and educational trends as developed by national professional educational organizations.

Edition	Year
1st	1923
2nd	1929
3rd	1940
4th	1953
5th	1964
6th	1973
7th	1982
8th	1989
9th	1996
10th	2003

Stanford 10 comprises a battery of thirteen test levels that assesses students from kindergarten through grade 12. *Stanford Early School Achievement Test* (SESAT) provides two test levels that assess the achievement of children in kindergarten and the first half of grade 1. The eight levels of *Stanford Achievement Test* measure the important instructional standards of the curricula from the second half of grade 1 through the end of grade 9. The three levels of *Stanford Test of Academic Skills* (TASK) are intended for use as measures of those basic skills in grades 9 through 12 requisite to continued academic training.

The first two test levels of Stanford 10 (SESAT 1 and SESAT 2) have only one test form each. All other test levels have two forms that are equivalent in both content and difficulty.

Ultimately, the validity of an intended interpretation of test scores relies on all the available evidence relevant to the technical quality of a testing system. This includes evidence of careful test construction; adequate score reliability; appropriate test administration and scoring; accurate score scaling, equating, and standard setting; and careful attention to fairness for all examinees.

TESTING WINDOW

Testing for the Spring Standardization Program of all levels and forms of Stanford 10 was from April 1, 2002, to April 26, 2002. Testing for the Fall Standardization Program was from September 9, 2002, to October 18, 2002.

OVERVIEW OF THIS TECHNICAL DATA REPORT

This *Technical Data Report* includes customary discussions of the test development process and associated supporting technical data. For Stanford 10, expanded descriptions are also included that address the testing of special populations, test structure and content, and additional research. The rest of this document is organized as follows:

- Part II describes the process of developing the Stanford 10 blueprint and specifications, item development, the National Item Tryout Program, procedures to eliminate bias, and the development and design of the final forms.

- Part III describes the Stanford 10 National Research Programs conducted in the spring and fall of 2002. The section describes the sampling procedures used for the various standardization and equating programs related to Stanford 10.

- Part IV explains the types of test scores and how to interpret them.

- Part V presents technical information on test scores and discusses the validity, fairness, and reliability of Stanford 10.

- Part VI presents information on testing special populations. Included are discussions of accommodations for students with disabilities, English language learners, and special editions of Stanford 10 for students who are blind, visually impaired, deaf, and hearing impaired.

- Part VII details the structure of Stanford 10, including the test levels, lengths, and forms. As further evidence of test validity, a thorough explanation of all content areas and the response processes required is also included.

- Part VIII summarizes several additional research studies, published and ongoing, related to Stanford 10.

Part II: The Test Development Process

The preparation of any test series involves extensive curriculum reviews and analyses as well as numerous large-scale research programs. This is an integral part of an achievement battery, which typically assesses performance in a variety of content areas across several grades.

Part II presents the rationale underlying the development of the content of Stanford 10. The important phases of the test development process are included: analyses of the curriculum, preparation of the test specifications, determination of the instructional standards to be measured, and the various review procedures.

RATIONALE/PURPOSE OF THE TEST

The process of producing a new edition of the *Stanford Achievement Test* Series involved the expertise and commitment of measurement experts, content and curriculum experts, psychometricians, and publishers, as well as the cooperation of teachers, thousands of school districts, and hundreds of thousands of students across the nation. Careful planning and implementation during each step of the development process ensured test validity and reliability.

Traditionally, the decision to launch a new edition of a major test series is based on changes in schools' curricula, changes in national assessment trends and methods, and the need for updated testing materials, normative information, and interpretive materials.

Although addressing changes in educational trends and requirements served as an initial catalyst for preparing the new test edition, other major factors furnishing impetus for development of Stanford 10 included the need to:

- maximize student engagement and involvement in the testing process;
- maintain student focus and motivation throughout the test;
- minimize any potential test navigation difficulties for students to maximize their ability to show what they know and can do;
- provide a seamless link between classroom instructional materials and testing materials utilizing full color and realistic illustrations (including computers and workbooks); and
- provide test content standards in a manner similar to the way they are taught.

REVIEW PROCESS

The development of Stanford 10 began with comprehensive reviews and careful analyses of the most recent editions of the major textbook series in every subject area, the most recent state and district school curricula and instructional standards, and the most important educational trends and directions as expressed by national professional organizations. See Appendix A for a list of the textbooks and other materials that were consulted during the development of Stanford 10.

In order to provide an accurate and objective measure of student achievement, Stanford 10 includes content that is closely aligned to standards-based national curricula. The list below presents a sampling of the academic standards and professional organizations consulted during the development of Stanford 10:

- National Council of Teachers of English and the International Reading Association, *Standards for the English Language Arts*

- National Council of Teachers of Mathematics, *Principles and Standards for School Mathematics*

- National Research Council, *National Science Education Standards*

- American Association for the Advancement of Science, *Benchmarks for Science Literacy*

- National Council for the Social Studies, *Expectations of Excellence: Curriculum Standards for Social Studies*

- National Assessment of Educational Progress (NAEP) test frameworks

- Academic standards for the states

Concurrent with these analyses was a careful review of the content, structure, and format of the *Stanford Achievement Test* Series, Ninth Edition (Stanford 9). That review determined which aspects of Stanford 9 would be retained or modified to meet the goals set for the Stanford 10. As a result, changes were made to the content areas and instructional standards to be assessed so that they aligned with current educational standards and curricula.

PREPARATION OF TEST BLUEPRINTS

The Stanford 10 test blueprints mapped the following key components:

- Number of test levels necessary for complete coverage across the elementary and secondary grades

- Test content areas

- Instructional standards coverage in each content area (i.e., reporting category)

- Articulation of the content and instructional standards across the grade levels

- Approximate number of items necessary for breadth of content and reliability of assessment

- Proportions of test items devoted to each of the two cognitive levels: Basic Understanding and Thinking Skills

The blueprints specified the development of new multiple-choice items for all Stanford 10 subtests, test levels, and forms. Each item was designed to measure up to four achievement parameters: (1) a content cluster; (2) a process cluster; (3) a cognitive level; and (4) an instructional standard.

For example, an Intermediate 2 Mathematics Problem Solving item might assess the content cluster "Patterns, Relationships, and Algebra," the process cluster "Mathematical Connections," the cognitive level "Thinking Skills," and the instructional standard "Translate problem situations into algebraic equations and expressions."

Content area specialists and editors at Harcourt, as well as distinguished test construction professionals throughout the nation, reviewed the test blueprints. Reviewers commented on all aspects of the blueprints, such as the content included at each test level, the grade level at which content areas were introduced or eliminated, and the proportion of test content devoted to an instructional standard. Reviewer input served as the basis for blueprint revisions.

For a detailed description of the content areas, process clusters, cognitive levels, and instructional standards assessed at each test level, refer to the Stanford 10 *Compendium of Instructional Standards*.

PREPARATION OF TEST ITEM SPECIFICATIONS

Once the test blueprints were finalized, test item specifications for Stanford 10 were developed. The test item specifications include the following information:

- Test item format
- Guidelines for subject matter and themes
- Content restrictions
- Item stem development practices
- Response option requirements
- Sample test items
- Conventions
- Procedures for aligning test items to the blueprint

All Stanford 10 test items have a multiple-choice format. A significant number of these were developed as enhanced multiple-choice items that elicit problem-solving processes rather than the mere recall of information.

Items assessing each instructional standard were written in sufficient numbers to allow for the production of twice as many complete test forms at each level as were actually required for the final Stanford 10 battery. With this strategy, items that did not function well in the classroom during the National Item Tryout Program were eliminated without adversely affecting the balance of content. In addition, this method allowed experimentation with different approaches to the assessment of a single instructional standard and the tryout of items of varying levels of complexity.

The final test item specifications became the framework that defined the test item writing process.

Test Item Writing and Review

Content area specialists and Harcourt-trained writers created pools of test items in their areas of expertise in accordance with finalized specifications. These pools of new items were developed specifically for Stanford 10.

The majority of individuals who wrote test items for Stanford 10 were practicing teachers from across the country with extensive experience in the various content areas. Test item writers were thoroughly trained on the principles of test item development and review procedures. They received detailed specifications for the content area for which they were writing, as well as lists of instructional standards and examples of both properly and improperly constructed test items.

As test items were written and received, each test item was submitted to internal screening processes at Harcourt that included examinations by:

- content experts, who reviewed each test item for alignment to specified instructional standards, cognitive levels, and processes;

- measurement experts, who reviewed each test item for adequate measurement properties; and

- editorial specialists, who screened each test item for grammatical and typographical errors.

At least four times as many items were written in each content area as would ultimately be needed for the final test forms. In addition, an even greater overage of items was developed for those content areas in which new types of items were being introduced. This overage allowed items that did not function well during the National Item Tryout Program or did not meet established psychometric criteria to be discarded.

Small local pilot studies were conducted prior to the National Item Tryout Program to determine how the new items would perform so that modifications could be made.

THE NATIONAL ITEM TRYOUT PROGRAM

Purpose

An item tryout program is critical to the development of any test because it provides information about the pool of items from which the final forms of the test will be constructed. The information provided by the Stanford 10 National Item Tryout Program included:

1. The appropriateness of the item format: How well does the item measure the particular instructional standard for which it was written?

2. The difficulty of the question: How many students in the tryout group responded correctly to the item?

3. The sensitivity of the item: How well does the item discriminate between students who score high on the test and those who score low?

4. The grade-to-grade progression in difficulty: For items tried out in different grades, did more students answer the question correctly at successively higher grades?

5. The functioning of the item options: How many students selected each option?

6. The suitability of test length: Are the number of items per subtest and recommended administration times satisfactory?

In addition to statistical information about individual items, information was collected from teachers and students concerning the appropriateness of the questions, the clarity of the directions, quality of the artwork, and other relevant information.

Construction and Administration of the Item Tryout Forms

Since it was not practical to administer a complete tryout battery to each student, tryout booklets were designed to assess only certain content areas in order to limit the administration time for each classroom. The tryout forms were assembled to mirror each other in terms of content and apparent difficulty level. Table 1 illustrates the scope of the item tryout program by test level, grade, number of forms, and content area.

11

Table 1. Scope of the National Item Tryout Program by Test Level, Grade, Form, and Content Area

Test Level	Grade	Number of Forms	Content Areas
SESAT 1	K	3 4 4	Sounds and Letters / Word Reading Mathematics / Environment Listening to Words and Stories
SESAT 2	K, 1	3 3 4 4	Sounds and Letters / Word Reading Sentence Reading Mathematics / Environment Listening to Words and Stories
Primary 1	1, 2	4	Word Study Skills Word Reading Sentence Reading Reading Comprehension / Language Form D Mathematics Problem Solving / Language Form A Mathematics Procedures Spelling Listening Environment
Primary 2	2, 3	4	Word Study Skills Reading Vocabulary Reading Comprehension / Language Form D Mathematics Problem Solving / Language Form A Mathematics Procedures Spelling Listening Environment
Primary 3	3, 4	4	Word Study Skills Reading Vocabulary Reading Comprehension / Language Form D Mathematics Problem Solving / Language Form A Mathematics Procedures Spelling Listening Science / Social Science
Intermediate 1	4, 5	4	Word Study Skills Reading Vocabulary Reading Comprehension / Language Form D Mathematics Problem Solving / Language Form A Mathematics Procedures Spelling Listening Science / Social Science

Table 1. Scope of the National Item Tryout Program by Test Level, Grade, Form, and Content Area (continued)

Test Level	Grade	Number of Forms	Content Areas
Intermediate 2	5, 6	4	Reading Vocabulary Reading Comprehension Mathematics Problem Solving / Language Form D Mathematics Procedures / Language Form A Spelling Listening Science / Social Science
Intermediate 3	6, 7	4	Reading Vocabulary Reading Comprehension Mathematics Problem Solving / Language Form D Mathematics Procedures / Language Form A Spelling Listening Science / Social Science
Advanced 1	7, 8	4	Reading Vocabulary Reading Comprehension Mathematics Problem Solving / Language Form D Mathematics Procedures / Language Form A Spelling Listening Science / Social Science
Advanced 2	8	4	Reading Vocabulary Reading Comprehension Mathematics Problem Solving / Language Form D Mathematics Procedures / Language Form A Spelling Listening Science / Social Science
TASK 1	9	4	Reading Vocabulary Reading Comprehension Mathematics / Language Form A Spelling Science / Social Science Language Form D
TASK 2	10	4	Reading Vocabulary Reading Comprehension Mathematics / Language Form A Spelling Science / Social Science Language Form D
TASK 3	11	4	Reading Vocabulary Reading Comprehension Mathematics / Language Form A Spelling Science / Social Science Language Form D

In addition to the new Stanford items, each tryout booklet included an appropriate subset of items from the same content area of the *Stanford Achievement Test* Series, Ninth Edition (Stanford 9). This allowed the sample of students in the Stanford 10 tryout to be compared with the Stanford 9 standardization sample in terms of achievement and ability so that the new items could be placed on a common scale of difficulty across all forms and levels of the battery. In each school district participating in the National Item Tryout Program, test booklets were randomly assigned by classroom to students in kindergarten through grade 11. Each booklet was administered to approximately 500 students per grade.

Description of the National Item Tryout Sample

The effectiveness of an item tryout program depends on the quantity, content, and quality of the test questions, as well as the characteristics of the students tested. The questions should be administered to students in numbers sufficient to yield stable data, and the students involved should represent demographically the population of students for whom the test is ultimately intended.

The Stanford 10 National Item Tryout Program took place in 1998, 1999, and 2000. The item tryout samples were chosen to be representative of the national school population. Altogether, a total of just over 170,000 students from 500 school districts in 42 states participated in the Stanford 10 National Item Tryout Program.

In addition, more than 10,000 teachers who administered the tryout editions contributed information on the match between test content and the local curriculum, the clarity and adequacy of the directions and administration times, the test layout, the item formats, and the artwork.

School districts were the sampling unit and were selected for inclusion in the National Item Tryout Program using a stratified random sampling technique. Geographic region, socioeconomic status (SES), urbanicity, and ethnicity were the stratification variables used to represent the diverse demographic characteristics of the national school population. The proportions of students in the different ethnic groups were observed during the sampling process to ensure that the ethnic groups in national school population were fairly represented.

Special needs students were included in the item tryout program. District test coordinators were instructed to include all students *except* those classified as Severely/Profoundly Mentally Disabled. Additionally, students not able to be tested under the prescribed conditions were exempted from the item tryout program.

Data Obtained from the Program

After testing, all tryout items were analyzed using both traditional item-analysis methods and Rasch model techniques, yielding the following kinds of information:

- The p-value, or percentage of students selecting the correct answer at each grade in which the item was administered

- The p-values for above-average, average, and below-average students, where the levels of ability were defined as the upper twenty-three percent, middle fifty-four percent, and lower twenty-three percent of the test score distribution

- The percentage of students in all categories (above-average, average, and below-average) selecting each distractor option or omitting the item entirely

- The biserial correlation coefficient, showing the correlation of the item with the total score on the test in which the item appeared

- The point biserial correlation coefficient, which provides an index of item discrimination and evidence that each item is functioning consistently with the overall test

- The Rasch model scaled score, a three-digit scaled score showing the item's difficulty in relation to the Stanford scaled score system

- The mean-square fit for the item, a statistical estimate of the match between the actual response and that predicted by the Rasch model

- Mantel-Haenszel bias analysis procedures were also implemented to determine whether items performed differentially for reference (majority) and focal (minority) groups

An item card was prepared containing all pertinent information for each selected tryout item. Information about an item as it performed at different grades was affixed to a single item card so that data could be compared across grades to determine the appropriate level of difficulty. Appendixes G, H, I, and J present item p-values, mean p-values, median biserial correlation coefficients, and median point biserial correlation coefficients.

Selected items from Stanford 9 were embedded in each Stanford 10 tryout test so that the difficulty of all new items could be expressed in terms of the existing Stanford scaled score system on a common scale of difficulty spanning all levels. For Pearson product-moment coefficients of correlation for corresponding Stanford 10 and Stanford 9 subtests, see Appendix L.

PROCEDURES TO ELIMINATE BIAS

At Harcourt, it is critically important that sufficient procedures be implemented during all phases of test development to protect against not only undesirable test content (e.g., bias) but also the presence of test items that functioned differently according to group

15

membership. Bias in testing "refers to construct-irrelevant components that result in systematically lower or higher scores for identifiable groups of examinees" (American Educational Research Association, 1999). Bias can result from a number of factors related to test content. Harcourt established procedures to eliminate testing bias during all stages of test development, including the way the items were constructed, the item review procedures, and the statistical analyses associated with the National Item Tryout Program and the National Research Programs (standardization).

Test content should neither be offensive to members of a particular group nor unfairly disadvantage the performance of a particular group because of extraneous factors irrelevant to the constructs (theoretical attributes) the test intends to measure. For this reason, it was critical throughout all phases of test development to minimize the presence of items that inadvertently reflected ethnic, gender, disability, socioeconomic status, cultural, or regional bias or stereotyping, or that included content that disadvantaged a group because of differences in culture or familiarity.

For example, genders were shown in both traditional and nontraditional roles. However, a single picture showing a man or woman cooking dinner cannot be considered stereotypical or non-stereotypical in and of itself; one must look at the roles in which people are depicted throughout the entire test.

Another factor considered when the final Stanford 10 forms were constructed was a balance of the frequency of the appearance of persons with obvious physical handicaps. Teachers of special populations and experts reviewed materials to ensure it was not biased against the deaf, hearing impaired, blind, visually impaired, or others with disabilities.

Harcourt also took into account the beliefs and sensitivities of religious groups regarding appropriate and inappropriate content for educational assessment. The items are designed to avoid areas of sensitivity to these groups.

Bias Review Advisory Panel

A major procedure to protect against undesirable test content involved ensuring that all test materials received sufficient review by the proper experts. The Stanford 10 Bias Review Advisory Panel comprised 20 members. Each Advisory Panel member is prominent in the educational community. Several of these members were chosen because of their views regarding the need to keep standardized achievement tests free of bias. The Advisory Panel reflected diverse backgrounds and represented many ethnicities; including African American, Hispanic, Native American, and Asian (see Table 2). In addition, the members came from diverse geographic locations including urban, suburban, and rural settings. Some were educators with experience and sensitivity towards disability issues. Eight of the members had served on other advisory panels for Harcourt.

Table 2. Stanford 10 Bias Review Advisory Panel Members

Name	Title	Affiliation
Kinley Alston	Professor	Trident Tech. College, Charleston, SC
Raul Aparicio	Professor of Mathematics	Blinn College, Brenham, TX
Anna Beck	Assistant Principal	Alexandria City Public Schools, VA
Dee Black	Senior Counsel	Home School Legal Defense Assoc., Purcellville, VA
Laura Bracken	Instructor/Author	Lewis-Clark State College, Lewiston, ID
Cecile Carter	Cultural Liaison	Caddo Nation, Oklahoma
Jeanie Crosby	Director	Vermont Adult Learning, St. Albans, VT
Shabbir Mansuri	Founding Director	Council on Islamic Education, Fountain Valley, CA
Sara Massey	Curriculum Specialist	Institute of Texan Cultures at UTSA, San Antonio, TX
Sonja Moore	Director of Distance Education	Virginia Commonwealth University, Richmond, VA
George Otero	Founder and Director	Las Palomas de Taos, Taos, NM
Graciela P. Rosenberg	Professor, Department of Curriculum & Instruction	University of Texas–Brownsville
Beverly Kelton Sheathelm	Professor of Reading	Retired
Albert Sinquah	Principal	Keams Canyon School, AZ
Karen Stewart	Associate Professor	College of the Mainland, Texas City, TX
Ivy Lin Syriotis	Assistant Principal	Chatsworth Avenue School, Larchmont, NY
Gilbert Wilkerson	President	Network of Black Home Schoolers, Richmond, VA
Vivian Williams	Systems Director for Learning and Development	Banner Health Systems, Phoenix, AZ
Samuel Yigzaw	Director of Curriculum & Instruction	Higher Ground Academy, St. Paul, MN
Gary F. Yung	Program Evaluator	San Francisco Unified School District

The Stanford 10 Bias Review Advisory Panel met during the spring of 2001 to review items for Stanford 10. This review occurred after completion of the National Item Tryout Program. Members received an orientation that outlined item review procedures in a workshop setting. The general bias categories covered were:

- gender;

- race/ethnicity;

- religion;

- geographic region;

- socioeconomic status (SES);

- level of English proficiency; and

- disability.

The Stanford 10 Bias Review Advisory Panel's training and guidance focused on the following specific forms of bias:

- Stereotyping: The demeaning/negative portrayal of a group using certain attributes or social roles

- Omission: The lack of representation of a group living in the United States

- Limited or Distorted Perspective: The presentation of only one view of an issue that affects more than one group

- Isolation/Unreality: The presentation of interactions between cultural groups that are inaccurate or unrealistic

- Bias in Language: The use of wording that offends or minimizes the contributions of a group or a culture

- Lack of Multicultural Perspective: The presentation of context related to one culture or ethnic group

- Disability: The use of items or wording that would be biased against persons with disabilities

Test items selected for training the Stanford 10 Bias Review Advisory Panel were examined to identify those that could be considered biased, and solutions to correct the bias problems were considered. These items served as examples to familiarize the advisory panel members with the specific forms of bias. Panelists reviewed the materials and made written comments on each item.

After the orientation, advisory panel members were divided into two groups according to level preference, race, and gender balance. The first group was responsible for reviewing the lower test levels of the Stanford 10 item tryout editions, while the second group was responsible for the upper levels. Once divided into the two groups, advisory panel members were assigned to particular tests based on their areas of interest, their educational backgrounds, and the special needs of a particular group representation.

An effort was made to divide test material equally among advisory panel members and to assign a particular upper- or lower-level subtest to the same individual. When an item appeared to be biased, it was flagged as containing a specific type of bias.

When the Stanford 10 final forms were constructed, great care was taken to balance the frequency and nature of the appearance of members of a particular racial/ethnic or gender group. This balance was regarded as an important means of reducing all forms of bias. For example, both genders are shown in both traditional and nontraditional roles.

During the development of Stanford 10, an item could be identified as potentially biased in any one of a number of ways:

- During initial item review by Stanford 10 assessment specialists, editors, and item writers

- During review by the Stanford 10 Bias Review Advisory Panel

- Following a review of the item analysis statistics from the National Item Tryout Program

- In response to teacher questionnaire comments from the National Item Tryout Program

- During generation of statistical data (Mantel-Haenszel or Rasch) from the National Item Tryout Program

When an item was flagged as being potentially biased, it was carefully reviewed by assessment specialists and appropriate Stanford 10 Bias Review Advisory Panel members. Items identified as biased were then immediately removed from consideration for inclusion in the final test forms. Some flagged items were retained in the item pool only if the reviewers agreed that there was no reason to assume bias in the item content or format and that removal of the item from the pool was unwarranted.

Most decisions regarding the rejection of a flawed item, particularly with respect to bias, were based on the available technical data. Any item the Stanford 10 Bias Review Advisory Panel deemed inappropriate was also removed from the Stanford 10 item pool. This was not the only review stage during which potentially biased items were rejected. During test construction and review by content experts (assessment specialists) and editorial staff, items not meeting test construction requirements were also removed. Items receiving negative comments from teachers as a result of the item tryout program were also eliminated.

Statistical Procedures and Protocols

It is a fact that *individuals* differ in the knowledge and skills they bring to a testing situation. Such differences, which may have an impact on achievement test results, may be the result of unequal educational opportunities or any number of environmental conditions. However, an achievement test is seriously flawed to the extent that it produces differential results that reflect anything other than true differences in achievement among examinees. In particular, items that show systematic *group* differences in performance can signal the presence of bias. Tests are particularly suspect if differential results occur among cohort groups, such that membership in a group indicates that examinees' scores differ from scores of examinees in another group. Therefore, in addition to the Stanford 10 Bias Review Advisory Panel's review discussed above, selected statistical procedures were used to identify items that did not fit predetermined psychometric specifications with respect to similarities and differences in group performance.

Prior to the selection of items for the final forms of Stanford 10, all items from the National Item Tryout Program were analyzed according to Mantel-Haenszel procedures, which examine differential item functioning between reference (majority) and focal (minority) groups, after matching the groups on test scores. Comparisons were made between Males and Females; Whites and African Americans; Whites and Hispanics; and Regular Education Students and Special Populations. An item was considered potentially biased if its Chi-square was greater than what would normally be expected by chance. Items showing differences greater than chance were flagged for review and possible exclusion from the final forms of the tests.

DEVELOPMENT OF THE FINAL FORMS

Items were selected and the final forms of Stanford 10 were assembled based on a combination of the following:

- Appropriate content fit to the test blueprint

- Appropriate difficulty for the intended grade and increase or decrease in difficulty for adjacent lower or higher grades

- Good discrimination between high scorers and low scorers (biserial correlation coefficient)

- Appropriate clarity and interest

- Absence of bias according to advisory panel and statistical procedures

- Good spread of students choosing each distractor

- Statistical properties of the individual items based on the results of the National Item Tryout Program

- Target mean item difficulties and the shapes of the percentage of items correct (*p*-value) distributions for the individual tests at every level

The foremost consideration for selection of items for the final forms was fit of content to the test blueprint. Indices of item difficulty and discrimination were used to balance the statistical properties of the test forms.

To eliminate any potential bias, every effort was made to counterbalance any items that favored one group over another with other items that favored the second group over the first. In this way, overall test scores would not be affected by the presence of items favoring one group over another. Care was taken to balance the frequency and nature of the appearance of members of a particular minority or gender group. For example, genders were shown in both traditional and non-traditional roles. The total composition of each form was checked again to eliminate clues from item to item, similarity of pictures, and repetition or overlap of content. Teachers of special populations and other experts reviewed materials to ensure there was no bias against students who are blind, visually impaired, deaf, or hearing impaired.

As an additional step in the item selection process at the Primary 1 through TASK 3 levels, a subset of items was selected for the abbreviated format of each test at each level. Thus, items for the full-length and abbreviated test formats were selected simultaneously.

The final forms were created to conform to the test blueprints, be appropriate in difficulty, be comparable in content and difficulty between test forms, and maintain continuity across levels. The cognitive levels were also built to be parallel in coverage and average item difficulty across forms.

Cognitive Levels of Test Items

In addition to the selection criteria already discussed, cognitive levels of test items were also considered. The test items selected for the final forms were balanced with respect to the cognitive level assessed—Basic Understanding or Thinking Skills.

1. **Basic Understanding** items measure the ability to recall or recognize factual information, to identify something that is explicitly stated, and to associate relevant aspects of specific content.

2. **Thinking Skills** items measure the ability to analyze and synthesize information; to classify and sequence information; to compare and contrast information; to evaluate information in order to determine cause and effect, fact and opinion, relevant and irrelevant; and to interpolate and/or extrapolate beyond information in order to draw conclusions, make predictions, and hypothesize.

Organization of Items within Subtests

Most achievement batteries arrange items within a subtest by order of difficulty, with the easiest item placed first and the most difficult item last. Although this is common practice, this way of ordering items can present a problem to many students, especially

21

those in special populations, who will typically respond to the first items on the subtest because they are easier but become increasingly frustrated as the items become more difficult. At some point in the subtest, these students may disengage from the testing process or lose motivation to continue. This may result in random marking of answer sheets or refusal to provide answers for the remainder of the items on the subtest. In either case, their test scores do not reflect their true level of achievement, e.g., students know and can do more than their performance shows.

In Stanford 10, items are arranged differently to facilitate student engagement in the testing process and test completion. Where appropriate and not prohibited due to other constraints (e.g., the passage-dependent requirement of items in the Reading Comprehension subtest), difficult items are interspersed throughout the subtest and are cushioned by easier items immediately preceding and following them. Therefore, the level of student frustration is dramatically reduced since difficult items are usually followed by items the students can answer successfully. This item configuration provides encouragement to students and increases their motivation to complete the test. Please see Appendix F for the student completion rates of each level and subtest.

Design of Final Forms

The Stanford 10 Full-Length Battery for kindergarten through grade 12 consists of 8,186 multiple-choice test questions across two forms. The number of items per subtest ranges from 30 to 54. The Abbreviated Battery consists of 20 to 30 items per subtest. The items contained in the Abbreviated Battery were selected to mirror the content and difficulty distribution of each subtest as defined by the Full-Length Battery. Table 3 on page 24 presents the subtests and recommended grade ranges for each level of the Stanford 10 Basic and Complete Batteries, including the number of items and recommended administration times for each subtest. Table 4 on page 25 presents the same information for the Stanford 10 Abbreviated Battery.

To maintain the focus on students, the Stanford 10 test booklets were designed to be easy to use, to help students complete the test, and to help students perform at their best. During the initial development stages of the Stanford 10, focus groups consisting of students and teachers were asked to describe what they wanted and needed in a standardized test. What was learned was that the booklets should be clean in appearance and contain a high degree of graphic imagery and illustration. However, these graphics should enhance the test without creating pages that are confusing to students. Also learned in related research was that the layout of the booklets should help students keep their place as they proceed through the test and that all information needed to provide an answer should be presented with the question.

The Stanford 10 test booklets were designed to address these needs. To help students keep their place on the page, items are presented in their own "boxes" on the pages, and each question number and the option letters are in a contrasting color. Each subtest is identified with its own unique icon, and all pages are numbered consecutively across the subtests. These features not only help students, they also allow teachers to quickly and easily see that students are on the proper page of the test booklet.

Illustrations accompany the Reading Comprehension passages and items to draw students into the passage. In the mathematics and Science tests, the extensive use of graphics helps minimize the amount of reading so that even those students who are not proficient readers understand the questions and can do their best.

The new design of Stanford 10 represents an important innovation in kindergarten through grade 12 test publishing. Harcourt believes that the use of full color and the crisp, clear navigation will maximize access to content for *all* students in ways never before available. By maintaining the integrity of the item format within a new page design, the psychometric integrity of scores will be maintained.

Table 3. Stanford 10 Multiple-Choice Scope and Sequence

Test Levels, Recommended Grade Ranges, Tests, and Administration Times

Test Levels	S1 Grade K.0-K.5 K	T	S2 Grade K.5-1.5 K	T	P1 Grade 1.5-2.5 K	T	P2 Grade 2.5-3.5 K	T	P3 Grade 3.5-4.5 K	T	I1 Grade 4.5-5.5 K	T	I2 Grade 5.5-6.5 K	T	I3 Grade 6.5-7.5 K	T	A1 Grade 7.5-8.5 K	T	A2 Grade 8.5-9.9 K	T	T1 Grade 9.0-9.9 K	T	T2 Grade 10.0-10.9 K	T	T3 Grade 11.0-12.9 K	T
Sounds and Letters	40	30	40	25																						
Word Study Skills					30	20	30	20	30	20	30	20														
Word Reading	30	15	30	25	30	25																				
Sentence Reading			30	30	30	30																				
Reading Vocabulary					40	40	30	20	30	20	30	20	30	20	30	20	30	20	30	20	30	20	30	20	30	20
Reading Comprehension							40	40	54	50	54	50	54	50	54	50	54	50	54	50	54	40	54	40	54	40
Total Reading	70	45	100	80	130	115	100	80	114	90	114	90	84	70	84	70	84	70	84	70	84	60	84	60	84	60
Mathematics	40	30	40	30																	50	50	50	50	50	50
Mathematics Problem Solving					42	50	44	50	46	50	48	50	48	50	48	50	48	50	48	50						
Mathematics Procedures					30	30	30	30	30	30	32	30	32	30	32	30	32	30	32	30						
Total Mathematics					72	80	74	80	76	80	80	80	80	80	80	80	80	80	80	80						
Language					40	40	48	45	48	45	48	45	48	45	48	45	48	45	48	45	48	40	48	40	48	40
Spelling			40	30	36	30	36	30	38	35	40	35	40	35	40	35	40	35	40	35	40	30	40	30	40	30
Listening to Words and Stories	40	30	40	30																						
Listening					40	30	40	30	40	30	40	30	40	30	40	30	40	30	40	30						
Environment	40	30	40	30	40	30	40	30																		
Science									40	25	40	25	40	25	40	25	40	25	40	25	40	25	40	25	40	25
Social Science									40	25	40	25	40	25	40	25	40	25	40	25	40	25	40	25	40	25
Basic Battery	150	105	180	140	318	295	298	265	316	280	322	280	292	260	292	260	292	260	292	260	222	180	222	180	222	180
Complete Battery	190	135	220	170	358	325	338	295	396	330	402	330	372	310	372	310	372	310	372	310	302	230	302	230	302	230
Total Testing Time	2 hrs. 15 min.		2 hrs. 50 min.		5 hrs. 25 min.		4 hrs. 55 min.		5 hrs. 30 min.		5 hrs. 30 min.		5 hrs. 10 min.		5 hrs. 10 min.		5 hrs. 10 min.		5 hrs. 10 min.		3 hrs. 50 min.		3 hrs. 50 min.		3 hrs. 50 min.	
Language Form D					40	40	40	40	45	45	48	45	48	45	48	45	48	45	48	45	48	40	48	40	48	40

K = No. of Items
T = Time in Minutes

24

Table 4. Stanford 10 Multiple-Choice Scope and Sequence, Abbreviated Battery

Test Levels, Recommended Grade Ranges, Tests, and Approximate Administration Times

Test Levels	P1 K	P1 T	P2 K	P2 T	P3 K	P3 T	I1 K	I1 T	I2 K	I2 T	I3 K	I3 T	A1 K	A1 T	A2 K	A2 T	T1 K	T1 T	T2 K	T2 T	T3 K	T3 T
Grade	1.5-2.5		2.5-3.5		3.5-4.5		4.5-5.5		5.5-6.5		6.5-7.5		7.5-8.5		8.5-9.9		9.0-9.9		10.0-10.9		11.0-12.9	
Word Study Skills	20	11	20	11	20	12	20	12														
Word Reading	20	17																				
Sentence Reading	20	20																				
Reading Vocabulary			20	14	20	14	20	14	20	14	20	14	20	14	20	14	20	14	20	14	20	14
Reading Comprehension	30	30	30	30	30	30	30	30	30	30	30	30	30	30	30	30	30	30	30	30	30	30
Total Reading	90	78	70	55	70	56	70	56	50	44	50	44	50	44	50	44	50	44	50	44	50	44
Mathematics																	30	30	30	30	30	30
Mathematics Problem Solving	30	34	30	34	30	33	30	33	30	31	30	31	30	30	30	29						
Mathematics Procedures	20	24	20	24	20	22	20	20	20	20	20	20	20	20	20	20						
Total Mathematics	50	58	50	58	50	55	50	53	50	51	50	51	50	50	50	49						
Language	30	28	30	28	30	28	30	28	30	28	30	28	30	28	30	28	30	25	30	25	30	25
Spelling	30	25	30	25	30	26	30	26	30	26	30	26	30	26	30	26	30	23	30	23	30	23
Environment	30	23	30	23																		
Science					30	19	30	19	30	19	30	19	30	19	30	19	30	19	30	19	30	19
Social Science					30	19	30	19	30	19	30	19	30	19	30	19	30	19	30	19	30	19
Abbreviated Battery	230	212	210	189	240	203	240	201	220	187	220	187	220	186	220	185	200	160	200	160	200	160
Total Testing Time	3 hrs.	32 min.	3 hrs.	9 min.	3 hrs.	23 min.	3 hrs.	21 min.	3 hrs.	7 min.	3 hrs.	7 min.	3 hrs.	6 min.	3 hrs.	5 min.	2 hrs.	40 min.	2 hrs.	40 min.	2 hrs.	40 min.
Language Form D	30	23	30	23	30	23	30	23	30	23	30	23	30	23	30	23	30	23	30	23	30	23

K = No. of Items

T = Time in Minutes

Part III: The National Research Programs

The National Research Program for the standardization of Stanford 10 took place during the spring and fall of 2002. The purpose of the National Research Programs were to provide the data used to (1) equate the levels and forms of the test series; (2) establish the statistical reliability and validity of the tests; and (3) develop normative information descriptive of achievement in schools nationwide.

Testing for the Spring Standardization Program of all levels and forms of Stanford 10 took place from April 1, 2002, to April 26, 2002. Testing for the Equating of Levels Program, Equating of Forms Program, and Equating of Stanford 10 to Stanford 9 took place from April 1, 2002, to May 24, 2002. Approximately 250,000 students from 650 school districts participated in the Spring Standardization Program, with another 85,000 students from 385 school districts participating in the spring equating programs. Some students participated in more than one program. See Appendix B for a list of school districts participating in the Stanford 10 Spring and Fall 2002 National Standardization Programs. The Stanford 10 *Spring Multilevel Norms Book* presents score conversion tables separately.

Testing for the Fall Standardization Program took place from September 9, 2002, to October 18, 2002. Testing for the Equating of Levels Program, Equating of Forms Program, and Equating of Stanford 10 to Stanford 9 took place from September 9, 2002, to November 1, 2002. Approximately 110,000 students participated in the Fall Standardization and Equating Programs. Some students participated in more than one program. The Stanford 10 *Fall 2002 Multilevel Norms Book* presents score conversion tables separately.

All students participating in the Stanford 10 Spring and Fall 2002 National Standardization Programs also completed the *Otis-Lennon School Ability Test*®, Eighth Edition (OLSAT® 8).

All forms of Stanford 10 were empirically standardized during the spring and fall standardization. All participating students were administered one of two groupings of *full-length* subtests. One grouping of subtests comprised:

- Total Reading (including the Sounds and Letters, Word Study Skills, Word Reading, Sentence Reading, Reading Vocabulary, and Reading Comprehension subtests, depending on test level);

- Language (traditional or comprehensive, at the Primary 1 through TASK 3 test levels); and

- Spelling (at the Primary 1 through TASK 3 test levels).

The other grouping comprised:

- Total Mathematics (including Mathematics, or Mathematics Problem Solving and Mathematics Procedures, depending on test level);

- Environment (at the SESAT 1 through Primary 2 test levels);

- Science (at the Primary 3 through TASK 3 test levels);

- Social Science (at the Primary 3 through TASK 3 test levels);

- Listening to Words and Stories (at the SESAT 1 and SESAT 2 test levels); and

- Listening (at the Primary 1 through Advanced 2 test levels).

All students were administered Stanford 10 subtests under *untimed* conditions with the exception of a special study of *timed* testing conditions in 150 classrooms nationwide at each grade level. The proposed schedule of approximate testing times in the Directions for Administering was used by test administrators for planning purposes only. The Directions for Administering recommends that test administrators allow a student to continue with a test as long as she or he is working productively.

Stanford 10 Practice Tests were administered to each student participating in the National Standardization Programs who was administered the Stanford 10 Primary 2 through Advanced 2 test level. Because of the administration of practice tests during standardization, Harcourt recommends their use when Stanford 10 is administered in state and local programs.

Table 5 shows the grades at which each level of the Stanford Series was standardized. Note that there was an intentional overlap of grades per test level to ensure a balance of easy to hard items for each level and grade-to-grade progression of difficulty.

Table 5. Grades at Which the Stanford Series Was Standardized, by Test Level

Grade / Test Level	K.1	K.8	1.1	1.8	2.1	2.8	3.1	3.8	4.1	4.8	5.1	5.8	6.1	6.8	7.1	7.8	8.1	8.8	9.1	9.8	10.1	10.8	11.1	11.8	12.1	12.8
SESAT 1	X	X																								
SESAT 2		X	X	X																						
Primary 1				X	X	X																				
Primary 2						X	X	X																		
Primary 3								X	X	X																
Intermediate 1										X	X	X														
Intermediate 2												X	X	X												
Intermediate 3														X	X	X										
Advanced 1																X	X	X								
Advanced 2																		X	X	X						
TASK 1																			X	X	X					
TASK 2																					X	X				
TASK 3																							X	X	X	X

SAMPLING PROCEDURES

School districts were selected for participation in the Stanford 10 National Research Programs according to demographic features that matched variables in the stratified cluster sampling design (classrooms served as clusters). The stratified sampling design was repeated in each state, i.e., within each state the samples were chosen to be representative of the national school population. All states, with the exception of Delaware, Wyoming, and the District of Columbia, were represented in the Stanford 10 National Research Programs. See Appendix B for a list of schools participating in the Stanford 10 National Standardization Program.

The stratification variables were geographic region (Northeast, South, Midwest, and West), socioeconomic status (SES), urbanicity, and ethnicity. Urbanicity was divided into three categories (urban, suburban, and rural), and socioeconomic status (SES) was classified as high, middle, or low. The SES index was a composite of median family income (in thousands of dollars) in the community and percentage of adults with high school diplomas.

Information relating to the stratification variables was obtained from the Census of Population and Housing (2000) and the National Center for Education Statistics (2000–2001). This information was used to evaluate, and weight where necessary, the Fall and Spring Standardization samples to approximate the desired proportional representation.

Table 6 shows the weighted percentages of students representing the various groups participating in the Spring and Fall 2002 National Standardization Programs as grouped by geographic region, socioeconomic status (SES), urbanicity, ethnicity, special condition, and non-public schools. National school enrollment percentages are shown for comparison purposes.

PARTICIPATION BY STUDENTS WITH DISABILITIES AND LIMITED ENGLISH PROFICIENT STUDENTS

Students receiving instruction as part of a regular education classroom who would normally test with other students in the regular classroom were asked to be part of the standardization sample. Schools were instructed to test all students except those classified as Severely/Profoundly Mentally Disabled or those who could not be tested under the prescribed standardization conditions. The percentage of special populations (students with disabilities and limited English proficient students) who participated in the standardization programs reflected the percentage of special education students routinely tested by participating schools, rather than the percentage of special education students attending these schools. Additional research, summarized in Part VI: Testing Special Populations, has been conducted on the effects of accommodations on the assessment of special populations.

Table 6. Demographic Characteristics of School Districts Participating in the *Stanford Achievement Test* Series, Tenth Edition, Standardization Programs

	School Enrollment Percentage of Total U.S.*	Percentage of Students in Spring Standardization Programs	Percentage of Students in Fall Standardization Programs
Geographic Region			
Northeast	19.7	13.7	21.8
Midwest	24.0	27.7	24.3
South	24.0	23.4	24.0
West	32.3	35.2	29.9
Socioeconomic Status (SES)			
Low	32.2	29.6	30.9
Middle	34.3	36.4	37.1
High	33.5	34.0	32.0
Urbanicity			
Urban	26.9	20.3	24.6
Suburban	48.1	49.9	48.2
Rural	25.0	29.8	27.2
Ethnicity (92.6% Reporting)			
Black or African American	15.1	14.0	14.7
Hispanic or Latino	15.0	14.5	13.5
White	65.2	65.1	64.9
Asian	3.8	3.4	3.8
Other	0.9	3.0	3.1
Special Condition			
Autism	---	0.1	0.1
Visual Impairment	---	0.1	0.2
Deaf-Blindness	---	0.0	0.0
Developmental Delay	---	0.1	0.1
Hearing Impairment	---	0.2	0.1
Orthopedic Impairments	---	0.1	0.1
Multiple Disabilities	---	0.0	0.0
Mental Retardation	---	0.2	0.2
Emotional Disturbance	---	0.3	0.3
Speech and Language Disorders	---	1.5	1.1
Specific Learning Disabilities	---	2.0	3.2
Other Health Care Needs	---	0.2	0.2
Traumatic Brain Injury	---	0.0	0.0
Nonpublic Schools			
Catholic	4.5	4.7	3.5
Private	4.7	4.3	4.8

*Census of Population and Housing (2000) and the National Center for Education Statistics, United States Department of Education, 2000–2001.

WEIGHTING PROCEDURES

The Stanford 10 standardization sampling methodology involved three steps:

1. Selection based on demographic data descriptive of the school districts

2. Description based on the previous step and on responses to a questionnaire distributed to participating districts

3. Statistical weighting of test scores after testing is completed, but before norms are derived, to effect final improvements in the sample

The purpose of this statistical weighting was to achieve a better approximation to national characteristics of those variables that show the highest relationship to test performance. The weighting procedure itself involved random deletion or duplication of complete student records until the desired sample characteristics were obtained.

EQUATING OF LEVELS PROGRAM: VERTICAL SCALING

The development of scaled score and normative information presented in the Stanford 10 *Multilevel Norms Books* began with the Equating of Levels Program. This program provided the elements to create the continuous score scale (vertical scale) that permits the interpretation of scores across levels of a test. To accomplish this, students in kindergarten through grade 11 completed two adjacent test levels. For each test in every content area that supported the program testing design, students were administered the on-grade level test and one level lower. Each dual-level test administration resulted in approximately 750 to 1,000 scores to develop this important component of the Stanford 10 multilevel assessment.

To control for test order and fatigue factors, a counterbalanced design was used to randomly administer the order of tests (lower level/higher level, higher level/lower level, and content grouping) to each participating classroom. One content grouping comprised Total Reading, Language, and Spelling. The other content grouping comprised Total Mathematics, Environment or Science and Social Science, and Listening to Words and Stories or Listening. Each classroom was randomly assigned to a content grouping in a manner that ensured the administration of all content areas within in a school district. To the extent possible, student participation in the Equating of Levels Program adhered to the demographic percentages displayed in Table 7.

The Pearson product-moment coefficients of correlation (r) that demonstrate relationships between subtests at adjacent levels can be found in Appendix N.

31

EQUATING OF FORMS PROGRAM

Test score information resulting from the Equating of Forms Program was used to develop scaled scores for each form of Stanford 10. The scaled scores indicate equivalent ability, e.g., a scaled score of 640 indicates the same achievement on Form A as it does on Form B. To establish this equivalence between forms, a different group similar in size and demographic composition to the Equating of Levels Program sample was administered the appropriate on-grade test level of Forms A and B, as well as the corresponding Form D or Form E Comprehensive Language subtest.

A testing design similar to that of the Equating of Levels Program was utilized. The test-content groupings and random assignments for classrooms were applied as in the Equating of Levels Program. Each student completed two forms of the assigned content grouping. The order of administration of the two forms was counterbalanced by classroom to obviate practice effects. To maintain the continuous vertical scale across forms, the scaling constants developed through the Equating of Levels Program were applied to test levels of each form. The information produced from the Equating of Levels and Equating of Forms Programs was used in the Standardization Programs to develop the normative information presented in the Stanford 10 *Multilevel Spring Norms* book and *Multilevel Fall Norms* book. Alternate forms reliability coefficients and related summary data for each level and subtest can be found in Appendix E.

EQUATING OF STANFORD 10 TO STANFORD 9

A major issue in the revision of an assessment series is the comparison of scores on the new edition with scores from the previous edition. That is, to what extent are scales of the two editions comparable? The Equating of Stanford 10 to Stanford 9 furnished this type of information.

A common-persons design similar to that of the Equating of Levels and Equating of Forms Programs was used to ensure that scaled score conversions from the Stanford 9 to Stanford 10 and those of Stanford 10 to Stanford 9 indicated equivalent achievement. A set of students different from the previous two programs was administered the appropriate on-grade test level from both Stanford 9 and Stanford 10. Approximately 1,000 students per grade participated. The content groupings, random assignments, and counterbalancing procedures as explained above were applied in this program. For Pearson product-moment coefficients of correlation for corresponding Stanford 10 and Stanford 9 subtests, see Appendix L.

SETTING PERFORMANCE STANDARDS

A recent trend in educational assessment involves providing information about students that goes beyond the norm-referenced information typically associated with achievement batteries. More and more, educators and parents want to know more than just how a student's performance compares with that of other students. Of growing interest is "what level of performance does a score represent?" Moreover, the *No Child Left Behind Act* of 2001 (NCLB) now provides and mandates that each state set achievement standards. Performance standards for Stanford 10 were determined by empirically equating Stanford 10 scaled scores with Stanford 9 scaled scores. Research linking the Stanford 10 scale to the Stanford 9 scale enabled a direct translation of performance standards for Stanford 10.

After Stanford 9 was standardized, approximately 200 teachers representing school districts from around the country were brought to a three-week series of standard-setting meetings, with each meeting lasting one week. The teachers were selected based on recommendations from principals, school or state department administrators, or other educators. They represented all content areas, grade levels, school district demographic variables, and ethnicities/cultures. Teachers were assigned to groups on the basis of their grade level/subject area expertise; each group consisted of 12 or 13 teachers, with 16 groups in all. The Stanford battery was divided into corresponding grade level/subject area combinations of subtests.

After a general orientation session that included training and practice, the teachers broke into small group sessions, where they were given the opportunity to ask further questions and receive further training. Then they were asked to actually take the tests that they would be evaluating. The standard setting was then accomplished through use of a modified Angoff procedure. Teachers were asked to make three independent judgments about each item in Form S of the Stanford 9 battery and to decide how students of various performance levels *should* perform on the item. For each multiple-choice item, the teachers were to judge what percentage of borderline students at the *Basic* level (Level 2) should answer the item correctly; what percentage of borderline students at the *Proficient* level (Level 3) should answer the item correctly; and what percentage of borderline students at the *Advanced* level (Level 4) should answer the item correctly. After the initial judgments were made, the data were compiled into frequency distributions, and the teachers received this feedback on their initial judgments. They were also provided the national p-value statistics for each item. The items and judgments were discussed, and teachers had the opportunity to revise their judgments, based on the discussion. The raw-score cut point for each performance level for each subtest was obtained by summing the ratings for all items in that subtest and averaging the sums across teachers.

Part IV: Types of Scores

The various types of scores utilized in Stanford 10 have different uses and yield different kinds of information. Since the underlying properties of these scores are not necessarily the same, the particular score type to be used to interpret test results depends on the purpose for which the test was administered. Types of scores that are frequently utilized in Stanford 10 are described below.

RAW SCORES

Raw scores are tied to a specific subtest and test content. A raw score refers to the number of test questions a student answered correctly, and its interpretation is limited to that set of questions. Because subtests differ in length, content, and difficulty, raw scores across subtests or test levels cannot be compared directly. Therefore, raw scores provide limited information about the relative performance of students.

It is not appropriate to use raw scores to compare performance over time or when different test levels have been administered. Instead, we convert raw scores into scaled scores, enabling the comparison of students' test scores with those of other students and the evaluation of changes in student performance across subtests and testing occasions. A scaled score can then be converted to one or more other derived scores allowing for further interpretation and evaluation of the test results.

SCALED SCORES

Scaled scores have the advantage of representing approximately equal units on a continuous scale. That is, a difference of 5 points between two students' scores represents the same amount of difference in achievement wherever it occurs on the scale. In addition, the Stanford scaled score system expresses student performance across all test levels of any given subtest on a single scale. For example, the Reading Comprehension subtest is linked across 11 test levels from Primary 1 through TASK 3, forming one continuous scale that makes it possible to compare scores from test level to test level. Scaled scores are especially suitable for comparing student performance in a particular subject area over time.

While scaled scores are comparable across test levels for the same subtest or total, they are not comparable from one content area to another or across subtests within a content domain total. For example, a scaled score on the Reading Comprehension subtest cannot be compared with a scaled score on the Spelling subtest; nor, can a scaled score on the Reading Vocabulary subtest be compared with a scaled score for Total Reading. Although these scaled scores may look similar, each subtest has its own scaled score system. For this reason, scaled scores cannot be used to develop score profiles across subtests.

Once a raw score earned on a particular subtest has been converted to its corresponding scaled score, the *test level* that was administered is no longer a concern. The scaled score can then be converted to other derived scores such as percentile ranks, stanines, and grade equivalents that correspond to the student's *grade level*.

Appendixes K and O demonstrate the validity of scaled scores as a measure of performance in a subject area over time. The mean scaled scores on each level and subtest for students participating in the 2002 National Standardization Program are presented in Appendix K. Appendix O presents scaled scores for subtests at key percentile ranks (10th, 25th, 50th, 75th, and 90th) across grade levels. These data demonstrate academic growth in a subject area over time.

INDIVIDUAL PERCENTILE RANKS

Percentile ranks range from a low of 1 to a high of 99, with 50 denoting average performance. Percentile ranks compare the relative standing of a student with students in a reference group who were in the same grade when they completed the same subtest at a comparable time of the year. For example, a percentile rank of 75 means that for a particular subtest the student performed as well as or better than 75% of the students in the reference group. The reference group may comprise a national or local sample of students and may represent a variety of population variables. Percentile ranks must always be interpreted with regard to the reference group from which they were derived.

Percentile ranks may be used to show the relative standing of students taking the Basic or Complete Batteries or to compare an individual student's performance across subtests within a battery. However, percentile ranks do not represent actual amounts of achievement. Furthermore, percentile ranks do not represent equal units along a scale. For example, the difference in achievement between percentile ranks 5 and 10 is not the same as the difference between percentile ranks 50 and 55. Percentile ranks do not represent equal units and their interpretation is limited to the reference group from which they were derived. For that reason, percentile ranks are best used for reporting scores when position within a reference group is of primary interest.

STANINES

Stanines range from a low of 1 to a high of 9, with 5 designating average performance. Stanines, like percentile ranks, indicate a student's relative standing when compared to a reference group. In contrast to percentile ranks, stanines represent approximately equal units of achievement. For example, the difference between stanines 2 and 4 represents about the same difference in achievement as the difference between stanines 5 and 7. Stanines are particularly useful for comparing or profiling a student's scores across subtests. Another benefit of using stanines is that broad performance categories can be easily identified. Usually, Stanines of 1, 2, and 3 indicate Below Average performance; 4, 5, and 6 Average; and 7, 8, and 9 Above Average. The relationship between stanines, percentile ranks, normal curve equivalents, and performance categories in a normally distributed set of scores is shown in Figure 1.

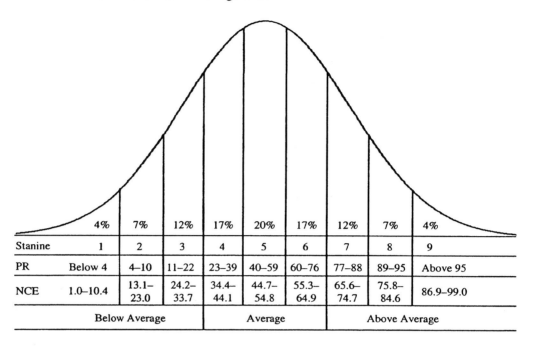

4%	7%	12%	17%	20%	17%	12%	7%	4%

Stanine	1	2	3	4	5	6	7	8	9
PR	Below 4	4–10	11–22	23–39	40–59	60–76	77–88	89–95	Above 95
NCE	1.0–10.4	13.1–23.0	24.2–33.7	34.4–44.1	44.7–54.8	55.3–64.9	65.6–74.7	75.8–84.6	86.9–99.0

Below Average	Average	Above Average

Figure 1. A Normal Distribution of Stanines, Percentile Ranks (PR), Normal Curve Equivalents (NCE), and Performance Categories

GRADE EQUIVALENTS

A grade equivalent is a score that represents an estimate of performance compared to the typical performance of students tested in a given month of the school year. Reported as a decimal number, the Stanford 10 grade equivalent scale ranges from K.0 (beginning kindergarten) to 12.9. Scores below K.0 or above 12.9 are designated as PK (pre-kindergarten) and PHS (post high school), respectively. The numeral to the left of the decimal point refers to a grade level. The numeral to the right of the decimal point represents one-tenth of the school year, or one school month. Table 7 shows the decimal relationships to months and weeks in school for the Stanford 10 grade equivalents. For example, a grade equivalent of 5.2 would represent the typical performance of the national sample of fifth graders taking Stanford 10 between the middle of October and the middle of November.

Table 7. **Decimal Relationships to Months and Weeks in School for Stanford 10 Grade Equivalents**

Month	Decimal	Weeks in School
August 16 – September 15	.0	0–4
September 16 – October 15	.1	5–8
October 16 – November 15	.2	9–12
November 16 – December 15	.3	13–16
December 16 – January 15	.4	17–20
January 16 – February 15	.5	21–24
February 16 – March 15	.6	25–28
March 16 – April 15	.7	29–32
April 16 – May 15	.8	33–36
May 16 – August 15	.9	37 +

The final Stanford 10 grade equivalent scale was derived from the performance of the Spring Standardization sample. Median scaled scores for each subtest and total were computed for students in the spring in each grade. These median scaled scores were plotted across grades, and a smooth line was fitted to the points. Grade equivalents corresponding to the median scaled score for each month within a school year were then read from the smoothed line.

Grade equivalents are used most effectively for comparing individual student performance with that of a normative sample. They can also be used to interpret the performance of groups of students. However, grade equivalents are frequently misinterpreted. When interpreting a grade equivalent, it is important to remember that grade equivalents estimate typical performance based on a standardization sample using median scaled scores across grade levels. If a fourth-grade student completed a test designed for fourth graders and received a grade equivalent of 6.3, this does not mean the fourth-grade student mastered sixth-grade work. Rather, this student's grade equivalent of 6.3 means that his or her score corresponds to the typical scores of sixth-grade students in the third month of the school year had they completed the same fourth-grade level test.

Like percentile ranks, grade equivalents represent unequal units. Therefore, a one-year difference in grade equivalents at one part of the scale is not the same as a one-year difference at another part of the scale. Further limitations of grade equivalents are that they are not comparable across subtests and across time periods. Although normal growth from the fall of one school year to the fall of the next is defined as one year (1.0) in grade equivalents, one year of growth is typical only for students who were tested in the fall of each year and who obtained average scores. Below-average students usually "grow" less than 1.0, and above-average students tend to show a pretest-posttest difference greater than 1.0.

NORMAL CURVE EQUIVALENTS

The normal curve equivalent is a normalized standard score with a mean of 50 and a standard deviation of 21.06. This score is most often used to enable test users to manipulate the test data algebraically. NCEs may be obtained by converting percentiles to normalized z-scores and making the transformation

$$NCE = 50 + 21.06z.$$

In contrast to percentile ranks, normal curve equivalents provide an equal-interval scale; thus, they should be used instead of percentile ranks when interpolating or averaging scores. Percentile ranks and normal curve equivalents have a direct, fixed relationship that is shown in Table 8, Normal Curve Equivalents Corresponding to Percentile Ranks, and Table 9, Percentile Ranks Corresponding to Normal Curve Equivalent Ranges. Table 8 provides NCEs corresponding to integer percentile ranks and is convenient to use when converting percentile ranks to NCEs. To convert NCEs to percentile ranks, Table 9 provides a percentile rank corresponding to ranges of NCEs.

Table 8. Normal Curve Equivalents Corresponding to Percentile Ranks

Percentile Rank	NCE	Percentile Rank	NCE	Percentile Rank	NCE	Percentile Rank	NCE
1	1.0	26	36.5	51	50.5	76	64.9
2	6.7	27	37.1	52	51.1	77	65.6
3	10.4	28	37.7	53	51.6	78	66.3
4	13.1	29	38.3	54	52.1	79	67.0
5	15.4	30	39.0	55	52.6	80	67.7
6	17.3	31	39.6	56	53.2	81	68.5
7	18.9	32	40.2	57	53.7	82	69.3
8	20.4	33	40.7	58	54.3	83	70.1
9	21.8	34	41.3	59	54.8	84	70.9
10	23.0	35	41.9	60	55.3	85	71.8
11	24.2	36	42.5	61	55.9	86	72.8
12	25.3	37	43.0	62	56.4	87	73.7
13	26.3	38	43.6	63	57.0	88	74.7
14	27.2	39	44.1	64	57.5	89	75.8
15	28.2	40	44.7	65	58.1	90	77.0
16	29.1	41	45.2	66	58.7	91	78.2
17	29.9	42	45.7	67	59.3	92	79.6
18	30.7	43	46.3	68	59.8	93	81.1
19	31.5	44	46.8	69	60.4	94	82.7
20	32.3	45	47.4	70	61.0	95	84.6
21	33.0	46	47.9	71	61.7	96	86.9
22	33.7	47	48.4	72	62.3	97	89.6
23	34.4	48	48.9	73	62.9	98	93.3
24	35.1	49	49.5	74	63.5	99	99.0
25	35.8	50	50.0	75	64.2		

Table 9. Percentile Ranks Corresponding to Normal Curve Equivalent Ranges

NCE Range	Percentile Rank	NCE Range	Percentile Rank	NCE Range	Percentile Rank	NCE Range	Percentile Rank
1.0–4.2	1	36.2–36.8	26	50.4–50.8	51	64.6–65.2	76
4.3–8.7	2	36.9–37.4	27	50.9–51.3	52	65.3–65.9	77
8.8–11.8	3	37.5–38.0	28	51.4–51.8	53	66.0–66.6	78
11.9–14.3	4	38.1–38.6	29	51.9–52.4	54	66.7–67.3	79
14.4–16.3	5	38.7–39.3	30	52.5–52.9	55	67.4–68.1	80
16.4–18.1	6	39.4–39.8	31	53.0–53.4	56	68.2–68.9	81
18.2–19.7	7	39.9–40.4	32	53.5–54.0	57	69.0–69.7	82
19.8–21.1	8	40.5–41.0	33	54.1–54.5	58	69.8–70.5	83
21.2–22.4	9	41.1–41.6	34	54.6–55.1	59	70.6–71.4	84
22.5–23.6	10	41.7–42.2	35	55.2–55.6	60	71.5–72.3	85
23.7–24.7	11	42.3–42.7	36	55.7–56.2	61	72.4–73.2	86
24.8–25.8	12	42.8–43.3	37	56.3–56.7	62	73.3–74.2	87
25.9–26.8	13	43.4–43.8	38	56.8–57.3	63	74.3–75.3	88
26.9–27.7	14	43.9–44.4	39	57.4–57.8	64	75.4–76.4	89
27.8–28.6	15	44.5–44.9	40	57.9–58.4	65	76.5–77.6	90
28.7–29.5	16	45.0–45.5	41	58.5–59.0	66	77.7–78.9	91
29.6–30.3	17	45.6–46.0	42	59.1–59.6	67	79.0–80.3	92
30.4–31.1	18	46.1–46.5	43	59.7–60.1	68	80.4–81.9	93
31.2–31.9	19	46.6–47.1	44	60.2–60.7	69	82.0–83.6	94
32.0–32.6	20	47.2–47.6	45	60.8–61.3	70	83.7–85.7	95
32.7–33.4	21	47.7–48.1	46	61.4–62.0	71	85.8–88.1	96
33.5–34.1	22	48.2–48.7	47	62.1–62.6	72	88.2–91.2	97
34.2–34.8	23	48.8–49.2	48	62.7–63.2	73	91.3–95.6	98
34.9–35.5	24	49.3–49.7	49	63.3–63.9	74	95.7–99.0	99
35.6–36.1	25	49.8–50.3	50	64.0–64.5	75		

ACHIEVEMENT/ABILITY COMPARISONS

Another type of score provided by the Stanford series is the Achievement/Ability Comparison (AAC). A Stanford 10 AAC shows how well a student has performed on a particular subtest or total in relation to other students at the same ability level as measured by the *Otis-Lennon School Ability Test*®, Eighth Edition (OLSAT® 8). An example of an Achievement /Ability Comparison table appears in Figure 2.

OLSAT Stanine	Low	Middle	High
1	Below 574	574–599	Above 599
2	Below 586	586–621	Above 621
3	Below 599	599–639	Above 639
4	Below 610	610–650	Above 650
5	Below 627	627–665	Above 665
6	Below 638	638–672	Above 672
7	Below 653	653–688	Above 688
8	Below 666	666–698	Above 698
9	Below 681	681–718	Above 718

Figure 2. Sample Achievement/Ability Comparison Scores for Total Reading for Fifth Graders Tested in the Spring

Figure 2 shows the Stanford scaled score ranges denoting Low, Middle, and High categories of achievement in Total Reading for each OLSAT grade stanine. Thus, if a student took Stanford 10 and OLSAT 8 in the spring and earned an OLSAT grade stanine of 6 and a Stanford scaled score of 660 in Total Reading, his or her achievement is "Middle" (average) in relation to the performance of a national sample of students who are in the same grade, at the same ability level, and tested at the same time of year.

For each grade tested during the national standardization programs, distributions of Stanford scaled scores on subtests and totals were generated separately by OLSAT grade stanine group. Within each OLSAT stanine group, the "High" category represents the top 23% of the Stanford scaled score distributions; the "Middle" category, the middle 54% of the Stanford distribution; and the "Low" category, the bottom 23% of the Stanford distribution. Therefore, AAC scores offer a normative comparison of students' achievement in relation to that of others of the same ability, while avoiding the difficulties associated with predicting achievement from ability measures.

See Appendix M for intercorrelations among Stanford 10 and the *Otis-Lennon School Ability Test*®, Eighth Edition (OLSAT® 8).

GROUP PERCENTILE RANKS AND STANINES

Percentile ranks and stanines are also provided for the interpretation of group performance on Stanford 10. Group percentile ranks and stanines are characterized by the same properties as those for the interpretation of individual student performance. Both types of scores indicate relative standing; the group norms provide information about a school group in relation to other similar groups of students in the same grade, taking the test at a comparable time.

It is important that group percentile ranks and stanines not be confused with those derived from the distribution of individual students' scores, since the group scores are far less variable than individual student scores. The larger the group, the smaller the variability. For example, a distribution of average scores for a district is less variable than a distribution of average scores for a school; and a distribution of the average scores of a school is considerably less variable than a distribution of scores of individuals.

Class, building, and district group percentile ranks and stanines are provided for each subtest and total, for use in fall and spring testing programs. These norms were computed from mid-interval cumulative percentage distributions of mean scaled scores earned by groups of students tested during the national standardization programs.

As with individual student percentile ranks and stanines, group percentile ranks and stanines must always be interpreted with reference to the groups from which they were derived. Group percentile ranks can be used to compare performance across subtests but have the same limitations as individual student percentile ranks; i.e., group percentile ranks do not represent equal units of ability. Group stanines, however, do represent approximately equal units of ability and are particularly useful for comparing scores across subtests in a stanine profile and for identifying broad performance categories.

CONTENT CLUSTER AND PROCESS CLUSTER PERFORMANCE CATEGORIES

An analysis of performance on the various Stanford 10 content clusters and subclusters can be useful in identifying students' strengths and weaknesses within a broader content area. An analysis of a student's performance on the process clusters can help in understanding other kinds of strengths and weaknesses. For example, if a student earns a Mathematics Procedures score that is lower than expected, a cluster analysis of this subtest will help determine whether the student's performance was consistently low for all of the mathematics operations or whether difficulties with, e.g., Computation in Context contributed to the low Mathematics Procedures score. Cluster analyses for a class can identify content areas in which the entire class may need instruction.

Stanford 10 provides an analysis of content clusters and process clusters for most subtests in the battery. For each level and form of Stanford 10, frequency distributions of individual raw scores on each content cluster and each process cluster are generated separately by grade, form, and time of year. The raw score interval corresponding to each stanine is then determined. Performance categories are established as Below Average (stanines 1, 2, and 3); Average (stanines 4, 5, and 6); and Above Average (stanines 7, 8, and 9). For individual students, then, a raw score on a particular content cluster or process cluster can be converted to a performance category (see Figure 1 on page 36), which provides a comparison of the student's performance with that of a national sample of students in the same grade taking the same test. For groups of students, the percent of students in each of the performance categories on particular content clusters or process clusters can be determined and used to identify instructional needs.

It should be noted that this approach to the interpretation of performance measured by a norm-referenced test is in contrast to the setting of criterion scores (e.g., "performance levels"). The norm-referenced approach is related to the actual performance of students in the nation rather than to stated expectations for these students.

p-VALUES

An item's *p*-value (i.e., item difficulty) is that percent of students in the reference group (standardization sample) who answered the item correctly. One use of *p*-values is to make comparisons of item difficulty values for a local group with those for the national sample of students. In evaluating the test performance of a school group, for example, the teacher or administrator may find it helpful to compare the group's *p*-values on items in a particular content cluster with the corresponding national *p*-values.

The Stanford 10 *Compendium of Instructional Standards* provides detailed descriptions of the content measured by every item on Stanford. Analysis of these content standards along with a group's *p*-values and the standardization sample's *p*-values can help teachers and administrators better understand their students' performance on Stanford 10. Appendix G presents by level and grade the *p*-values of individual Stanford 10 items for each subtest on forms A/D and B/E. Appendix H presents by level and grade the mean *p*-values for each subtest on forms A, B, D, and E.

PERFORMANCE STANDARDS

Performance standards are criterion-referenced scores that were obtained through the standard-setting procedure described earlier in this report. A Performance Standard represents a level of mastery based on the judgment of teachers and provides information about what students should know and be able to do.

To maintain consistency and comparability of performance standards across Stanford editions, the performance standards established for Stanford 10 were the result of a comprehensive Stanford 9 to Stanford 10 equating study. Thus, performance standards achieved on Stanford 10 may be interpreted as representing the same performance standard that would have been achieved on Stanford 9.

Four levels of performance standards for Stanford 10 are available routinely:

Level 4, Advanced, signifies superior performance beyond grade level mastery. At the high school levels, students achieving at this level show readiness for advanced academic courses, advanced technical training, or career-oriented employment.

Level 3, Proficient, represents solid academic performance, indicating that students are prepared for the next grade. In high school, this level reflects competency in a body of subject-matter knowledge and skills that prepares students for responsible adulthood and productive work.

Level 2, Basic, denotes partial mastery of the knowledge and skills that are fundamental for satisfactory work.

Level 1, Below Basic, indicates little or no mastery of fundamental knowledge and skills.

For those who wish to have only three levels of performance, special arrangements can be made, in which case, Level 1 and Level 2 are combined.

Appendix P presents performance levels corresponding to the scaled scores for Stanford 10 levels and subtests. Appendix Q presents by level and subtest the percentage of students from the 2002 National Standardization Program in each performance level.

Part V: Technical Information on Test Scores

VALIDITY

The current edition of the *Standards for Educational and Psychological Testing* (American Educational Research Association [AERA], American Psychological Association, and National Council on Measurement in Education, 1999) provides the most authoritative consensus of opinion on the meaning of validity in the field:

> "Validity refers to the degree to which evidence and theory support the interpretations of test scores entailed in the use of tests ... [and] is, therefore, the most fundamental consideration in developing and evaluating tests" (p. 9).

Harcourt psychometricians and test development experts are aware of the evolution of the concept of validity and have reviewed the earliest conceptualization from Lindquist (1942), Gulliksen (1950), Cronbach and Meehl (1955), Ebel (1961) through Guion (1980), and all the various editions of the *Standards for Educational and Psychological Testing*. Harcourt has also looked at the movement toward an integrated view of validity (Messick, 1989) and ongoing discussions through the 1990s.

Harcourt's definition of validity is based on the *Standards for Educational and Psychological Testing*. Hence, Harcourt views validity as an integrated and unifying concept as it relates to the development and evaluation of Stanford 10. This view calls for descriptions of validity evidence that in the past would have been characterized as "types" of validity, e.g., content validity, criterion-related validity, or construct validity. This view also calls for evidence based on the consequences of test use and interpretation.

Stanford 10 supports the 24 validity-related standards (1.1 through 1.24) set forth in the *Standards for Educational and Psychological Testing*. Harcourt's judgments about test validity are based primarily on the following sources of evidence of validity:

1. Test content

2. Response processes

3. Internal structure

4. Relationships to other variables

5. Convergent and discriminant analysis

6. Test criterion relationships

7. Consequences of testing

45

According to the *Standards for Educational and Psychological Testing*, evidence of validity can be based on *test content*, i.e., "...an analysis of the relationship between a test's content and the construct it is intended to measure" (p. 11); on *response processes*, i.e., "...the fit between the construct and the detailed nature of performance or response actually engaged in by examinees" (p. 12); on *internal structure*, i.e., "...the degree to which the relationships among test items and test components conform to the construct on which the proposed test score interpretations are based" (p. 13); on *relationships to other variables*, i.e., "Analyses of the relationship of test scores to variables external to the test..." (p. 13); and on *consequences of testing*, i.e., "...the incorporation of the intended and unintended consequences of test use..." (p. 16). The Standards further state "...the validity of an intended interpretation of test scores relies on all the available evidence relevant to the technical quality of a testing system" (p. 17).

Validity of the Abbreviated Form

When forms are being assembled for standardization, the items that will comprise the Abbreviated form are actually selected first. Using item data obtained from field-testing and the test blueprint, items are selected to represent the targeted mean and spread of item difficulty and to ensure that every test objective is covered. After the items for the Abbreviated form are selected, additional items are selected for the full-length form, still keeping the mean and spread of item difficulties within tolerance limits and adding additional items to provide deeper coverage at the objective level. Since the items for the abbreviated form are selected first, it is clear that the abbreviated form is not simply an afterthought but a shorter test measuring the same core of achievement measured by the full-length form.

Probably the key question in the discussion of Stanford 10 Abbreviated form concerns how it relates to norms and technical information developed for the full-length forms. Central to this is the use of item response theory (IRT), specifically the Rasch model, to create the scales and norms upon which Stanford 10 is based.

Stanford 10 full-length and Stanford 10 Abbreviated are both expressed on the same underlying ability scale. Although the relationship of raw score to ability may differ from one test form to another, the relationship of ability (scaled score) to percentile rank is the same. There is in essence a single norm set which applies equally to any Stanford 10 form linked to the underlying Stanford 10 scale.

Thus, any information that pertains to norms for the Stanford 10 full-length test applies equally to Stanford 10 Abbreviated. Because the abbreviated form is a core subset of items on the full-length form, all of the validity information for the full-length form applies equally to the abbreviated form. The only real difference is that since the abbreviated form has fewer items, it does not measure with quite the same precision as the full-length test due to the slightly lower reliability.

Integrating the Validity Evidence

Harcourt has integrated the various types of evidence of validity into a framework that supports the intended interpretation of Stanford 10 test scores for specific uses. This framework encompasses evidence from previous studies as well as new Stanford 10-related studies. As Harcourt continues to study and document evidence of validity, the information may be used to identify areas needing further study, redefine directions for administering tests, or revise constructs.

FAIRNESS

The primary and overarching purpose behind the development of Stanford 10 is to provide an assessment system that accurately represents what a student knows and can do. This is possible only if each and every student to whom Stanford 10 is administered is fully engaged, motivated, and given the opportunity to complete the test to the best of his or her ability.

From making certain that content accurately represents what students are taught, to following procedures, to minimizing sources of bias in test items, to employing the statistical assessment of differential item functioning, Harcourt has diligently applied industry standard procedures to the development of Stanford 10.

PRIMARY 1 SENTENCE READING

One significant departure in Stanford 10 from previous editions involved the Sentence Reading subtest. In previous editions, the Sentence Reading subtest was offered at the SESAT 2 level only, where it assessed students in the second half of kindergarten transitioning from the simpler Word Reading subtest to the more complex Reading Comprehension subtest. In response to requests from customers for an instrument that assessed Sentence Reading in the first-grade also, the Sentence Reading subtest was added to the Primary 1 test level of Stanford 10. This change also allowed Stanford 10 to better assess all students who, under the *No Child Left Behind Act* of 2001 (NCLB), must demonstrate the ability to read by the third grade.

At the first-grade level, the Sentence Reading subtest will prove to be inordinately easy for students with high overall reading ability. This was demonstrated during the standardization of Stanford 10, when a relatively large number of first-grade students earned perfect raw scores on the Sentence Reading subtest. This resulted in lower percentile ranks for all first-grade students on this subtest.

A student's percentile rank is the percentage of students in the norm group who scored below the student's score. Therefore, if a high percentage of students earn perfect scores on a particular subtest, the percentage of students earning below a perfect score is that much lower. If only 85 percent of the students earned less than a perfect score, the percentile rank for the student with the perfect score is only 85. This is considerably lower than a percentile rank in the high 90s, which would be expected for a student with a close to perfect score.

This is not an unusual occurrence and is most common with relatively short subtests (30 items or less) in the early grades. It happens because the growth rate for students at these young ages is very steep. A test that is designed to be appropriate for a particular grade level and time of year—say, the spring of first grade—may be too difficult for an earlier group (fall of first grade) and too easy for a later group (fall of second grade).

As a result, it may not be possible to properly evaluate and rank the higher achieving students based on the results of the Sentence Reading subtest alone. This will be more easily accomplished by using the students' Total Reading score. At the SESAT 1 through Primary 3 test levels, there are at least two Total Reading score points that will yield percentile ranks of 95 or higher. If desired, schools can request Total Reading scores that do not include Sentence Reading.

RELIABILITY AND ERRORS OF MEASUREMENT

The reliability of a test is reflected in evidence of test accuracy, precision and consistency. Numeric indices provide quantified estimates of reliability. These indices may denote the consistency of scores as a form of correlation coefficient (coefficient *alpha*, alternate forms correlations, etc.) or as estimates of the amount of error in a given test score (standard error of measurement).

Test reliability is an essential first condition to support validity but is not sufficient by itself. Reliability data must be used in conjunction with other sources of validity evidence. A test may demonstrate excellent reliability but show little validity for the intended use when other sources of validity evidence are reviewed. A test reporting lower reliability indices may demonstrate superior qualities from other sources of validity evidence. A test is validated by "the degree to which all the accumulated evidence supports the intended interpretation of test scores" (AERA, et al., 1999, p. 11).

Indices of internal consistency and alternate-forms reliability as well as standard errors of measurement for Stanford 10 are explained below.

48

Internal Consistency

The reliability coefficient that demonstrates internal consistency emphasizes the consistency of test performance from item to item. This is accomplished by subdividing a test into portions, typically halves, and correlating the scores from each portion. To overcome the possibility of non-equivalent portions, the Kuder-Richardson Formula 20 (KR20) is used to generate the KR20 reliability coefficient.

$$KR_{20} = \left(\frac{n}{n-1}\right)\left(\frac{SD_t^2 - \sum p_i q_i}{SD_t^2}\right)$$

Where, n =   the number of items in the test,

           SD_t =   the standard deviation of the test scores,

           p_i =   the proportion of correct item responses, and

           q_i =   the proportion of incorrect item responses.

This formula averages all possible half-test correlations. KR20 reliability coefficients are presented in Appendix C for each subtest across every grade level for the full-length and abbreviated formats of Stanford 10. These data were obtained from the Spring and Fall Standardization Samples. The KR20 reliability coefficients show that Stanford 10 is reliable based on a high degree of internal consistency.

Alternate-Forms Reliability

The reliability coefficient that demonstrates alternate-forms reliability compares equivalent forms of a test in a rigorous measure of test precision. Each form of Stanford 10 comprises different but equivalent test content. Each student participating in the Equating of Forms Program was administered both forms of the applicable subtest groupings. (See Part III: The National Research Program for additional information.) By using this type of testing design, the resulting alternate-forms reliability coefficients provide an excellent measure of the equivalency of forms. Alternate-forms reliability coefficients and related summary data for each level and subtest can be found in Appendix E.

Classification Probabilities

The values presented in Appendix R demonstrate the accuracy and consistency with which the performance level boundaries presented in Appendix P classify a student's achievement into a performance level. Classification accuracy is the measure of agreement between the classification obtained from an assessment's results and a hypothetical classification that would be obtained if all possible forms of the assessment could be administered to the student. Classification consistency is the measure of agreement between the classification obtained from an assessment's results and the classification that would be obtained from a different form of the assessment of equivalent level and content. Appendix R presents by assessment level and grade level the accuracy and consistency of the performance level boundaries for each subtest.

Standard Error of Measurement

A standard error of measurement (SEM) provides information regarding the degree to which chance fluctuation in test scores can be expected. An SEM represents inconsistencies occurring in repeated observations of obtained test scores around a student's true test score, which is assumed according to classical test theory to remain constant across repeated measurements of the same trait. For example, an SEM of 3 raw score units means that chance fluctuations within 3 points of the student's "true" test score can be expected roughly two-thirds of the time. SEM values are presented with the reliability coefficient data in Appendixes C and E.

The SEM values reported were calculated using the standard deviation of observed scores and the test reliability coefficient under the assumptions of classical test theory.

$$SEM = s_x\sqrt{1 - r_{xx}}$$

Where,
$\quad$ SEM $\quad$ = $\quad$ standard error of measurement,

$\quad\quad$ s_x $\quad$ = $\quad$ standard deviation of observed scores, and

$\quad\quad$ r_{xx} $\quad$ = $\quad$ test reliability coefficient.

Conditional Standard Errors of Measurement

The standard error of measurement (SEM) value for a particular subtest and test level is not the same at all score levels; it is conditional upon the specific scaled score level. The SEM values at the top and bottom of the scaled score range for a given subtest and test level are typically larger than those near the middle of the range. This means that scores earned by students in the middle of the range are more accurate than scores that are at the high and low ends of the scaled score range. Standard errors of measurement as a function of scaled score for subtests and totals, are presented in Appendix D.

50

Part VI: Testing Special Populations

OVERVIEW

Harcourt is committed to making Stanford 10 as accessible as possible to all students. Harcourt conducted several studies to ensure that Stanford 10 measures the same academic knowledge and skills for all students regardless of gender, disability, ethnicity, or level of English proficiency. Special studies conducted by Harcourt have demonstrated that Stanford 10 scores do not substantially and systematically underestimate or overestimate the knowledge or skills of members of any particular group.

An important feature of Harcourt's test development process is that during the item review process, items are submitted to a Special Populations Advisory Group representing minority and disabled groups who screened the items in terms of appropriateness of content. This group included several prominent national researchers and psychometricians who focused on the issues regarding the assessment of special populations.

In terms of special populations as well as regular test-takers, Harcourt's Stanford assessment products were developed to be inclusive and accountable. These characteristics are the marks of a *universally designed assessment* that is accessible and valid for the widest range of students including students with disabilities and with limited English proficiency.

ACCOMMODATIONS FOR STUDENTS WITH DISABILITIES

Requirements for Inclusion

Requirements for including all students with disabilities (SWD) in assessments stem from a number of federal laws, including Section 504 of the *Rehabilitation Act* of 1973; Title II of the *Americans with Disabilities Act* of 1990 (ADA); Title I of the *Elementary and Secondary Education Act* (ESEA); and the *Individuals with Disabilities Education Act* of 1997 (IDEA). More recently, the *No Child Left Behind Act* of 2001 (NCLB), which amended Title I of the ESEA, requires that students with disabilities be provided accommodations, where appropriate, and if documented on either the Individualized Education Plan (IEP) or Section 504 plan. Since assessment is often associated with direct individual benefits such as promotion and graduation, and is an integral part of accountability systems, Harcourt examined closely the accommodations allowed on Stanford 10.

Defining Accommodations

According to Tindal & Fuchs (1999), "Accommodations are changes in standardized assessment conditions to 'level the playing field' for students by removing the construct-irrelevant variance created by their disabilities. Valid accommodations produce scores for students with disabilities that measure the same attributes as standard assessments measured in non-disabled students" (p. 7).

Harcourt uses the accommodations taxonomy listed below, which was developed by the National Center on Educational Outcomes (NCEO), University of Minnesota. Harcourt combines the categories of Timing and Scheduling, where NCEO keeps them separate.

Timing/Scheduling	Changes to when the assessment is given
Setting/Administration	Changes to where the assessment is given
Presentation Format	Changes to how the assessment is given
Response Format	Changes to how a student responds to the assessment
Other	Use of dictionaries/word lists/glossaries

It is important to note that in the recent evolution of assessment terminology in psychometrics, the use of "accommodations" defines changes in format, response, setting, timing, or scheduling that do not alter in any significant way what the test measures or the comparability of scores (Phillips, 1993). In contrast, "modifications" refers to where changes in the assessment alter what the test is supposed to measure or the comparability of scores (e.g., Braille in some accountability systems) (Phillips, 1993).

Harcourt's Policy on Accommodations for Students with Disabilities

As shown in Table 10, students using accommodations marked under "standard administration" can receive norm-referenced scores that are considered to be valid and can be aggregated with those of other students.

Table 10. Accommodations for Students with Disabilities

Accommodation	Standard Administration	Non-standard Administration
Timing/Scheduling		
• Breaks between subtests	x	
• Time of day most beneficial to student	x	
• Frequent breaks within a subtest	x	
Setting/Administration		
• Test in a small group with Special Ed. teacher	x	
• Test individually with Special Ed. teacher	x	
• Test in regular classroom	x	
• Home/hospital setting	x	
• Environmental modifications: Special lighting, adaptive furniture, noise buffers, carrels, special seating	x	
• Sign language (ASL, cued speech) for directions	x	
Presentation Format		
• Large Print (18 point text)	x	
• Repeating directions	x	
• Simplifying directions	x	
• Visual aids (magnifiers, templates)	x	
• Audio amplification equipment	x	
• Calculator/talking calculator use allowed for Mathematics Problem Solving subtest, grades 4 and up (disable device's programming capability)	x	
• Audio recordings/audio (except decoding and reading comprehension)	x	
• Abacus for visually impaired (VI) students	x	
• Braille		x
Response Format		
• Visual aids (graph paper, templates, rulers)	x	
• Special pencil, pen, pencil grip	x	
• Auditory aids	x	
• Braille		x
Other		
• Augmentative, assistive, or adaptive technology	(contact local DOE)	

Harcourt recognizes that some students with disabilities require the use of accommodations when our assessments are administered. Often, the conditions under which accommodations are used differ from those present when the test was standardized. These differences, in some cases, reach a level sufficient to jeopardize the validity of interpretations. However, based on available evidence, most of the accommodations listed in the table above are considered to be "incidental to the construct intended to be measured by the test" (AERA, et al., 1999, p. 101).

ACCOMMODATIONS FOR LIMITED ENGLISH PROFICIENT (LEP) STUDENTS

Background

Title I of the *Elementary and Secondary Education Act* (ESEA) was amended by the *No Child Left Behind Act* of 2001 (NCLB). Under NCLB, all students are to be included in the measurement of progress toward state achievement standards. In order to evaluate the progress made by schools toward these standards, states must disaggregate and report the performance of limited English proficient (LEP) students, as well as students with disabilities and disadvantaged students. Furthermore, states must compare the performance of these groups to the performance of the general population and report on the findings. States must also disaggregate and report the performance of LEP students within different ethnic groups.

Harcourt's Policy on Accommodations for LEP Students

Harcourt's position is that:

- the use of accommodations for LEP students is a decision that must be made on an individual basis;

- the accommodation to be used should have been properly documented in the district's LEP Plan; and

- the accommodation should have been used in the classroom assessment window or since the original LEP Plan was written, whichever is earlier.

As shown in Table 11, students using accommodations marked under "standard administration" can receive norm-referenced scores that are considered to be valid and can be aggregated with those of other students.

Table 11. Accommodations for Limited English Proficient (LEP) Students

Accommodations	Standard Administration	Nonstandard Administration
Timing/Scheduling		
• Breaks between subtests	x	
• Time of day most beneficial to student	x	
• Frequent breaks within a subtest	x	
Setting/Administration		
• Test in a small group	x	
• Test individually	x	
• Environmental modifications: Location with minimal distractions, preferential seating, noise buffers	x	
Presentation Format		
• Repeating directions	x	
• Simplifying directions	x	
• Calculator use allowed for Mathematics Problem Solving subtest, grades 4 and up (disable device's programming capability)	x	
• Items read aloud to student/audio (except decoding and reading comprehension)	x	
Response Format		
• Visual aids (graph paper, templates, rulers)	x	
• Oral response in native language interpreted by school		x
Other		
• Use of dictionaries (without definitions)	x	
• Use of word lists/glossaries	x	

SPECIAL EDITIONS OF STANFORD 10

Braille Edition

Stanford 10 is available in a Braille edition. Braille tests will differ when compared to the regular-print edition of Stanford 10. Some test items may have been revised, but care was taken to not change the construct being measured. In other instances where responses are critically dependent on art, text, or numbers that cannot be Brailled, items were omitted and Rasch model techniques were then used to provide "adjusted" norms based on the remaining Stanford items.

The Braille materials required a quality review different than that performed for the regular-print materials. All transcribers and proofreaders were certified by the Library of Congress in the codes approved by the Braille Authority of North America (BANA). The Braille editions were produced by the Texas Education Agency's Education Service Center Region 4 located in Houston. The Braille test materials were reviewed twice by certified Braille proofreaders.

Large-Print Editions

Large-print editions of Stanford 10 and OLSAT® 8 are available in black and white and do not differ in content from the regular-print editions. Some graphics have been adjusted based on the recommendation of a panel of experts in consultation with Harcourt assessment specialists.

Answer documents are also available in a large-print format. These materials may be purchased by special arrangement through Harcourt Assessment, Inc.

Editions for the Deaf and Hearing Impaired

Special Stanford 10 materials have been prepared for students who are deaf and hearing impaired. There are two basic components: (1) screening tests for students who are deaf and hearing impaired; and (2) special norms by age for students who are deaf and hearing impaired. The screening tests, which contain reading and mathematics items, are used to identify the proper test level of Stanford 10 to be administered to these students. The Stanford 10 subtests themselves are not changed. The special norming studies were conducted by the Gallaudet Research Institute, Gallaudet University, Washington, D.C., and the Harcourt Assessment Research Group.

Part VII: Test Structure and Content

STANFORD 10 TEST STRUCTURE

Stanford 10 consists of a number of multiple-choice subtests and totals covering a spectrum of content areas and test levels. Table 3 on page 24 presents the scope and sequence of the Stanford 10 Basic and Complete Batteries. Table 4 on page 25 presents the scope and sequence of the Stanford 10 Abbreviated Battery.

Test Levels

The 13 test levels of Stanford 10 are configured as follows:

Stanford Early School Achievement Test (SESAT)

- SESAT 1
- SESAT 2

Stanford Achievement Test

- Primary 1, 2, and 3
- Intermediate 1, 2, and 3
- Advanced 1 and 2

Stanford Test of Academic Skills (TASK)

- TASK 1, 2, and 3

Each test level is recommended for use at specified grade ranges as shown in Tables 3 and 4.

Test Lengths

Stanford 10 subtests are available in full-length and abbreviated formats. The abbreviated format comprises a subset of items selected from the corresponding full-length test. The number of items in each full-length subtest ranges from 30 to 54. The number of items in each abbreviated subtest is either 20 or 30, depending on the subtest. Tables 3 and 4 on pages 24 and 25 show the number of test items included in each full-length and abbreviated subtest and totals across the test levels.

Test Forms

Forms A and D are identical with the exception of the format of the Language test. The Form A Language subtest uses a traditional approach that measures language proficiency through the assessment of language mechanics and expression. The Form D Language subtest uses a comprehensive approach that assesses language proficiency through actual writing processes such as prewriting, composing, and editing.

Form A assesses each of the 13 test levels from SESAT 1 through TASK 3. Form D assesses the Primary 1 through TASK 3 test levels.

Forms B and E are equivalent in content and difficulty to Forms A and D, respectively. Forms B and E contain unique test items and assess the Primary 1 through TASK 3 test levels.

STANFORD 10 CONTENT

One of the most important sources of evidence of validity for a nationally norm-referenced achievement test series is the test content. How well the test items represent the emphasis of subject matter and learning processes taught and assessed in classrooms nationwide is a vital element in judging the validity of test score interpretations. From the beginning of Stanford 10 development, Harcourt staff conducted comprehensive reviews of current national instructional standards, state and local curricula, and widely used textbooks to ensure that Stanford 10 accurately reflected instruction and assessment practices in classrooms today. Harcourt staff also reviewed the literature on criteria for alignment of expectations, standards, and assessments. A list of textbooks and other materials consulted during Stanford 10 development is presented as Appendix A.

The extensive reviews aided Harcourt's assessment specialists with their choices of content to include in Stanford 10. Care was taken to develop test items that assessed content which is important both to teach and to measure. Test items were further focused to reflect the specific constructs of each content domain. This was accomplished by writing test items to mirror instructional processes in classrooms; utilizing vocabulary, to the extent possible, specific to the subject area; and minimizing the amount of reading required to respond to test items in content domains other than reading. The descriptions that follow delineate the constructs of major content domains included in Stanford 10.

The academic content areas assessed by the Stanford series encompass the knowledge and skills taught in the nation's highest-achieving schools. Stanford 10 offers a selection of subtests and totals that assess Total Reading, Total Mathematics, Spelling, Language, Environment, Science, Social Science, Listening to Words and Stories, and Listening.

Total Reading comprises the following different subtests depending on test level:
- Sounds and Letters
- Word Study Skills
- Word Reading
- Sentence Reading
- Reading Vocabulary
- Reading Comprehension

Total Mathematics consists of a single Mathematics subtest at the SESAT and TASK test levels and comprises the following two subtests at the Primary 1 through Advanced 2 test levels:

- Mathematics Problem Solving

- Mathematics Procedures

Stanford 10 Reading

The goal of reading instruction is to help students learn to construct meaning with various texts in a variety of situations. Definitions of content domains are determined to correspond with instructional practice. However, before students can read connected discourse with fluency and good comprehension, they are taught strategies that will help them decode words they do not recognize on sight, thus allowing them to continue building their reading vocabularies. A comprehensive and balanced reading assessment includes measures in all three areas—decoding, vocabulary, and comprehension—at the appropriate grades.

In the elementary school grades, the focus of instruction shifts to comprehension of printed discourse and, to a lesser degree, vocabulary acquisition and strategies while keeping an emphasis on oral language development, basic story structure, conventions of print, decoding, and word recognition that are critical for beginning readers.

The six subtests that assess Total Reading at the various test levels of Stanford 10 are described in more detail below.

Sounds and Letters

Before students can read connected discourse with fluency and good comprehension, they need to develop a variety of prereading skills. The Sounds and Letters subtest measures those important early reading skills that form the basis for constructing meaning with text. These early reading skills are assessed by requiring students to demonstrate:

- Phonological Awareness, the ability to distinguish between auditory likenesses and differences. Students are asked to match two words that begin with the same sounds or that end with the same sounds;

- Orthographic Awareness, the ability to determine the distinctive characteristics of given visual elements; and

- Alphabetic Principles, the ability to recognize letters and match letters with the sounds those letters represent.

Sounds and Letters is assessed at the SESAT 1 and SESAT 2 test levels.

Word Reading

As students begin formal reading instruction, they are expected to apply their newly learned decoding skills to the words already existing in their listening/speaking vocabularies. The Word Reading subtest focuses on appropriate word recognition development, which is assessed by requiring students to demonstrate the ability to identify:

- Printed Word to Spoken Word, or the printed name for a picture of an object after the name has been pronounced;

- Printed Word to Picture, or the printed name for a picture of an object;

- Multiple Printed Words to Picture, or two or more printed words associated with a given picture; and

- Dictated Word to Printed Word, or a printed memorized word that has been pronounced.

Word Reading is assessed at the SESAT 1, SESAT 2, and Primary 1 test levels.

Sentence Reading

The goal of reading instruction is the development of the reading and thinking skills that enable students to comprehend connected discourse. By the middle of kindergarten, simple sentence structures are generally familiar to most students. The Sentence Reading subtest measures students' ability to comprehend single, simple sentences and two related sentences. These sentence skills are assessed as students demonstrate their ability to comprehend:

- Predictable Text, or printed predictable sentences;

- Onset-Rime, or simple printed sentences with decodable onset-rime;

- Simple Sentence, or printed simple sentences; and

- Two Simple Sentences, or two related printed sentences.

Sentence Reading is assessed at the SESAT 2 and Primary 1 test levels.

Word Study Skills

The Word Study Skills subtest continues to measure those important reading skills that form the basis for constructing meaning with text. These early elementary level skills are assessed by requiring students to demonstrate:

- Structural Analysis, or the ability to recognize within words the structural elements required for decoding;

- Phonetic Analysis-Consonant Sounds, or the ability to relate consonant sounds to their most common spellings; and

- Phonetic Analysis-Vowel Sounds, or the ability to relate vowel sounds to their most common spellings.

Word Study Skills is assessed at the Primary 1 through Intermediate 1 test levels.

Reading Vocabulary

Students with extensive vocabulary knowledge and strategies for acquiring and refining new vocabulary tend to be efficient and effective at comprehending text. In a conceptually based reading program, vocabulary is taught by utilizing vocabulary within meaningful contexts to associate words with vivid and memorable concepts rather than rote memorization of definitions.

The Reading Vocabulary subtest focuses on reading vocabulary development appropriate for each test level:

- Synonyms assess a student's general word knowledge. The tested word appears within a sentence, not in isolation. However, the sentence does not supply context clues to the meaning of the word.

- Multiple-Meaning Words are assessed by challenging a student's skill to choose the appropriate meaning of a word for the particular context in which it appears.

- Context Clues require students to use clues supplied by other parts of a sentence to determine the meaning of an unknown word.

Reading Vocabulary is assessed at the Primary 2 through TASK 3 test levels.

Reading Comprehension

The goal of reading instruction is the development of the reading and thinking skills that enable students to comprehend connected discourse. The Reading Comprehension subtest is composed of increasingly complex selections of conceptually appropriate text, each accompanied by questions. The Reading Comprehension subtest aligns with the National Assessment of Educational Progress (NAEP) and assesses students' reading achievement within the framework of three types of text:

- Literary material typically read for enjoyment such as contemporary fiction, folktales, humor, poetry, and historical fiction

- Informational and expository material with content from the natural, physical, and social sciences, as well as other nonfiction general information materials found in grade-appropriate textbooks

- Functional material typically encountered in everyday life, both in and outside school, that is designed to help the reader perform a task, including directions, forms, advertisements, and labels

Many of the Stanford 10 literary and informational passages were written by award-winning authors of children's and young people's literature. The accompanying illustrations, many of which were created by well-known illustrators of children's publications, help the students focus attention, recall and activate prior knowledge, and set purposes for reading. The Stanford 10 Reading Comprehension selections closely resemble the kinds of materials that students read in school and in everyday life. The selections reflect literature-based curricula taught in most classrooms today. The passages include a variety of topics and diverse cultural themes appealing to students of varying backgrounds, experiential levels, and interests.

At the Primary 1 test level, the Reading Comprehension subtest assesses the following content clusters:

- Two-Sentence Stories, where comprehension is demonstrated by identifying the picture described by the story

- Cloze, where comprehension of explicit and implicit information in short reading selections is demonstrated by completing sentences presented in modified cloze format

- Short Passages with Questions, where comprehension of explicit and implicit information in short reading selections is demonstrated by answering questions about the passages

At the Primary 2 test level and above, the Reading Comprehension subtest consists entirely of increasingly complex literary, informational, and functional reading passages appropriate for the test level followed by the number of multiple-choice test items that query the reading selection. Nine reading selections are included in each Reading Comprehension subtest at the Primary 2 through the Advanced 2 test levels (grades 2 through 8). At the TASK 1 through TASK 3 test levels (grades 9 through 11), six selections are included in each subtest.

The Reading Comprehension subtest also assesses the following cognitive processes:

- Initial Understanding, or the ability to comprehend explicitly stated details or relationships in a variety of reading selections

- Interpretation, or the ability to extend meaning, infer relationships, and form interpretations based on explicit and implicit information in the text

- Critical Analysis, or the ability to analyze and evaluate explicit and implicit information and relationships in a variety of reading selections

- Strategies, or the ability to recognize text characteristics and structures and select and apply appropriate reader strategies in a given situation

Stanford 10 Mathematics

When developing the Stanford 10 mathematics subtests, careful attention was paid to the National Council of Teachers of Mathematics (NCTM) *Principles and Standards for School Mathematics* (2000), which emphasizes the necessity of problem solving as the focus of school mathematics.

At each of the Primary 1 (grade 1) through Advanced 2 (grade 8) test levels, two mathematics subtests are included: Mathematics Problem Solving and Mathematics Procedures. A single subtest, Mathematics, is included at the SESAT test levels (early and late kindergarten) and the TASK test levels (grades 9, 10, and 11/12).

The Stanford 10 mathematics subtests go beyond merely assessing skills at a basic level by emphasizing:

- the use of logical and mathematical reasoning;

- the employment of communication skills to recognize alternative forms of equivalent values and alternative representations of data; and

- the implementation of non-routine problem-solving strategies.

Stanford 10 was designed to encourage students to think and to enable them to demonstrate the extent to which their mathematics instructional programs have empowered them. The subtests are also a fair and practical assessment of meaningful, valuable mathematics. Mathematics is most meaningful to students when it is presented in a relevant context. Therefore, many Stanford 10 mathematics items are contextualized, which also can show the connection between mathematics and other content areas.

Using Rulers, Reference Sheets, and Calculators

Students demonstrate actual performance through the use of both customary and metric rulers as they take the Mathematics Problem Solving and Mathematics subtests. Mathematics reference sheets for the Advanced and TASK levels contain commonly used formulas.

Calculator use is an option when the Mathematics Problem Solving and Mathematics subtests are administered, beginning at the Intermediate 1 test level (grade 4) up through the TASK test levels (high school).

In addition to incorporating the use of rulers, reference sheets, and calculators into the assessment, Stanford 10 mathematics subtests reference common classroom instructional practices and tools including thermometers and manipulatives such as base-10 blocks and tangrams.

Content and Format of the Mathematics Subtests

The Stanford 10 mathematics subtests assess the breadth of mathematical content recommended by the NCTM *Principles and Standards* (2000) including number theory, geometry, algebra, data analysis, and probability.

It is not possible to capture which strategy a student uses to select a test item response during the administration of a multiple-choice test. Therefore, each problem in the mathematics subtests was constructed so that one or more of the classic problem-solving strategies would be effective. In addition, regardless of the strategy selected, reasoning skills are required to arrive at a solution. Some test items in both the Mathematics Problem Solving and Mathematics Procedures subtests require or encourage students to guess and check, work backwards, make a list, analyze data, make predictions, and validate conclusions. Other test items are designed to encourage students to draw pictures or to construct diagrams or models in order to find solutions. Also, distracter options are typically based on errors students commonly make.

Mathematics Problem Solving and Mathematics

The items in the Mathematics Problem Solving (grades 1–9) and Mathematics (kindergarten and grades 9–12) subtests assess student proficiency with the fundamental concepts and processes of mathematical problem solving in keeping with the NCTM *Principles and Standards* (2000). Students will encounter situations that encourage them to use virtually every problem-solving strategy appropriate to their grade level. Although some Mathematics Problem Solving and Mathematics test items are easy, many require understanding of concepts and procedures far beyond recall or simple use-of-knowledge levels. These items generally call for careful analysis, synthesis of information, attention to detail, and the selection and use of appropriate strategies.

Stanford 10 distinguishes between mathematical concepts and mathematical vocabulary. For example, a Mathematics Problem Solving and Mathematics test item may require students to recognize the pattern in a factor tree that is associated with a prime number in order to answer the question rather than asking students to simply "select the prime number from the numbers listed below." This example also illustrates assessment of thinking skills as a process required to respond to mathematics test items.

Many of the Mathematics Problem Solving and Mathematics test items are contextualized so that students relate to them and become engaged in the test. In Stanford 10, however, "problem context" is not equivalent to "word problem." Nearly all problems are accompanied by illustrations that clarify, restate, or complete the verbal portion of the item. Reading included in the mathematics items is on or below grade level, and, to the greatest extent possible, Stanford 10 Mathematics Problem Solving and Mathematics test items were developed so that student reading ability would not impact performance.

Mathematics Procedures

The Mathematics Procedures subtest (grades 1–9) stresses the importance of being able to successfully apply the computational procedures of mathematics. Computational proficiency plays a valuable role in establishing the foundations of effective and fluent problem solving (National Council of Teachers of Mathematics, 2000). Because of the computational construct being assessed by this subtest, calculator use is *not* allowed during the administration of the Mathematics Procedures subtest.

The Mathematics Procedures subtest includes some traditional computation items that use symbolic notation rather than a context. However, about half of the items are enhanced by relevant context and engaging artwork. Natural and varied contexts were purposely selected for the computation items. For example, problems with fractions and decimals were integrated within realistic situations or scenarios that students are likely to recognize. In this way, Stanford 10 provides educators with meaningful information that allows a comparison between students' abilities to perform arithmetic algorithms when indicated symbolically and their abilities to *select* and *apply* the appropriate algorithms to a problem presented in a textual context.

Stanford 10 Language

The Stanford 10 Language subtest is available in two different formats—traditional Language and Comprehensive Language. The Form A Language subtest uses a traditional approach that measures word and sentence level skills to whole paper skills in mechanics (capitalization, punctuation, and usage) and expression (sentence structure, pre-writing, and content and organization). The Form D Language subtest uses a comprehensive approach that assesses language proficiency through techniques that support actual instruction, including prewriting, composing, and editing processes.

Stanford 10 Spelling

Stanford 10 assesses Spelling beginning with the Primary 1 test level (grade 1). Like other language arts skills, spelling is best taught and assessed in context. Beginning at the Primary 2 test level, each Spelling item consists of a sentence having three underlined words. Students must recognize and mark the misspelled word. Starting at the Primary 3 test level (grade 3), a fourth "No mistake" option is included. The Spelling items include the most commonly found misspellings in student writing. Because research indicates that seeing spelling errors in print tends to reinforce them, each item contains only one misspelled word rather than one correctly spelled word that students must find among the incorrect ones.

Stanford 10 Environment

The Stanford 10 Environment subtest is a teacher-dictated test that encompasses kindergarten through grade two. Generally, students in these grades have been exposed to the natural and social sciences in informal ways. Students in this age range acquire concepts in these disciplines through an integrated curriculum as well as through their daily activities. The Environment subtest assesses a combination of the natural sciences and the social sciences and mirrors the accepted instructional standards appropriate for children at these ages.

Stanford 10 Science

Science is the search for a verifiable understanding of how the universe works. The Stanford 10 Science subtest reflects the types of activities that are valued in science instruction and assessment. Mirroring the philosophy presented in the *National Science Education Standards* (National Research Council, 1996), the Science subtest de-emphasizes specific vocabulary and emphasizes the unifying themes and concepts of science. The same criteria used in *Benchmarks for Science Literacy* (American Association for the Advancement of Science, 1993) were employed to determine the importance of conceptual information and to aid with inclusion decisions for Stanford 10. Important science content integrates concepts that have strong predictive power for future learning, apply to many situations, guide observations, encourage questioning, and represent organizing principles.

Each science test item supports high standards of student achievement in processing science information. Test items allow students to use reasoning skills and apply understanding to reach answers rather than having to recall memorized, detailed facts and information. Students may be asked to apply an understanding of the concept directly to a situation, interpret data, draw conclusions, predict events, or utilize other skills in tandem with their understanding.

Design of the Science Subtest

The Stanford 10 Science subtest is designed to assess understanding of life sciences, physical sciences, Earth and space sciences, and the nature of science. The test content for the areas of life sciences, physical sciences, and Earth and space sciences targets concepts unique to each; the test content for the nature of science area focuses on the basic skills of science and an understanding of the strengths and limitations of science.

Test items developed within each of the science areas incorporate the cognitive processes required for both learning science and applying it to novel contexts: Basic Understanding and Thinking Skills. Test items written to assess basic understanding reflect the "big ideas" of science, direct translation of data, and terminology that is commonly encountered in newspapers and magazines. Test items assessing thinking skills are designed to allow students to use many skills in concert with their science knowledge. Students who are curious about the world will find many of these items intriguing and thought-provoking. Since students may potentially learn from test items, the information presented is accurate and as realistic as possible. For example, actual experiments are cited, and students may be asked to evaluate the experimental design, data, or conclusions.

Many of the habits of mind outlined in *Science for All Americans* (American Association for the Advancement of Science, 1989) form the foundation of the reasoning skills that students are asked to use throughout the Science subtest. These skills include estimating, making calculations, seeking patterns, making observations, recognizing cause and effect, reading standard instruments, and drawing conclusions.

Most of the science test items are accompanied by graphic images and illustrations, which serve several purposes. First, the use of graphics removes the need for students to complete extensive reading to respond to test items. Text included in the science test items is on or below grade level, and reading ability should minimally impact students' ability to answer a test item. Second, information is provided with the items to reduce the use of recall and recognition. Finally, visual representation of science data has become an important part of daily life and classroom instruction. The ability to read and interpret data correctly is essential for the development of lifelong learners.

Stanford 10 Social Science

Social science education is increasingly steering away from the rote memorization of facts, dates, and names and moving toward the teaching of real-world survival skills that help develop students into informed voters, more knowledgeable participants in the national experience, wiser consumers, thoughtful decision-makers in the political spectrum, and persons more aware of the nature of what society is, where it came from, and where it is going.

The Stanford 10 Social Science subtest addresses major social science themes at appropriate levels. Questions are based on the most current practice and research in social science education, making it the most comprehensive and rigorous social studies test available today. The Stanford 10 Social Science subtest includes grade-appropriate content and aligns with *Expectations of Excellence: Curriculum Standards for Social Studies* (National Council for the Social Studies, 1994) as well as *National Standards for History* (National Center for History in the Schools, 1996), *Geography for Life: National Geography Standards* (National Geographic Research and Exploration, 1994), *Voluntary National Content Standards in Economics* (National Council on Economics Education, 1997), as well as *National Standards for Civics and Government* (U.S. Department of Education, 1994). Stanford 10 test items represent the most current thinking in the content areas of history, geography, political science, and economics.

History

The *National Standards for History* (National Center for History in the Schools, 1996, p. 1) states "knowledge of history is the precondition of political intelligence." Our nation's multicultural background makes it imperative that students develop a common understanding of our democratic heritage. Stanford 10 history test items were developed to assess broad trends, issues, and themes in United States and world history rather than rote memorization of dates and facts.

Geography

Stanford 10 geography test items are based on the five themes of geography (location, place, human/environment interaction, movement, and regions) and the six essential elements outlined in *Geography for Life: National Geography Standards* (National Geographic Research and Exploration, 1994): (1) The World in Spatial Terms; (2) Places and Regions; (3) Physical Systems; (4) Human Systems; (5) Environment and Society; and (6) The Uses of Geography.

Economics

Voluntary National Content Standards in Economics (National Council on Economics Education, 1997, p. V) states that economic understanding is important so that students "will be better-informed workers, consumers and producers, savers and investors, and most important, citizens." These standards provide the foundation for Stanford 10 economic item development. Both macroeconomic and microeconomic concepts are assessed.

Political Science

These test items assess a basic understanding of the United States system of government, with an emphasis on the rights and responsibilities that accompany citizenship and the role of the citizen as a member of a community, a nation, and the world. At the upper grade levels, comparative government systems are also assessed.

Design of the Social Science Subtest

At all test levels, the test items on the Stanford 10 Social Science subtests are drawn from up-to-date and real-life information to measure students' understanding of history, geography, economics, and political science. These content areas are integrated within the test to assess students' understanding of the interrelationships between them. Students must use the skills they will need to succeed in tomorrow's competitive environment. Some of those skills include:

• reading maps to determine cause and effect;

• reading charts to gather, organize, and use information; and

• understanding the meaning of political cartoons.

The Stanford 10 Social Science subtest primarily measures students' thinking skills. Students must use both knowledge and process skills to analyze, evaluate, draw conclusions, make a distinction, or interpret a cause and effect relationship.

Stanford 10 Listening

The Stanford tradition of assessing listening skills continues in Stanford 10. Although listening instruction is not a separate subject in most schools, it is the mode of language

68

by which vocabulary acquisition first occurs, as well as the means by which early learning in all subject areas is acquired. Educators are recognizing the importance of listening in instruction and in everyday life. In recognition of this, Stanford 10 assesses both listening vocabulary and listening comprehension at the SESAT 1 through Advanced 2 test levels.

Design of the Listening Subtest

Listening Vocabulary

The listening vocabulary section comprises sentences that the teacher reads to the class. Each sentence is accompanied by a question about the meaning of one of the words included in the sentence. Students' test booklets display the answer choices, which are also read aloud by the teacher. The vocabulary tested represents a carefully chosen sample of the types of words that students at each grade are likely to encounter in everyday life and in school.

Listening Comprehension

This section of the Listening subtest includes dictated selections that reflect the kinds of real-life listening materials that students encounter both in and out of the classroom. Three types of dictated selections correspond to the types of texts used in the Reading Comprehension subtests:

- Literary selections are stories and poems.

- Informational selections cover topics such as science, history, current events, geography, health, and government.

- Functional selections include step-by-step directions, advertisements, and announcements.

Since younger children are most familiar with stories, a greater emphasis is placed on literary selections at the lower levels. Beginning at the Primary 3 (grade 3) test level, students are encouraged to take notes as the selections are read to them. Many of the informational and functional pieces mirror the types of listening situations these students normally encounter.

As in reading comprehension, listening comprehension measures how well students form an initial understanding, interpret information, and analyze it critically.

Additional Test Content Information

Detailed information about the content and structure of Stanford 10 is presented in the *Compendium of Instructional Standards*, available from Harcourt. The compendium presents the content cluster, process cluster, and cognitive level measured by each Stanford 10 item. Also included in the compendium are the instructional standards measured within each subtest and cluster.

Part VIII: Additional Research

This section presents information concerning additional research related to the development of the Tenth Edition of the Stanford series. Additional studies beyond the scope of the research programs implemented as part of the National Item Tryout Program and the National Research Program are described in this section.

TEST AND ANSWER DOCUMENT DESIGN AND LAYOUT

Harcourt drew from a variety of sources to create and validate its innovative test and answer document design for Stanford 10.

- Reviews were performed of pertinent literature and empirical research conducted around the world. The literature review related to test design and answer document design and layout was limited to the concept of maximum legibility, i.e., the capability of being deciphered with ease.

- Lessons learned from assessment programs conducted in key client states were applied.

- Within these state programs, cognitive labs were conducted to try out modified designs.

- A panel of expert reviewers, including in-house specialists, evaluated the content and graphic design of the test and answer document prototypes. Each test level was reviewed for style, layout, and appropriateness. Care was taken to ensure that the design for Stanford 10 was appropriate for all students in terms of content, format, and administration.

Harcourt utilized focus groups and cognitive labs from 1996 to 2000 to review prototype test and answer document designs. The layout of the tests and answer documents, and the use of white space, borders, and larger print, appeared to assist students in responding to test items. These design features were incorporated into a pilot test conducted in a key client state during May 2000.

The tests and answer document designs were revised based on the results of the pilot test. In the fall of 2000, the revised materials were field tested in Texas. In response to an accompanying survey question about whether the layout of the tests and answer documents was appropriate, 97% of teachers participating agreed.

COLOR BLINDNESS

Color Vision

Color vision is determined by the discrimination of three qualities of color: *hue* (such as red vs. green), *saturation* (pure vs. blended colors), and *brightness* (vibrant vs. dull reflection of light) (Arditi, 1999a). The color blind person is unable to distinguish between hues that appear very different to those with normal color perception. In other words, having a color vision deficit means that the ability to discriminate hue, saturation, and brightness is reduced. To accommodate test users with these deficits, Harcourt has developed assessments with more dramatic color contrast to address each of these three qualities of color.

Research on Color Blindness

To better understand color blindness, Harcourt began a literature review on the normal perception of color, beginning with research related to the psychophysics of reading. Legge, Pelli, Rubin, and Schleske (1985) studied how those with normal vision read. Legge, Rubin, and Luebker (1987) documented the importance of color contrast in normal vision. Other research conducted by Lighthouse International (2000) and the American Printing House for the Blind (D. Willis, personal communication, 2000, and B. Henderson, personal communication, 2001, 2002) was also used to shape the use of effective color contrast for Stanford 10 as well as Harcourt's own reviews of the material.

Harcourt's Review

Harcourt engaged six nationally known experts in the field of visual impairment. All were provided with the background articles referenced in this report, a web designer's color card, a rating instrument, and directions. Each expert was asked to review every item at every level of Stanford 10 for imagery that would have an effect on students with color vision deficiencies. The reviewers were asked to identify issues by item number and to offer suggestions for correcting any potential color problems.

The responses were returned to Harcourt and then summarized and analyzed. Assessment specialists reviewed the recommendations to evaluate any effects the changes might have on content. The most significant color issues identified were the item-number icons for the SESAT, Primary 2, and Advanced 1 level test booklets, which were quickly corrected.

Reviewers had different perceptions on some items in the reading, mathematics, and science domains. Using the Delphi Method, the reviewers were asked to look at the items again and rank ways to correct the items. Three rounds were utilized. Finally, four of the expert reviewers assembled to conduct a final review of the items, some of which had been revised.

As a result of this entire review process, Harcourt is confident that the colors utilized in Stanford 10 meet the needs of the vast majority of students with color deficiencies. For students with visual impairments and color deficiencies, the large-print version is available on non-glare paper in black and white.

STUDIES ON SCALED SCORE FUNCTIONING

Two special studies examined the need to develop additional scaled score information.

First, during the Spring 2002 Standardization Program, a sample of approximately 1,000 third-grade students took the Primary 3 test level using separate answer documents. This study resulted in two separate norms sets: one for students at the Primary 3 test level who marked answers in their test booklets and another for those who used separate answer documents. See the Stanford 10 *Multilevel Spring Norms* and *Multilevel Fall Norms* books for separate norms established for these groups.

Second, Harcourt investigated the possible effects of calculator use on student performance on the Mathematics Problem Solving and Mathematics subtests. Calculator use has over the years become increasingly integrated into mathematics instruction and testing. At the same time, a curricular emphasis on mathematics problem solving over computation has increased. As it relates to standardized testing today, the question is: "Do students who use calculators during testing have a differential advantage over those who do not use calculators?"

To answer this question, a sample of students was chosen from schools in which calculators were used in daily mathematics instruction. During the Spring 2002 Standardization Program, 1,000 students from each of grades 4 through 12 were administered the Mathematics Problem Solving or Mathematics subtest, depending on test level. It is important to note that students participating in all other Stanford 10 National Research Programs did *not* use calculators. Based on analyses conducted on the resulting data, Harcourt concluded that the observed score differences between calculator users and non-users were not significant and did not warrant the development of separate score conversion tables.

Part IX: Document References

American Association for the Advancement of Science. (1993). *Benchmarks for science literacy.* New York, NY: Oxford University Press.

American Association for the Advancement of Science. (1987). *Science for all Americans.* New York, NY: Oxford University Press.

American Educational Research Association (AERA), American Psychological Association (APA), & National Council on Measurement in Education (NCME). (1999). *Standards for educational and psychological testing.* Washington DC: Author.

Americans with Disabilities Act of 1990, 42 U.S.C. 12101 *et seq.*

Arditi, A. (1999a). *Effective color contrast: Designing for people with partial sight and congenital color deficiencies.* New York: Lighthouse International. Retrieved September 21, 2001, from the World Wide Web: http://www.lighthouse.org/color_contrast.htm

Cronbach, L. J., & Meehl, P. E. (1955). Construct validity in psychological tests. *Psychological Bulletin, 52,* 281–302.

Ebel, R.L. (1961). Must all tests be valid? *American Psychologist, 16,* 640–647.

Elementary and Secondary Education Act of 1965, 20 U.S.C. 6301 *et seq.*

Guion, R. M. (1980). On trinitarian doctrines of validity. *Professional Psychology, 11,* 385–398.

Gulliksen, H. (1950). *Theory of mental tests.* New York, NY: Wiley.

Individuals with Disabilities Education Act of 1997. 20 U.S.C. 1400 *et seq.*

Legge, G. E., Pelli, D. G., Rubin, G. S., & Schleske, M. M. (1985). Psychophysics of reading. I. Normal vision. *Vision Research, 25,* 239–252.

Legge, G. E., Rubin, G. S., & Luebker, A. (1987). Psychophysics of reading. V. The role of contrast in normal vision. *Vision Research, 27,* 1165–1171.

Lighthouse International's "Big Type is Good Business" Campaign. (2000).

Lindquist, E. F. (1942). *A first course in statistics: Their use and interpretation in education and psychology* (Rev. ed.). Boston, MA: Houghton Mifflin.

Messick, S. (1989). Meaning and values in test validation: The science and ethics of assessment. *Educational Researcher, 18*(2), 5–11.

National Center for History in the Schools. (1996). *National standards for history.* Los Angeles, CA: University of California.

National Center on Educational Outcomes. (2001). *Crosswalk of Title I and IDEA assessment and accountability provisions for students with disabilities.* Minneapolis, MN: University of Minnesota.

National Council for the Social Studies. (1994). *Expectations of excellence: Curriculum standards for social studies.* Washington, DC: Author.

National Council of Teachers of Mathematics. (2000). *Principles and standards for school mathematics.* Reston, VA: Author.

National Council on Economics Education. (1997). *Voluntary national content standards in economics*. New York, NY: Author.

National Geographic Research and Exploration. (1994). *Geography for life: National geography standards*. Washington DC: Author.

National Research Council. (1996). *National science education standards*. Washington DC: National Academy Press.

No Child Left Behind Act of 2001, Pub. L. No. 107–110, § 1, 115 Stat. 1425 (2002).

Phillips, S.E. (1993). *Legal implications of high-stakes assessments: What states should know*. Oak Brook, IL. North Central Regional Laboratory.

Rehabilitation Act of 1973, 29 U.S.C. 701 *et seq.*

U.S. Department of Education. (1994). *National standards for civics and government*. Washington DC: Author.

Appendix A:

Textbooks and Other Materials Consulted During Stanford 10 Development

Appendix A

Textbooks and Other Materials Consulted During Stanford 10 Development

NATIONAL AND STATE STANDARDS

American Association for the Advancement of Science. (1993). *Benchmarks for science literacy* (Project 2061). New York, NY: Oxford University Press.

Baker, E. L., Freeman, M., & Clayton, S. (1991). Cognitive assessment of history for large-scale testing. In M. C. Wittrock & E. L. Baker (Eds.), *Testing and cognition*, (pp. 131–153). Englewood Cliffs, NJ: Prentice-Hall.

Blank, R. K., Pechman, E. M., & Goldstein, D. (1996). *State mathematics and science standards, frameworks, and student assessments: What is the status of development in the 50 states?* Washington DC: Council of Chief State School Officers.

Board of Education, Commonwealth of Virginia. (1995). *Standards of learning for Virginia public schools*. Richmond, VA: Author.

California Department of Education. (1990). *Academic content standards for California public schools: Kindergarten through grade 12*. Sacramento, CA: Author.

Center for Civic Education. (1994). *National standards for civics and government*. Calabasas, CA: Author.

Consortium for Policy Research in Education. (1991). *Putting the pieces together: Systemic school reform*. (CPRE Policy Briefs). New Brunswick, NJ: State University of New Jersey.

Council of Chief State School Officers. (1997). *Mathematics and science content standards and curriculum frameworks*. Washington DC: Author.

Economics America: National Council on Economic Education. (1997). *Voluntary national content standards*. New York, NY: Author.

Florida Department of Education. (1996). *Florida curriculum*. Tallahassee, FL: Author.

Geography Education Standards Project. (1994). *Geography for life: National geography standards*. Washington DC: Author.

Illinois Academic Standards Project. (1996). *Illinois academic standards for public review and comment: English language arts and mathematics* (Vol. 1, State Goals 1–10). Preliminary draft of unpublished manuscript. Springfield, IL: Author.

Kendall, J. S., Snyder, C., Schintgen, M. A., Wahlquist, A. M., & Marzano, R. J. (1999). *A distillation of subject matter content for the subject areas of language arts, mathematics, and science*. Retrieved from: http://www.mcrel.org

Massachusetts Department of Education. (1999). *Massachusetts curriculum frameworks*. Retrieved from: http://www.doe.mass.edu/frameworks/current.html

McKnight, C., Britton, E. D., Valverde, G. A., & Schmidt, W. H. (1992). *Survey of mathematics and science opportunities: Document analysis manual* (Research report series No. 42). East Lansing, MI: Third International Mathematics and Science Study, Michigan State University.

Michigan State Board of Education. (1996). *Michigan framework for social studies education: Content standards*. Lansing, MI: Author.

National Assessment of Educational Progress Civics Consensus Project. (n.d.). *Civics framework for the 1998 national assessment of educational progress.* Washington DC: National Assessment Governing Board.

National Assessment of Educational Progress Geography Consensus Project. (n.d.). *Geography framework for the 1994 national assessment of educational progress.* Washington DC: National Assessment Governing Board.

National Assessment of Educational Progress Mathematics Consensus Project. (n.d.). *Mathematics framework for the 1996 and 2000 national assessment of educational progress.* Washington DC: National Assessment Governing Board.

National Assessment of Educational Progress Reading Consensus Project. (1990). *Reading assessment framework for the 1992 national assessment of educational progress.* Washington DC: National Assessment Governing Board.

National Assessment of Educational Progress Reading Consensus Project. (1990). *Reading assessment and exercise specifications for the 1992 national assessment of educational progress.* Washington DC: National Assessment Governing Board.

National Assessment of Educational Progress Science Consensus Project. (n.d.). *Science framework for the 1996 national assessment of educational progress.* Washington DC: National Assessment Governing Board.

National Assessment of Educational Progress Science Consensus Project. (1993). *Science assessment and exercise specifications for the 1994 national assessment of educational progress.* Washington DC: National Assessment Governing Board.

National Assessment of Educational Progress U.S. History Consensus Project. (n.d.). *U.S. history framework for the 1994 national assessment of education progress,* Washington DC: National Assessment Governing Board.

National Center for History in the Schools. (1996). *National Standards for History.* Los Angeles, CA: Author.

National Council of Teachers of English and the International Reading Association. (1995, October). *Standards for the English Language Arts.* Draft of unpublished manuscript.

National Council for the Social Studies. (1994). *Expectations of excellence: Curriculum standards for social studies.* Washington DC: Author.

National Council of Teachers of Mathematics. (1989). *Curriculum and evaluation standards for school mathematics.* Reston, VA: Author.

National Council of Teachers of Mathematics. (1991). *Professional standards for teaching mathematics.* Reston, VA: Author.

National Council of Teachers of Mathematics. (2000). *Principles and standards for school mathematics.* Reston, VA: Author.

National Research Council. (1996). *National science education standards.* Washington DC: National Academy Press.

New Standards. (1997). *Performance standards: English language arts, mathematics, science, applied learning* (Vol. 1, Elementary school). Washington DC: National Center on Education and the Economy.

Appendix A

New Standards. (1997). *Performance standards: English language arts, mathematics, science, applied learning* (Vol. 2, Middle School). Washington DC: National Center on Education and the Economy.

New Standards. (1997). *Performance standards: English language arts, mathematics, science, applied learning* (Vol. 3, High School). Washington DC: National Center on Education and the Economy.

Newman, F. M. (1993). Beyond common sense in educational restructuring: The issues of content and linkage. *Educational Researcher, 22*(2), 4–13.

Roeber, E. D. (1996). *Review of the Oregon content and performance standards* (National Standards Review Team Rep.). Salem, OR: Oregon Department of Education.

Schmidt, W. H., & McKnight, C. (1995). Surveying educational opportunity in mathematics and science: An international perspective. *Educational Evaluation and Policy Analysis, 3,* 337–353.

South Carolina Department of Education. (1996). *South Carolina science academic achievement standards*. Draft of unpublished manuscript.

Stein, M. K., Grover, B. W., & Henningsen, M. (1996). Building student capacity for mathematical thinking and reasoning: An analysis of mathematical tasks used in reform classrooms. *American Educational Research Journal, 33*(2), 455–488.

Teachers of English to Speakers of Other Languages, Inc. (TESOL) (1997). *ESL standards for pre-K–12 students*. Alexandria, VA: Author.

Texas Education Agency. (1998). *Texas essential knowledge and skills*. Austin, TX: Author.

U.S. Department of Education. (n.d.). *Mathematics framework for the 1996 and 2000 national assessment of educational progress*. Washington DC: Author.

Webb, N. L. (1997). *Criteria for alignment of expectations and assessments in mathematics and science education* (Research Monograph No. 8). Washington DC: Council of Chief State School Officers.

Webb, N. L. (1997). *Determining alignment of expectations and assessments in mathematics and science education* (Brief Vol. 1, No. 2). Washington DC: National Institute for Special Education.

Webb, N. L. (1999). *Alignment of science and mathematics standards and assessments in four states* (Research Monograph No.18). Washington DC: Council of Chief State School Officers.

MATHEMATICS

Addison-Wesley mathematics (Grades K–2). (1993). Menlo Park, CA: Addison-Wesley.

Advanced math for Christian schools. (1989). Greenville, SC: Bob Jones University Press.

Anytime math, grades K–2. (1995). Orlando, FL: Harcourt Brace.

Assessing higher order thinking in mathematics. (1990). Washington DC: American Association for the Advancement of Science.

Curriculum frameworks for mathematics and science (TIMSS Mongraph No. 1). (1997). Vancouver, Canada: Pacific Educational Press.

Heath geometry (Teacher's ed.). (1998). Evanston, IL: McDougal Littell.

HRW advanced algebra, Texas (Teacher's ed.). (1997). Austin, TX: Holt, Rinehart & Winston.

HRW algebra, Texas (Teacher's ed.). (1997). Austin, TX: Holt, Rinehart & Winston.

HRW geometry, Texas (Teacher's ed.). (1997). Austin, TX: Holt, Rinehart & Winston.

Math advantage, elementary series for grades K–5. (1998). Orlando, FL: Harcourt Brace.

Math advantage, middle school series for grades 6–8. (1998). Orlando, FL: Harcourt Brace.

Mathematics applications and connections, grades 6–8. (1999). New York, NY: Glencoe/McGraw-Hill.

Mathematics, grades 4 and 5. (2001). Parsippany, NJ: Silver Burdett Ginn.

Mathematics plus, grades 4 and 5. (1994). Orlando, FL: Harcourt Brace Jovanovich.

Mathematics unlimited, grade 4. (1991). Orlando, FL: Harcourt Brace Jovanovich.

Middle school math, grades 6–8. (1999). Menlo Park, CA: Scott Foresman.

NCTM handbook of research on mathematics teaching and learning. (1992). New York, NY: Simon & Schuster Macmillan.

Pre-algebra: A transition to algebra. (1992). Menlo Park, CA: Addison-Wesley.

TIMSS mathematics items released set for population 1 (seventh and eighth grades). (n.d.). Amsterdam, The Netherlands: International Association for the Evaluation of Educational Achievement (IEA).

TIMSS mathematics items released set for population 2 (third and fourth grades). (1997). Amsterdam, The Netherlands: International Association for the Evaluation of Educational Achievement (IEA).

TIMSS released item set for the final year of secondary school: Mathematics and science literacy, advanced mathematics, physics. (n.d.). Amsterdam, The Netherlands: International Association for the Evaluation of Educational Achievement (IEA).

SCIENCE

Biology: Principles and exploration. (1998). Austin, TX: Holt, Rinehart & Winston.

Biology: The study of life (4th ed.). (1991). Englewood Cliffs, NJ: Prentice Hall.

Chemistry (4th ed.). (1997). Boston, MA: Zumdahl; Houghton Mifflin.

Conceptual physics (2nd ed.). (1992). Menlo Park, CA: Addison-Wesley.

Earth science. (1994). Austin, TX: Holt, Rinehart & Winston.

Modern chemistry. (1993). Austin, TX: Holt, Rinehart & Winston.

Modern physics. (1992). Austin, TX: Trinklein; Holt, Rinehart & Winston.

Physical science. (1997). New York, NY: Glencoe/McGraw-Hill.

Science. (2000). Orlando, FL: Harcourt School Publishers.

Science. (2000). New York, NY: McGraw-Hill.

Science horizons. (1991). Parsippany, NJ: Silver Burdett and Ginn.

Appendix A

Science plus. (1995). Austin, TX: Holt, Rinehart & Winston.

Science, modules a, b, and c. (2000). Glenview, IL: Scott Foresman.

ENGLISH LANGUAGE ARTS

Afflerbach, P. (2000). *Scott, Foresman reading.* Glenview, IL: Scott Foresman.

Beers, J. W., Cramer; R. L., & Hammond, W. D. (1998). *Everyday spelling.* Glenview, IL: Scott Foresman.

BK English: Communication skills in the new millennium. (2001). Austin, TX: Barrett-Kendall Publishing.

Carroll, J. A., et al. (2001). *Writing and grammar: communication in action.* Upper Saddle River, NJ: Prentice Hall.

Chin, B. A. (2002). *Glencoe literature: The reader's choice.* New York, NY: Glencoe/McGraw-Hill.

Cook, G. E. (1998). *McGraw-Hill spelling.* New York, NY: McGraw-Hill.

Dickson, S., et al. (1998). *Sing, spell, read & write: a total language arts curriculum.* St. Petersburg, FL: International Learning Systems.

DLM early childhood program. (1995). Columbus, OH: SRA/McGraw-Hill.

Elements of literature. Introductory course. (2000). Austin, TX: Holt, Rinehart & Winston.

Farr, R. C.; Strickland, D. S. (2001). *Collections, a Harcourt reading/language arts program.* Orlando, FL: Harcourt.

Farr, R. C., Strickland, D. S., Beck, I. L. (2002). *Harcourt language.* Orlando, FL: Harcourt.

Hasbrouck, J. E. (2001). *McGraw-Hill language arts.* New York, NY: Macmillan/McGraw-Hill.

Houghton Mifflin English. (2001). Boston; Morris Plains, NJ: Houghton Mifflin.

Houghton Mifflin spelling and vocabulary. (1998). Boston, MA: Houghton Mifflin.

Language network: Grammar, writing, communication. (2001). Evanston, IL: McDougal Littell.

The language of literature. (2001). Evanston, IL: McDougal Littell.

Literature and the language arts: Discovering literature. (2001). St. Paul, MN: EMC/Paradigm.

Macmillan/McGraw-Hill reading. (1991). New York, NY: Macmillan/McGraw-Hill.

Moats, L., et al. (1998). *Scholastic spelling.* New York, NY: Scholastic.

Odell, L., & Warriner, J. E. (2001). *Elements of language. Introductory course.* Austin, TX: Holt, Rinehart & Winston.

Prentice Hall literature: Timeless voices, timeless themes. (2000). Upper Saddle River, NJ: Prentice Hall.

Rebecca Sitton's spelling sourcebook series, grades 1–4, 1996–97. Scottsdale, AZ: Egger.

Scholastic early childhood workshop. (1995). New York, NY: Scholastic.

Shurley, B., Wetsell, R. K. (2000). *The Shurley method, English made easy.* Cabot, AR: Shurley Instructional Materials.

Simmons, L., & Calvert, L. (1996) *Saxon phonics K: An incremental development.* Norman, OK: Saxon Publishers.

Spell It—Write!. (1998). Columbus, OH: Zaner-Bloser.

SRA journeys. (2000). Columbus, OH: SRA/McGraw-Hill.

SRA open court reading. (2000). Columbus, OH: SRA/McGraw-Hill.

Tomorrow's promise spelling. (1997). San Diego, CA: Jostens Learning.

Touchphonics: The manipulative multi-sensory phonics system. (1995). Newport Beach, CA: Touchphonics Reading Systems.

Writer's choice: Grammar and composition. (1996). New York, NY: Glencoe/McGraw-Hill.

SOCIAL SCIENCE

Adventures in time and place. (1997). New York, NY: Macmillan/McGraw-Hill.

The Americans. (1998). Evanston, IL: McDougal Littell.

The American nation. (1998). Upper Saddle River, NJ: Prentice Hall.

American pageant. (1998). Evanston, IL: McDougal Littell.

America will be. (1994). Boston, MA: Houghton Mifflin.

Civics: Responsibilities and citizenship. (1996). New York, NY: Glencoe.

Geography: People and places in a changing world. (1995). St. Paul, MN: West Publishing.

Geography: the world and its people. (1996). New York, NY: Glencoe/McGraw-Hill.

Harcourt Brace social studies. (1997). Orlando, FL: Harcourt Brace.

Holt economics. (1999). Austin, TX: Holt, Rinehart & Winston.

Magruder's American government. (1998). Evanston, IL: McDougal Littell.

Prentice Hall world geography: A global perspective. (1995). Englewood Cliffs, NJ: Prentice Hall.

Regions and resources. (1997). Parsippany, NJ: Silver Burdett Ginn.

The story of America. (1992). Austin, TX: Holt, Rinehart & Winston.

World history: People and nations. (1993). Austin, TX: Holt, Rinehart & Winston.

World history: Perspectives on the past. (1990). Lexington, Mass: DC Heath.

World history: The human experience. (1999). New York, NY: Glencoe/McGraw-Hill.

SPECIAL POPULATION TESTING ACCOMMODATIONS

Accommodations (In-house study). (1999). San Antonio, TX: Harcourt Educational Measurement.

Case, B. (2003). *Accommodations on Stanford 10 for limited English proficient (LEP) students.* San Antonio, TX: Harcourt Educational Measurement.

Case, B. (2003). *Accommodations on Stanford 10 for students with disabilities.* San Antonio, TX: Harcourt Educational Measurement.

Appendix A

Gallaudet Research Institute. (1996). *Norms booklet for deaf and hard-of-hearing students.* Washington DC: Gallaudet University.

Messick, S. (1989). Meaning and values in test validation: The science and ethics of assessment. *Educational Researcher, 18*(2), 5–11.

Norms booklet for Braille edition. (1996). Louisville, KY: American Printing House for the Blind.

Phillips, S.E. (1993), High-stakes testing accommodations: Validity versus disabled rights. *Applied Measurement in Education, 7*(2), 93–120.

Salvia, J., & Ysseldyke, J. E. (2003). *Assessment* (9th ed.). Boston, MA: Houghton Mifflin.

Special groups—Comparative analysis (In-house study). (2000). San Antonio, TX: Harcourt Educational Measurement.

Thurlow, M., Quenemoen, R., Thompson, S., & Lehr, C. (2001). *Principles and characteristics of inclusive assessment and accountability systems* (Synthesis Report No. 40). Minneapolis, MN: University of Minnesota, National Center on Educational Outcomes.

Tindal, G., Fuchs, L. (1999). *A summary of research on tests changes: An empirical basis for defining accommodations.* Lexington, KY: Mid-South Regional Resource Center.

Tindal, G., & Haladyna, T. M. (Eds.) (2002). *Large-scale assessment programs for all students: Validity, technical adequacy, and implementation.* Mahwah, NJ: Lawrence Erlbaum Associates.

Tindal, G., Heath, B., Hollenback, K., Almond, P., & Harness, M. (1998). Accommodating students with disabilities on large-scale tests: An empirical study of student response and test administration demands. *Exceptional Children, 64*(4), 439–450.

Appendix B:

List of School Districts Participating in the Stanford 10 Spring and Fall Standardization Programs

Appendix B

List of School Districts Participating in the Stanford 10 Spring and Fall Standardization Programs

Midwest

Illinois

Armstrong Township High School District 225	F
Buncombe Consolidated School District 43	F
Burnham School District 154-155	F
Cambridge Community Unit School District 227	F
Carrollton Community Unit School District 1	Sp-F
Christian Fellowship School	Sp
Community Consolidated School District 168	Sp
Iroquois County Community Unit School District 9	Sp
Iroquois West Community Consolidated School District 10	F
Maroa-Forsyth Community Unit School District 2	Sp
Nelson Public School District 8	Sp
Odell Community School District	F
Pleasant Hill Community Unit School District 3	F
Prairieview Elementary School District	F
Round Lake Area School District 116	Sp
Saint Libory Consolidated School District 30	F
Woodlawn Community Consolidated School District 4	Sp
Zion Lutheran School	Sp

Indiana

Baptist Academy	Sp
Barr-Reeve Community Schools, Inc.	Sp
Chesterton Montessori School	Sp
Lakeland Christian School	F
North Miami Community School	Sp
Northside Montessori School	Sp
Saint John's Lutheran School	Sp
Saint Peter Immanuel Lutheran School	F
Spencer-Owen Community School District	Sp
Taylor Community School Corporation	Sp
Trinity Lutheran School	Sp

Iowa

BCLUW Community School District	F
Central Iowa Christian Academy	Sp
Clinton Community School District	Sp-F
Grundy Center Community School District	Sp
Mount Olive Lutheran School	Sp
Rock Valley Christian School	Sp
Tri-County Community School District	Sp

Kansas

Abilene Baptist Academy	F
Lawrence Unified School District 4	Sp
Saint John's Military	F
Tonganoxie Unified District 46	F

Michigan

Akron-Fairgrove School District	Sp-F
Alba School District	Sp
Baptist Park School	Sp
Barry County Christian School	Sp
Bethany Christian School	Sp
Bethesda Christian School	F
Capitol City Baptist School	Sp
Cedar Crest Academy	F
Cross Lutheran School	Sp
Deford Christian Academy	Sp
Diocese of Lansing Schools	Sp
Faithway Christian Academy	Sp
Family Altar Christian School	Sp
Free Soil Community School District	Sp
Genesee Christian Academy	F
Grand Rapids Public School District	F
Grass Lake Community School District	Sp
Heritage Christian Academy	Sp
Heritage School	Sp
Immanuel Lutheran School	Sp
Inkster Public School District	Sp
Ionia Nazarene Christian School	Sp
Ionia Public School District	Sp
Kaleva Norman Dickson School District	F
Lincoln Park Public School District	Sp
Macomb Christian School	Sp
Marysville Public School District	F
Melvindale-North Allen Park School District	Sp
Mount Calvary Lutheran School	Sp
Mount Clemens Montessori School	Sp
Mount Hope Lutheran School	F
New Buffalo Area School District	Sp
Otsego Baptist Academy	Sp
Saint John Lutheran School	Sp
Saint Paul Lutheran School	Sp
Saint Stephen Lutheran School	Sp
Saint Thomas Lutheran School	Sp
Skeels Northern Christian School	Sp
Sterling Christian School	Sp
The Montessori School	Sp
Tri County Area School District	Sp
Trinity Lutheran School	Sp
United Christian School	Sp
Wesleyan Christian Academy	Sp
West Highland Christian Academy	F
White Pigeon Community School District	Sp

Minnesota

Christian Life School	Sp
Cross of Christ Lutheran School	Sp
East Central Independent School District	Sp
Fulda Independent School District 505	F
Lester Prairie Independent School District	Sp
Litchfield Independent School District	F
Lynd School District 415	F

List of School Districts Participating in the Stanford 10 Spring and Fall Standardization Programs (continued)

Shakopee Independent School District	Sp	Minnewaukan School District 5	F
Victory Christian Academy	F	Taylor School District 3	Sp
Woodcrest Baptist Academy	F	Twin Buttes School District 17	Sp
		United School District 7	Sp-F
Missouri			
Adair Consolidated School District R-2	F	*Ohio*	
Clarksburg C2 School District	F	Akron City School District	Sp
Crawford County School District	Sp	Austintown Local School	Sp
Diocese of Jefferson Schools	Sp	Central Baptist School	F
El Dorado Springs School District	Sp	East Cleveland City School District	Sp
Gorin School District R3	F	East Muskingum Local School District	Sp-F
Grain Valley School District R-5	F	Edon-Northwest Local School District	F
Hardin-Central School District C-2	Sp	Heath City School District	Sp
Hermitage School District R-4	F	Huron City School District	Sp
Kearney School District R-1	Sp	Mentor Christian School	Sp
Keytesville School District R-3	Sp	Miamisburg City School District	Sp
Laclede County C-5 School District	Sp	Monroeville Local School District	Sp
Lexington School District R-5	F	North College Hill City School District	F
Liberal School District R-2	Sp	Put-In-Bay Local School District	Sp
Lutie School District R-6	F	South Point Local School District	Sp
Macks Creek School District R-5	Sp		
Meadville School District R-4	F	*South Dakota*	
North Daviess School District R-3	F	Bristol School District 18-1	Sp-F
Northwestern R-1 School District	Sp	Geddes Community School District 11-2	Sp
Osborn School District R-0	F	Haakon School District 27-1	F
Perry County School District 32	Sp	Marion School District 60-3	Sp
Phelps County School District R-3	F	McCook Central School District 43-7	Sp
Sarcoxie School District R-2	Sp	Memorial Christian School	Sp
Sherwood School District R-8	Sp	Montrose School District 43-2	Sp
Tri City Christian School	Sp	Pierre School District 32-2	Sp
		Plankinton School District 1-1	Sp
Nebraska		Platte Christian School	Sp
Bingham Public School District	Sp	Rosholt School District 54-4	Sp
Cambridge Public Schools	Sp	Stanley County School District 57-1	F
Centura Public School District 100	Sp	Stickney School District 1-2	Sp
Cherry County School District	Sp	White Lake School District 1-3	Sp
Christ Lutheran School	Sp	Wood School District 47-2	Sp
Dawson County School District 16	F		
Diocese of Lincoln Schools	F	*Wisconsin*	
Holt County School District 10	Sp	Diocese of Madison Schools	Sp
Holt County School District 60	F	Diocese of La Crosse Schools	Sp
Kearney Public School District	Sp	Elm Grove Lutheran School	Sp
Nebraska Christian School	F	Faith Community Christian School	Sp
Plattsmouth Community School District 1	F	Fontana Joint School District 8	Sp
Trinity Lutheran School	Sp	Hales Corners Lutheran School	Sp
Wynot Public School District	Sp	Ladysmith-Hawkins School District	Sp
		Maranatha Academy	F
North Dakota		Northeast Wisconsin Lutheran School	Sp
Bakker School District 10	Sp	Oakhill Christian School	F
Dodge School District 8	Sp	Pilgrim Lutheran School	F
Edinburg School District	F	Saint John's Lutheran School	Sp-F
Eureka School District 106	Sp-F	Utica Christian School	F
Golden Valley School District 20	Sp		

Appendix B

Northeast

Connecticut

Colebrook School District	F
Derby Public School District	Sp
Thomaston Public School District	Sp

Maine

Diocese of Portland Schools	F
Eastport-Union School District 104	Sp
Eggemoggin Christian Academy	Sp-F
Greater Portland Christian School	F
Maine School Administrative District #32	F
Maine School Administrative District #53	F
Maine School Administrative District #59	F

Maryland

Bethel Christian Academy	Sp
Carroll Christian Academy	Sp
Chestertown Christian Academy	F
Grace Bible Baptist Church School	Sp
Saint John's Christian School	Sp-F
Saint Paul's Lutheran School	Sp
The Tome School	Sp

Massachusetts

Baptist Temple Christian Academy	Sp
Bellingham School District	F
Springfield School District	F
Wachusett Hills Christian School	F
Webster School District	F

New Hampshire

Calvary Christian School	Sp
Hampstead Academy	Sp-F
Landaff School District SAU #3	F
Stratham School District SAU #16	Sp
Wakefield School District	Sp

New Jersey

Cheder School	Sp
Delaware Township School District	F
Diocese of Metuchen Schools	Sp
King's Kids Academy	Sp
Lakewood Cheder Girls School	F
Little Ferry School District	F
Metuchen Christian Academy	Sp
Park Bible Academy	F
Somerdale Public School District	Sp
Stafford Township School District	F
Trocki Hebrew Academy	Sp
Wall Township School District	F

New York

Beth Jacob Elementary School	Sp

Faith Fellowship Christian School	F
Highland Elementary School	Sp
Interlaken Christian School	F
Letchworth Central School District	F
New York City Community School District	F
Raquette Lake Union Free School District	F
Saint John the Evangelist School	Sp
Saint Mark's Lutheran School	Sp
Saint Peter's Lutheran School	Sp
West Seneca Christian School	Sp

Pennsylvania

Allegheny Valley School District	Sp
Butler Area School District	Sp
Calvary Temple Christian Academy	Sp
Cambria County Christian School	Sp-F
Career Connections Charter School	F
Cedar Grove Christian Academy	Sp
Central Cambria School District	Sp
Crusaders for Christ Academy	Sp
Diocese of Allentown Schools	Sp-F
Faith Christian Academy	F
Faith Christian School	Sp
Gateway School District	Sp
Gettysburg Area School District	Sp
Girard Alliance Christian Academy	Sp
Heritage Christian School	Sp
High Point Baptist Academy	F
Living Word Academy	Sp
Midwest School District	F
Mohawk Area School District	F
Penn Cambria School District	Sp
Salisbury-Elk Lick School District	F
Smethport Area School District	F
South Hills Christian School	F
Summit Baptist Academy	Sp
Sunbury Christian Academy	Sp
Zion Academy	Sp

Rhode Island

Barrington Town School District	Sp
Bristol Warren Region School District	Sp
Glocester School District	F

Vermont

Austine School for the Deaf	Sp
Diocese of Burlington Schools	F
Roxbury School District	Sp-F
Waterville School District	Sp

Virginia

Augusta County School District	Sp-F
Calvary Christian School	Sp

List of School Districts Participating in the Stanford 10 Spring and Fall Standardization Programs (continued)

Holloman's Child Development	Sp	Leslie Public School District 23	Sp
Norfolk Christian School Center	Sp	Lockesburg School District 16	Sp
Norfolk Christian Middle School	Sp	Lynn Public School District	Sp
Northside Christian School	Sp	Magnolia Public School District	Sp
Poquoson City Public Schools	F	Manila School District	Sp
Ryan Academy of Norfolk	F	Mansfield School District	F
Westover Christian Academy	Sp	Marmaduke School District	Sp
		Mayflower Public School District	Sp
West Virginia		Nettleton School District	Sp
Braxton County District School	Sp-F	Norphlet Public School District	F
Clay County School District	F	Ouachita Hills Academy	F
Diocese of Wheeling-Charleston Schools	F	Ouachita School District	F
Grant County School District	Sp	Ozark Public School District	Sp
Heritage Christian Academy	F	Palestine Wheatley School District	Sp
Morgantown Christian Academy	F	Piggott School District 52	Sp
Pocahontas County School District	F	Prescott Public School District	Sp
Roane County School District	Sp	Quitman School District 21	F
		Rural Special School District	Sp
		Saint Joe Public School District	Sp
Southeast		Smackover School District	Sp
Alabama		Southside School District 2	Sp
Albertville City School District	Sp	Taylor School District	Sp
Archdiocese of Mobile Schools	Sp	Walnut Ridge School District	Sp
Calvary Baptist Academy	F	Williford School District 39	Sp
Dothan City School District	Sp-F		
East Alabama Christian Academy	F	*Florida*	
Florence Christian Academy	Sp	Beaches Chapel School	Sp
Gadsden City School District	F	Duval County School District	F
Garywood Christian School	F	Faith Christian Academy	F
Greene County School District	Sp	First Christian Church Day School	Sp
Guntersville City School District	Sp	Fort Lauderdale Preparatory School	Sp
Huntsville Achievement School	Sp	Grace Lutheran School	F
Jacksonville City School District	Sp	Harvest Time Christian School	F
Magnolia Springs Christian School	Sp	Heritage Christian School	Sp
Piedmont City School District	Sp	Heritage Preparatory	Sp
The Country Day School	Sp	Jefferson County School District	Sp
The Lakeside School	Sp	Maitland Montessori School	Sp
		Martin County School District	Sp
Arkansas		North Florida Christian School	F
Alpena Public School District	Sp	Orange County Public Schools	Sp
Bay School District 29	Sp	Our Savior Lutheran School	Sp
Bearden Public School District	F	Palm Beach County School District	Sp-F
Beebe Public School District	Sp	Panama City Christian School	F
DeSoto School	F	Park Avenue Baptist School	F
DeValls Bluff School District	Sp	Princeton Christian School	F
East Poinsett County School District	Sp	Santa Rosa Christian School	F
Faith Christian School	Sp	Sumter County School District	Sp
Fayetteville School District	Sp-F	Victory Christian Academy	Sp
Fouke School District 15	Sp-F		
Fourche Valley School District 13	Sp	*Georgia*	
Genoa Central School District 1	Sp	Appling Christian Academy	Sp
Gurdon Public School District	Sp	Baldwin County School District	F
Holly Grove Public School District	Sp	Barrow County School District	Sp
Lee County School District 1	Sp	Catoosa County School District	Sp

Appendix B

List of School Districts Participating in the Stanford 10 Spring and Fall Standardization Programs (continued)

Christian Lutheran School	Sp	Simpson County School District	Sp
Curtis Baptist School	Sp-F	The Indianola Academy	Sp
Forsyth County School District	Sp	Tri-County Academy	Sp
Glennville Christian Academy	Sp-F	Union County School District	Sp
Maranatha Christian Academy	Sp	University Christian School	Sp
Meriwether County School District	Sp	Winston Academy	Sp
Pathway Christian Academy	F		
Peach County School District	Sp-F	*North Carolina*	
Shiloh Hills Christian School	Sp	Alamance Christian School	Sp
Southland Academy	Sp	Gaston Christian School	Sp
Twiggs County School District	Sp	Grace Christian School	F
		Myrtle Grove Christian School	Sp
Kentucky		Pender County School District	F
Bethany Christian School	Sp	Victory Christian Center School	Sp
Eliahu Academy	Sp		
Garrard County School District	Sp	*South Carolina*	
Jackson Independent School District	F	Bob Jones Elementary School	F
Landmark Christian Academy	Sp	Jefferson Davis Academy	Sp
Lewis County School	Sp	Laurens County School District 55	Sp
Painstville Independent School District	Sp	Wilson Hall School	Sp
Pikeville Independent School District	Sp		
Robertson County School District	F	*Tennessee*	
		Bornblum Solomon School	F
Louisiana		Central Baptist School	F
Avoyelles Parish School District	Sp	Coffee County School District	Sp
Boutee Christian Academy	Sp	Covington City School District	Sp-F
Catahoula Parish School District	Sp	Dickson County School District	Sp-F
Community Christian Academy	Sp	Haywood County School District	Sp
Diocese of Houma-Thibodaux Schools	F	Hollow Rock Bruceton School District	Sp-F
Diocese of Lake Charles Schools	Sp	Lewis County School District	Sp
Faith Christian Academy	Sp	Pioneer Christian Academy	Sp
John Curtis Christian School	Sp	Pleasant View Christian School	F
Lake Castle Private School	Sp	Putman County School District	Sp
Ouachita Parish School District	F	Rhea County School District	Sp
Trafton Academy	Sp	Robertson County School District	Sp-F
		Thrifthaven Baptist Academy	Sp
Mississippi		Tullahoma City School District	Sp
Alcorn County School District	Sp	Van Buren County School District	Sp
Calvary Christian School	Sp	Volunteer Christian Academy	F
Durant Public School District	Sp		
Emmanuel Christian School	Sp	**West**	
Greenville Christian School	Sp	*Alaska*	
Leake County School District	Sp	Far North Christian School	F
Marshall Academy	Sp		
North Panola Consolidated School District	Sp	*Arizona*	
North Pike School District	Sp	Avondale Elementary School District 44	F
Noxubee County School District	Sp-F	Bicentennial Union High School District	F
Ocean Springs School District	Sp	Buckeye Elementary School District 33	F
Parklane Academy	Sp-F	Canon School District 50	Sp-F
Pontotoc County School District	Sp	Cave Creek Unified School District	Sp
Prentiss County School District	Sp-F	Coolidge Unified School District	F
Presbyterian Day School	Sp	Crane School District 13	F
Quitman County School District	Sp	Diocese of Tucson Schools	Sp

List of School Districts Participating in the Stanford 10 Spring and Fall Standardization Programs (continued)

Diocese of Phoenix Schools	Sp	Redemption Academy	Sp
Gila Bend Unified School District 24	F		
Golf Links Christian Academy	F	*Idaho*	
Lake Havasu Unified School District 1	Sp	Kimberly School District 414	Sp-F
Laveen Elementary School District 59	F	Marsing Joint School District 363	F
Mammoth-San Manuel Unified School District 8	F		
Miami Area Unified School District 40	Sp	*Montana*	
Naco School District 23	Sp	Auchard Creek School District 27	Sp
Oracle Elementary School District 2	F	Birch Creek Colony School	Sp
Paradise Valley Unified District 69	Sp-F	Bonner School District 14	F
Scottsdale Unified School District 48	F	Havre Christian School	Sp
Shiloh Christian School	F	Hutley Project School District 24	Sp
Tolleson Elementary School District 17	Sp-F	Paradise School District 8	F
Topock School District 12	Sp	Raynesford School District 49	Sp
Tri-City Christian Academy	Sp	Springdale School District 63-56	Sp
Unity School of Creative Learning	Sp	Swan Valley School District 33	Sp
William K. Eaton School	Sp	Trinity Lutheran School	Sp
		Wolf Creek School District 13	Sp
California			
Buena Vista Elementary School District	F	*Nevada*	
Cornerstone Academy	F	Diocese of Reno Schools	Sp
Corning Union Elementary School District	Sp	Paradise Christian Academy	Sp
Country Day Montessori School	F	Ruby Mountain Christian School	Sp
Fort Sage Unified School District	Sp		
Franklin McKinley School District	F	*New Mexico*	
Irvine Unified School District	F	Valley Christian Academy	Sp
Krouzian Zekarian Elementary/Middle School	F		
Linns Valley-Poso Flat Union School District	F	*Oklahoma*	
Los Angeles Unified School District	Sp	Aline Cleo Independent School District 4	F
Southern Kern Unified School District	F	Allen Bowden School District 35	F
New Hope Elementary School District	Sp-F	Anadarko School District I 20	Sp
Saint George's Academy	F	Anderson School District 52	F
Saint Stephen's Episcopal School	F	Berryhill Independent School District 10	Sp
San Lucas Union Elementary School District	Sp	Bishop School District 49	Sp
South Bay Lutheran High School	F	Boone-Apache Independent School District	Sp
Sunnyside Union Elementary School District	Sp	Bowring School District 7	Sp
Tracy Joint Unified School District	F	Butner Independent School District 15	Sp-F
Vallejo City Unified School District	Sp	Byng-Francis Elementary School District	Sp
Weimar Academy	F	Calumet Independent School District 76	Sp
Western Avenue Baptist School	F	Cameron Independent School District 17	Sp-F
Westside Elementary School District	Sp	Chelsea Independent School District 3	Sp
		Chisholm Public Schools	Sp
Colorado		Christian Heritage Academy	Sp
Adams County District No. 1	Sp	Coalgate Independent School District 1	Sp
Bethlehem Lutheran School	Sp	Coleman Independent School District 35	Sp
Boulder Country Day	Sp	Collinsville School District 6	F
Faith Baptist School	Sp	Crescent Independent School District 2	Sp
Humanex Academy	Sp	Crutcho School District 74	Sp
Miami/Yoder School District 60 JT	Sp	Cyril Independent School District 64	Sp
Riverview Christian School	F	Diocese of Tulsa Schools	Sp
Trinity Lutheran School	Sp	Drummond Independent School District 85	Sp
		Duke Independent School District 14	Sp
Hawaii		Duncan Independent School District I-1	Sp
Navy Hale Keiki School	Sp	Emmanuel Christian School	F

Appendix B

Flower Mound School District 48	Sp	Diocese of Tyler Schools	F
Geronimo Independent School District 4	Sp	Frankston Independent School District	Sp
Guthrie Independent School District 1	Sp	Glen Oaks Elementary School	Sp
Gypsy School District 12	F	Grace Lutheran School	Sp
Healdton Independent School District 55	F	Hidalgo Independent School District	Sp
Heavener Independent School District I-3	Sp	Ingram Independent School District	F
Hulbert Independent School District 16	Sp	Lake Dallas Independent School District	F
Kansas Public Schools	Sp	Lometa Independent School District	Sp
Ketchum Independent School District 6	F	Lydia Patterson Institute	Sp
Laverne Independent School District I-1	Sp	Maud Independent School District	Sp
Leach School District 14	F	Memorial Lutheran School	Sp
Lindsay Independent School District 9	Sp	Mount Enterprise Independent School District	Sp
Lukfata School District 9	Sp	Nederland Independent School District	F
Macomb Independent School District 4	Sp	New Summerfield School District	Sp
Marlow Independent School District 3	F	Odessa Christian School	Sp
Mid Del Christian School	F	Pearland Independent School District	Sp
Millwood Independent School District 37	Sp	Pharr-San Juan-Alamo Indep. School District	Sp
Owasso Independent School District 11	F	Rains Independent School District	Sp
Peavine School District 19	Sp	Ranger Independent School District	Sp
Pleasant Grove School District 5	Sp	Redeemer Lutheran School	F
Porter Consolidated	Sp-F	S & S Consolidated Independent School District	Sp
Ripley Independent School District I-3	Sp	Saint Andrew's School	Sp
Ryal School District 3	Sp	Saint James Day School	Sp
Sandy Point School District 4	Sp	Saint Martin's Lutheran Day School	Sp
Schulter Independent School District 6	F	San Antonio Christian Schools	F
Sequoyah Independent School District 6	Sp-F	Spring Hill Independent School District	F
Taloga Independent School District 10	Sp	Strawn Independent School District	F
Trinity Christian School	F	Throckmorton Independent School District	Sp
Westville Independent School District 11	Sp	Trinity Lutheran School	Sp
Zion School District 28	Sp	Utopia Independent School District	F
		Uvalde Consolidated Independent School District	F
Oregon		Valley Grande Academy	Sp
Faith Bible Christian School	Sp	Valley View Independent School District	Sp
Kingsview Christian School	F	Vega Independent School District	Sp
Milo Adventist Academy	Sp	Veribest Independent School District	F
Portland Christian Elementary School	Sp	Wells Independent School District	Sp
Portland Lutheran School	Sp	White Oak Independent School District	F
		Wolfe City Independent School District	F
Texas			
Albany Independent School District	Sp	*Utah*	
Arp Independent School District	Sp-F	Berean Baptist Christian Academy	F
Beeville Independent School District	Sp	Children's Classic	Sp
Bethel Temple Christian School	Sp	Intermountain Christian School	F
Boerne Independent School District	F	Ogden City School District	Sp
Bonham Independent School District	Sp	Saint Paul Lutheran School	Sp
Burleson Adventist School	Sp	Wayne Community School District	Sp
Central Baptist Academy	Sp		
Dallas Independent School District	F	*Washington*	
Diocese of Dallas Schools	Sp	Buena Vista SDA School	F
Diocese of Fort Worth Schools	Sp	North Seattle Christian School	Sp
Diocese of Galveston-Houston Schools	Sp	Raymond School District 116	F
Diocese of Victoria Schools	Sp	Riverside School District 416	Sp
Diocese of Corpus Christi Schools	Sp	Saint Matthew Lutheran School	Sp
Diocese of El Paso Schools	Sp	Tonasket School District 404	Sp

Appendix C:

Kuder-Richardson Formula 20 (KR20) Reliability Coefficients, Standard Errors of Measurement, and Related Summary Data for the Full-Length and Abbreviated Batteries for the Stanford 10 Standardization Samples

Appendix C

Table C–1.   KR20 Reliability Coefficients, Standard Errors of Measurement, and
             Related Summary Data for SESAT 2, Spring, Grade K, Full-Length*

Subtest	Number of Items	Form A				
		N	Mean	SD	SEM	KR20
Total Reading	100	2082	68.7	16.3	4.01	0.94
Sounds and Letters	40	2176	33.3	5.7	2.09	0.87
Word Reading	30	2337	18.0	6.3	2.40	0.85
Sentence Reading	30	2356	16.6	6.8	2.33	0.88
Mathematics	40	2404	29.2	7.1	2.36	0.89
Environment	40	2411	28.1	5.3	2.52	0.77
Listening to Words and Stories	40	2402	24.2	6.8	2.71	0.84

* In terms of raw scores

Table C–2. KR20 Reliability Coefficients, Standard Errors of Measurement, and Related Summary Data for Primary 1, Spring, Grade 1, Full-Length*

Subtest	Number of Items	Form A					Form B				
		N	Mean	SD	SEM	KR20	N	Mean	SD	SEM	KR20
Total Reading	130	3320	100.2	21.8	4.03	0.97	3258	99.5	22.6	4.10	0.97
Word Study Skills	30	3493	21.1	5.2	2.02	0.85	3443	21.3	4.8	2.06	0.81
Word Reading	30	3527	23.8	6.6	1.84	0.92	3486	23.4	6.7	1.90	0.92
Sentence Reading	30	3481	25.1	5.1	1.74	0.89	3397	25.1	5.8	1.70	0.92
Reading Comprehension	40	3392	29.4	7.8	2.34	0.91	3330	29.2	7.8	2.37	0.91
Total Mathematics	72	3525	46.9	12.4	3.55	0.92	3500	46.3	12.7	3.55	0.92
Mathematics Problem Solving	42	3543	29.2	7.6	2.62	0.88	3546	29.1	7.5	2.62	0.88
Mathematics Procedures	30	3555	17.6	5.9	2.35	0.84	3518	17.1	6.3	2.35	0.86
Language	40	2422	25.2	7.4	2.76	0.86	3013	25.0	7.7	2.75	0.87
Spelling	36	3547	24.5	7.9	2.36	0.91	3543	25.0	7.6	2.34	0.90
Environment	40	3515	30.1	5.2	2.44	0.78	3519	29.7	4.7	2.39	0.75
Listening	40	3402	25.8	7.4	2.72	0.87	3459	25.7	7.2	2.73	0.86

Subtest	Number of Items	Form D					Form E				
		N	Mean	SD	SEM	KR20	N	Mean	SD	SEM	KR20
Language	40	1380	27.2	6.5	2.61	0.84	1214	24.6	6.8	2.74	0.84

* In terms of raw scores

Table C–3.   KR20 Reliability Coefficients, Standard Errors of Measurement, and Related Summary Data for Primary 2, Spring, Grade 2, Full-Length*

Subtest	Number of Items	Form A					Form B				
		N	Mean	SD	SEM	KR20	N	Mean	SD	SEM	KR20
Total Reading	100	3443	70.5	17.6	3.84	0.95	2972	70.6	17.9	3.83	0.95
Word Study Skills	30	3542	21.5	4.6	1.94	0.82	3042	21.5	4.7	1.94	0.83
Reading Vocabulary	30	3527	22.5	6.7	1.98	0.91	3031	22.4	6.3	2.00	0.90
Reading Comprehension	40	3558	26.2	8.3	2.56	0.91	3057	26.6	8.7	2.55	0.91
Total Mathematics	74	3546	51.1	12.8	3.47	0.93	3060	53.0	12.8	3.37	0.93
Mathematics Problem Solving	44	3568	29.8	7.9	2.70	0.88	3075	31.0	8.0	2.60	0.89
Mathematics Procedures	30	3579	21.4	6.0	2.14	0.87	3082	22.0	5.8	2.09	0.87
Language	48	2768	36.5	8.3	2.61	0.90	2075	35.4	8.5	2.75	0.90
Spelling	36	3595	28.5	6.7	2.14	0.90	3069	29.1	6.7	2.06	0.91
Environment	40	3564	29.6	4.8	2.50	0.73	3008	27.9	5.2	2.54	0.76
Listening	40	3555	25.4	6.8	2.68	0.85	2994	27.5	6.6	2.62	0.84

Subtest	Number of Items	Form D					Form E				
		N	Mean	SD	SEM	KR20	N	Mean	SD	SEM	KR20
Language	40	1102	28.1	5.9	2.59	0.81	1304	28.7	6.6	2.58	0.85

* In terms of raw scores

94

Table C–4. KR20 Reliability Coefficients, Standard Errors of Measurement, and Related Summary Data for Primary 3, Spring, Grade 3, Full-Length*

Subtest	Number of Items	Form A					Form B				
		N	Mean	SD	SEM	KR20	N	Mean	SD	SEM	KR20
Total Reading	114	2110	76.0	21.0	4.38	0.96	2395	75.5	20.7	4.40	0.96
Word Study Skills	30	2147	20.4	5.5	2.15	0.85	2424	20.2	5.6	2.23	0.84
Reading Vocabulary	30	2148	20.2	6.3	2.24	0.87	2433	21.3	6.1	2.15	0.88
Reading Comprehension	54	2160	35.1	11.2	3.03	0.93	2455	33.9	11.1	3.06	0.92
Total Mathematics	76	2161	48.9	15.4	3.63	0.94	2454	49.9	14.4	3.59	0.94
Mathematics Problem Solving	46	2170	30.5	9.5	2.80	0.91	2463	32.3	8.7	2.68	0.91
Mathematics Procedures	30	2177	18.3	7.0	2.25	0.90	2473	17.6	6.8	2.33	0.88
Language	48	1805	31.4	9.8	2.89	0.91	2062	30.7	9.2	2.90	0.90
Spelling	38	2159	21.3	7.7	2.69	0.88	2458	22.3	7.5	2.68	0.87
Science	40	2158	24.8	7.1	2.69	0.86	2445	25.1	7.1	2.70	0.86
Social Science	40	2147	23.7	8.3	2.69	0.89	2426	27.0	7.8	2.57	0.89
Listening	40	2141	27.3	6.7	2.65	0.84	2417	26.1	6.4	2.68	0.82

Subtest	Number of Items	Form D					Form E				
		N	Mean	SD	SEM	KR20	N	Mean	SD	SEM	KR20
Language	45	669	24.3	8.6	2.98	0.88	638	23.5	8.1	3.02	0.86
Prewriting	14	673	8.0	2.9	1.63	0.69	646	7.7	2.9	1.66	0.67
Composing	16	670	7.8	3.6	1.81	0.74	644	7.5	3.2	1.83	0.66
Editing	15	671	8.5	3.2	1.71	0.71	640	8.1	3.2	1.71	0.72

* In terms of raw scores

Appendix C

Table C–5.   KR20 Reliability Coefficients, Standard Errors of Measurement, and Related Summary Data for Intermediate 1, Spring, Grade 4, Full-Length*

Subtest	Number of Items	Form A					Form B				
		N	Mean	SD	SEM	KR20	N	Mean	SD	SEM	KR20
Total Reading	114	2876	72.1	21.3	4.49	0.96	3518	73.5	22.8	4.44	0.96
Word Study Skills	30	2939	19.2	6.4	2.21	0.88	3575	19.2	6.5	2.26	0.88
Reading Vocabulary	30	2922	20.6	6.6	2.21	0.89	3564	21.2	7.1	2.11	0.91
Reading Comprehension	54	2963	31.8	10.7	3.15	0.91	3593	32.7	11.6	3.10	0.93
Total Mathematics	80	2947	47.3	15.6	3.82	0.94	3567	49.7	16.8	3.74	0.95
Mathematics Problem Solving	48	2971	27.7	9.3	2.93	0.90	3591	30.0	10.0	2.88	0.92
Mathematics Procedures	32	2961	19.7	7.3	2.39	0.89	3585	19.6	7.8	2.34	0.91
Language	48	2488	30.1	9.8	2.97	0.91	2228	29.4	10.3	2.96	0.92
Language Mechanics	24	2488	14.9	5.0	2.09	0.83	2228	15.1	5.3	2.04	0.85
Language Expression	24	2488	15.2	5.5	2.08	0.86	2228	14.3	5.7	2.11	0.86
Spelling	40	2911	23.4	8.0	2.72	0.89	3527	23.9	8.3	2.68	0.90
Science	40	2888	24.7	7.3	2.76	0.86	3506	23.9	7.8	2.70	0.88
Social Science	40	2879	23.1	8.3	2.77	0.89	3506	24.7	8.8	2.67	0.91
Listening	40	2875	24.3	6.5	2.76	0.82	3486	24.1	7.1	2.72	0.85

Subtest	Number of Items	Form D					Form E				
		N	Mean	SD	SEM	KR20	N	Mean	SD	SEM	KR20
Language	48	334	26.2	8.7	3.11	0.87	1218	28.7	9.3	3.03	0.89
Prewriting	12	336	7.1	2.5	1.54	0.62	1222	8.0	2.8	1.42	0.75
Composing	18	336	9.4	4.0	1.89	0.78	1219	10.0	3.7	1.93	0.72
Editing	18	336	9.6	3.6	1.89	0.73	1223	10.8	3.8	1.84	0.76

* In terms of raw scores

Table C–6. KR20 Reliability Coefficients, Standard Errors of Measurement, and Related Summary Data for Intermediate 2 Spring, Grade 5, Full-Length*

Subtest	Number of Items	Form A					Form B				
		N	Mean	SD	SEM	KR20	N	Mean	SD	SEM	KR20
Total Reading	84	3055	53.1	15.7	3.89	0.94	2704	54.6	16.2	3.85	0.94
Reading Vocabulary	30	3087	21.5	5.9	2.19	0.86	2707	21.1	6.0	2.21	0.86
Reading Comprehension	54	3132	31.4	10.7	3.18	0.91	2751	33.2	11.2	3.12	0.92
Total Mathematics	80	3129	46.3	15.3	3.89	0.94	2718	43.1	14.9	3.90	0.93
Mathematics Problem Solving	48	3141	27.9	9.6	3.01	0.90	2737	24.9	9.0	3.06	0.89
Mathematics Procedures	32	3153	18.3	6.8	2.41	0.88	2728	18.2	6.9	2.38	0.88
Language	48	2332	31.9	9.3	2.88	0.90	2324	29.2	9.5	3.00	0.90
Language Mechanics	24	2332	16.6	4.6	1.97	0.82	2324	15.2	4.9	2.05	0.82
Language Expression	24	2332	15.2	5.3	2.07	0.85	2324	14.0	5.3	2.17	0.84
Spelling	40	3005	25.3	8.1	2.67	0.89	2692	26.2	7.5	2.67	0.88
Science	40	2992	21.7	7.0	2.85	0.84	2678	22.6	7.5	2.79	0.86
Social Science	40	2975	23.3	8.1	2.75	0.88	2671	21.8	7.7	2.79	0.87
Listening	40	2977	24.7	6.3	2.74	0.81	2676	24.4	6.4	2.77	0.81

Subtest	Number of Items	Form D					Form E				
		N	Mean	SD	SEM	KR20	N	Mean	SD	SEM	KR20
Language	48	611	23.1	9.1	3.09	0.88	342	30.3	8.9	2.96	0.89
Prewriting	12	613	6.8	2.6	1.54	0.66	343	8.9	2.5	1.33	0.71
Composing	18	614	7.9	3.8	1.89	0.76	343	10.5	3.6	1.87	0.73
Editing	18	615	8.4	3.6	1.89	0.73	342	10.9	3.8	1.87	0.76

* In terms of raw scores

97

Table C–7. KR20 Reliability Coefficients, Standard Errors of Measurement, and Related Summary Data for Intermediate 3, Spring, Grade 6, Full-Length*

Subtest	Number of Items	Form A					Form B				
		N	Mean	SD	SEM	KR20	N	Mean	SD	SEM	KR20
Total Reading	84	3020	57.3	15.1	3.74	0.94	3472	54.5	15.5	3.81	0.94
Reading Vocabulary	30	3048	22.3	5.3	2.15	0.84	3491	21.7	5.7	2.15	0.86
Reading Comprehension	54	3053	34.9	10.7	3.03	0.92	3500	32.8	10.7	3.12	0.92
Total Mathematics	80	3046	43.4	16.3	3.93	0.94	3484	42.9	15.6	3.95	0.94
Mathematics Problem Solving	48	3072	25.6	9.9	3.06	0.91	3507	25.4	9.5	3.10	0.89
Mathematics Procedures	32	3053	17.7	7.4	2.42	0.89	3504	17.4	7.1	2.40	0.88
Language	48	2620	32.5	9.0	2.87	0.90	2468	32.5	9.5	2.87	0.91
Language Mechanics	24	2620	16.4	4.4	2.00	0.79	2468	16.5	4.7	2.00	0.82
Language Expression	24	2620	16.1	5.2	2.03	0.85	2468	16.0	5.4	2.04	0.86
Spelling	40	3059	26.2	7.6	2.65	0.88	3487	25.0	7.9	2.68	0.89
Science	40	3056	25.1	6.4	2.75	0.82	3482	23.3	6.8	2.83	0.83
Social Science	40	3046	22.9	7.4	2.78	0.86	3483	21.9	7.8	2.80	0.87
Listening	40	3022	25.7	6.7	2.71	0.84	3468	25.2	7.0	2.68	0.85

Subtest	Number of Items	Form D					Form E				
		N	Mean	SD	SEM	KR20	N	Mean	SD	SEM	KR20
Language	48	338	29.5	8.8	3.02	0.88	973	25.1	9.8	3.06	0.90
Prewriting	12	340	7.7	2.4	1.48	0.62	974	6.9	2.8	1.50	0.71
Composing	18	342	10.3	3.9	1.89	0.77	975	8.6	4.0	1.90	0.78
Editing	18	340	11.4	3.8	1.81	0.77	975	9.6	4.0	1.85	0.79

* In terms of raw scores

Table C–8. KR20 Reliability Coefficients, Standard Errors of Measurement, and Related Summary Data for Advanced 1, Spring, Grade 7, Full-Length*

Subtest	Number of Items	Form A					Form B				
		N	Mean	SD	SEM	KR20	N	Mean	SD	SEM	KR20
Total Reading	84	2307	51.7	17.1	3.86	0.95	2697	54.1	16.2	3.81	0.95
Reading Vocabulary	30	2316	19.8	6.9	2.22	0.90	2710	20.3	6.3	2.19	0.88
Reading Comprehension	54	2345	31.7	11.0	3.13	0.92	2760	33.6	10.8	3.09	0.92
Total Mathematics	80	2320	38.6	16.7	3.92	0.95	2696	41.2	16.6	3.93	0.94
Mathematics Problem Solving	48	2348	22.6	9.6	3.07	0.90	2751	24.8	10.1	3.04	0.91
Mathematics Procedures	32	2331	15.9	8.0	2.39	0.91	2724	16.2	7.7	2.42	0.90
Language	48	1640	31.1	10.3	2.89	0.92	1884	30.0	10.5	2.95	0.92
Language Mechanics	24	1640	15.9	4.9	2.02	0.83	1883	15.7	5.2	2.04	0.85
Language Expression	24	1640	15.1	5.9	2.04	0.88	1884	14.4	5.9	2.10	0.87
Spelling	40	2332	24.5	7.4	2.71	0.87	2753	23.8	7.2	2.71	0.86
Science	40	2337	23.0	7.5	2.78	0.86	2648	22.9	7.1	2.77	0.85
Social Science	40	2310	19.9	7.5	2.87	0.85	2646	19.7	7.2	2.87	0.84
Listening	40	2208	25.1	6.9	2.74	0.84	2623	25.0	7.2	2.73	0.86

Subtest	Number of Items	Form D					Form E				
		N	Mean	SD	SEM	KR20	N	Mean	SD	SEM	KR20
Language	48	582	23.8	9.6	3.08	0.90	696	27.1	9.0	3.08	0.88
Prewriting	12	582	6.3	2.7	1.50	0.70	698	6.5	2.3	1.54	0.54
Composing	18	585	8.5	3.9	1.91	0.76	707	10.3	3.8	1.90	0.75
Editing	18	585	8.9	3.9	1.89	0.77	707	10.0	4.2	1.84	0.81

* In terms of raw scores

Table C–9. KR20 Reliability Coefficients, Standard Errors of Measurement, and Related Summary Data for Advanced 2, Spring, Grade 8, Full-Length*

Subtest	Number of Items	Form A					Form B				
		N	Mean	SD	SEM	KR20	N	Mean	SD	SEM	KR20
Total Reading	84	1924	54.7	16.2	3.86	0.94	1752	52.9	16.2	3.86	0.94
Reading Vocabulary	30	1935	21.1	6.0	2.20	0.87	1771	20.7	5.9	2.21	0.86
Reading Comprehension	54	1937	33.5	11.1	3.14	0.92	1904	32.4	10.9	3.15	0.92
Total Mathematics	80	1824	37.1	15.9	4.00	0.94	1848	37.8	16.9	3.95	0.95
Mathematics Problem Solving	48	1855	22.2	9.4	3.10	0.89	1905	22.2	10.0	3.07	0.91
Mathematics Procedures	32	1849	14.8	7.4	2.48	0.89	1868	15.3	7.8	2.43	0.90
Language	48	1626	30.7	10.2	2.91	0.92	1047	31.6	10.0	2.84	0.92
Language Mechanics	24	1626	15.8	4.8	2.04	0.82	1047	16.4	5.0	1.97	0.85
Language Expression	24	1626	14.9	6.1	2.04	0.89	1047	15.2	5.5	2.03	0.87
Spelling	40	1927	25.0	7.1	2.66	0.86	1750	25.4	7.4	2.62	0.88
Science	40	1949	20.7	6.9	2.86	0.83	1880	20.2	6.8	2.86	0.82
Social Science	40	1872	18.6	7.6	2.85	0.86	1864	18.2	7.7	2.86	0.86
Listening	40	1816	24.0	6.4	2.75	0.82	1698	22.8	7.3	2.82	0.85

Subtest	Number of Items	Form D					Form E				
		N	Mean	SD	SEM	KR20	N	Mean	SD	SEM	KR20
Language	48	276	34.0	8.3	2.76	0.89	675	27.7	9.7	3.02	0.90
Prewriting	12	276	9.7	2.3	1.22	0.71	675	8.0	2.6	1.45	0.70
Composing	18	276	12.0	3.6	1.76	0.76	675	9.2	3.6	1.91	0.72
Editing	18	276	12.2	3.4	1.73	0.73	675	10.5	4.6	1.81	0.85

* In terms of raw scores

Table C–10. KR20 Reliability Coefficients, Standard Errors of Measurement, and Related Summary Data for Advanced 2, Spring, Grade 9, Full-Length*

Subtest	Number of Items	Form A					Form B			
		N	Mean	SD	SEM	KR20	N	Mean	SD	SEM	KR20
Total Reading	84	1498	59.2	14.8	3.72	0.94	1826	56.8	15.4	3.75	0.94
Reading Vocabulary	30	1498	22.0	6.3	2.11	0.89	1854	21.9	6.1	2.09	0.88
Reading Comprehension	54	1515	37.0	10.1	3.00	0.91	1853	34.5	10.7	3.07	0.92
Total Mathematics	80	1470	43.6	16.2	3.93	0.94	1828	39.7	17.4	3.93	0.95
Mathematics Problem Solving	48	1495	26.1	10.0	3.02	0.91	1861	23.6	10.6	3.04	0.92
Mathematics Procedures	32	1481	17.4	7.4	2.45	0.89	1848	16.0	7.9	2.42	0.91
Language	48	881	33.6	10.0	2.75	0.92	1088	31.8	10.8	2.79	0.93
Language Mechanics	24	881	16.8	4.8	1.95	0.83	1088	16.1	5.5	1.96	0.87
Language Expression	24	881	16.8	5.8	1.91	0.89	1088	15.7	5.8	1.97	0.88
Spelling	40	1515	26.8	7.0	2.54	0.87	1865	25.4	7.3	2.66	0.87
Science	40	1500	22.4	7.1	2.79	0.85	1862	21.6	7.3	2.83	0.85
Social Science	40	1499	20.5	8.5	2.81	0.89	1851	19.2	8.0	2.84	0.87
Listening	40	1418	25.9	6.4	2.68	0.83	1814	24.3	7.2	2.77	0.85

Subtest	Number of Items	Form D					Form E			
		N	Mean	SD	SEM	KR20	N	Mean	SD	SEM	KR20
Language	48	634	30.7	10.4	2.87	0.92	754	29.6	9.3	2.96	0.90
Prewriting	12	634	8.6	3.0	1.33	0.80	757	8.7	2.5	1.36	0.71
Composing	18	638	10.9	4.2	1.80	0.82	757	9.9	3.5	1.91	0.71
Editing	18	638	11.2	4.2	1.76	0.82	754	11.0	4.5	1.77	0.84

* In terms of raw scores

Table C-11. KR20 Reliability Coefficients, Standard Errors of Measurement, and Related Summary Data for TASK 1, Spring, Grade 9, Full-Length*

Subtest	Number of Items	Form A					Form B				
		N	Mean	SD	SEM	KR20	N	Mean	SD	SEM	KR20
Total Reading	84	1715	54.3	14.8	3.82	0.93	1116	56.8	15.8	3.68	0.95
Reading Vocabulary	30	1721	21.6	5.7	2.12	0.86	1116	20.6	5.6	2.19	0.85
Reading Comprehension	54	1721	32.6	10.1	3.15	0.90	1121	36.2	11.0	2.93	0.93
Mathematics	50	1733	21.8	8.2	3.12	0.86	1092	22.8	8.6	3.14	0.87
Language	48	913	30.1	10.5	2.87	0.93	672	31.4	11.2	2.77	0.94
Language Mechanics	24	913	15.1	4.9	2.04	0.83	672	16.1	5.2	1.94	0.86
Language Expression	24	913	15.0	6.2	1.99	0.90	672	15.3	6.7	1.93	0.92
Spelling	40	1711	24.6	7.6	2.67	0.88	1087	23.9	7.6	2.73	0.87
Science	40	1689	20.8	6.7	2.80	0.82	1104	20.4	7.4	2.82	0.85
Social Science	40	1681	18.4	6.9	2.91	0.82	1120	18.2	7.2	2.86	0.84

Subtest	Number of Items	Form D					Form E				
		N	Mean	SD	SEM	KR20	N	Mean	SD	SEM	KR20
Language	48	774	31.7	8.7	2.91	0.89	419	27.1	10.2	3.01	0.91
Prewriting	12	777	7.9	2.6	1.47	0.68	419	7.3	2.4	1.50	0.62
Composing	18	779	12.4	3.9	1.72	0.80	421	10.0	4.5	1.83	0.84
Editing	18	776	11.3	3.4	1.80	0.73	421	9.8	4.1	1.85	0.80

* In terms of raw scores

Table C–12. KR20 Reliability Coefficients, Standard Errors of Measurement, and Related Summary Data for TASK 2, Spring, Grade 10, Full-Length*

Subtest	Number of Items	Form A					Form B				
		N	Mean	SD	SEM	KR20	N	Mean	SD	SEM	KR20
Total Reading	84	2426	58.0	14.1	3.69	0.93	2729	54.5	15.2	3.77	0.94
Reading Vocabulary	30	2451	22.5	5.0	2.04	0.84	2736	21.7	5.3	2.11	0.84
Reading Comprehension	54	2443	35.5	10.2	3.03	0.91	2744	32.7	11.0	3.10	0.92
Mathematics	50	2337	22.6	8.5	3.16	0.86	2670	20.6	8.3	3.15	0.86
Language	48	1767	30.4	9.5	2.94	0.90	1830	31.2	9.8	2.88	0.91
Language Mechanics	24	1767	14.3	4.7	2.11	0.80	1830	15.4	4.8	2.03	0.82
Language Expression	24	1767	16.1	5.4	2.03	0.86	1830	15.8	5.6	2.01	0.87
Spelling	40	2457	25.7	7.8	2.63	0.89	2678	25.5	7.5	2.63	0.88
Science	40	2343	22.2	7.3	2.79	0.86	2616	20.0	6.9	2.84	0.83
Social Science	40	2339	17.1	6.3	2.90	0.79	2612	16.0	6.2	2.89	0.78

Subtest	Number of Items	Form D					Form E				
		N	Mean	SD	SEM	KR20	N	Mean	SD	SEM	KR20
Language	48	657	31.5	8.9	2.88	0.90	790	27.6	9.0	3.00	0.89
Prewriting	12	659	8.6	2.4	1.38	0.67	792	7.9	2.6	1.40	0.70
Composing	18	657	11.3	3.6	1.80	0.75	791	9.6	3.6	1.89	0.72
Editing	18	659	11.5	3.8	1.77	0.79	793	10.1	4.0	1.84	0.79

* In terms of raw scores

Table C–13. KR20 Reliability Coefficients, Standard Errors of Measurement, and Related Summary Data for TASK 3, Spring, Grade 11, Full-Length*

Subtest	Number of Items	Form A					Form B				
		N	Mean	SD	SEM	KR20	N	Mean	SD	SEM	KR20
Total Reading	84	1992	53.2	18.0	3.77	0.96	1650	53.6	17.5	3.77	0.95
Reading Vocabulary	30	1996	21.2	6.9	2.05	0.91	1663	21.1	6.0	2.14	0.87
Reading Comprehension	54	1996	31.9	12.1	3.12	0.93	1651	32.5	12.6	3.06	0.94
Mathematics	50	1956	20.5	8.5	3.20	0.86	1626	18.0	7.5	3.14	0.82
Language	48	1384	27.4	11.2	2.95	0.93	894	26.0	10.2	3.05	0.91
Language Mechanics	24	1384	13.4	5.4	2.09	0.85	894	12.7	4.8	2.16	0.80
Language Expression	24	1381	14.0	6.3	2.05	0.90	894	13.3	6.1	2.10	0.88
Spelling	40	1939	23.9	8.8	2.61	0.91	1588	24.6	8.5	2.65	0.90
Science	40	1952	17.9	6.9	2.84	0.83	1564	17.9	6.8	2.83	0.83
Social Science	40	1962	15.3	6.5	2.86	0.81	1569	14.2	6.0	2.84	0.77

Subtest	Number of Items	Form D					Form E				
		N	Mean	SD	SEM	KR20	N	Mean	SD	SEM	KR20
Language	48	509	27.9	10.8	2.93	0.93	641	30.1	10.0	2.92	0.92
Prewriting	12	511	7.6	3.0	1.40	0.79	643	7.7	2.7	1.44	0.73
Composing	18	512	10.1	4.3	1.83	0.82	645	11.3	4.3	1.78	0.83
Editing	18	510	10.1	4.3	1.80	0.82	643	11.0	3.9	1.80	0.79

* In terms of raw scores

Table C-14. KR20 Reliability Coefficients, Standard Errors of Measurement, and Related Summary Data for TASK 3, Spring, Grade 12, Full-Length*

Subtest	Number of Items	Form A					Form B				
		N	Mean	SD	SEM	KR20	N	Mean	SD	SEM	KR20
Total Reading	84	2029	52.4	18.0	3.78	0.96	1297	49.1	20.3	3.80	0.97
Reading Vocabulary	30	2054	21.8	6.6	2.05	0.90	1299	19.4	7.3	2.18	0.91
Reading Comprehension	54	2046	30.5	12.7	3.11	0.94	1302	29.7	13.9	3.06	0.95
Mathematics	50	2045	19.5	8.2	3.19	0.85	1282	16.9	7.9	3.11	0.85
Language	48	1460	26.4	11.6	2.96	0.94	697	25.1	11.2	2.99	0.93
Language Mechanics	24	1459	13.2	5.6	2.11	0.86	697	12.0	5.4	2.15	0.84
Language Expression	24	1460	13.2	6.7	2.04	0.91	697	13.1	6.4	2.06	0.90
Spelling	40	2052	24.7	8.9	2.58	0.92	1226	23.1	9.8	2.61	0.93
Science	40	2019	17.5	7.0	2.85	0.84	1205	16.4	7.5	2.80	0.86
Social Science	40	2002	14.7	6.5	2.85	0.81	1209	13.7	6.3	2.82	0.80

Subtest	Number of Items	Form D					Form E				
		N	Mean	SD	SEM	KR20	N	Mean	SD	SEM	KR20
Language	48	528	29.7	10.0	2.89	0.92	472	25.0	11.5	2.96	0.93
Prewriting	12	538	8.1	2.9	1.38	0.77	474	6.5	3.0	1.46	0.76
Composing	18	531	10.9	4.1	1.80	0.80	477	9.2	4.6	1.84	0.84
Editing	18	528	10.6	4.1	1.77	0.81	477	9.1	4.7	1.79	0.86

* In terms of raw scores

Table C–15. KR20 Reliability Coefficients, Standard Errors of Measurement, and Related Summary Data for Primary 1, Spring, Grade 1, Abbreviated*

Subtest	Number of Items	Form A					Form B				
		N	Mean	SD	SEM	KR20	N	Mean	SD	SEM	KR20
Total Reading	90	3320	69.5	15.5	3.34	0.95	3258	68.3	16.0	3.47	0.95
Word Study Skills	20	3493	14.1	3.7	1.64	0.80	3443	14.2	3.3	1.74	0.72
Word Reading	20	3527	16.0	4.5	1.46	0.89	3486	15.4	4.7	1.56	0.89
Sentence Reading	20	3481	16.7	3.5	1.42	0.84	3397	16.6	4.0	1.41	0.88
Reading Comprehension	30	3392	22.1	6.0	2.03	0.88	3330	21.7	5.9	2.07	0.88
Total Mathematics	50	3525	32.7	8.4	2.94	0.88	3500	32.8	8.8	2.92	0.89
Mathematics Problem Solving	30	3543	21.1	5.4	2.20	0.83	3546	21.0	5.4	2.19	0.83
Mathematics Procedures	20	3555	11.6	3.9	1.92	0.76	3518	11.8	4.2	1.89	0.80
Language	30	2422	18.9	5.6	2.39	0.82	3013	18.5	5.8	2.38	0.83
Spelling	30	3547	20.8	6.6	2.13	0.89	3543	20.9	6.5	2.12	0.89
Environment	30	3515	22.6	4.0	2.14	0.71	3519	22.5	3.7	2.04	0.69
Listening	30	3402	19.5	5.8	2.32	0.84	3459	19.1	5.4	2.37	0.81

Subtest	Number of Items	Form D					Form E				
		N	Mean	SD	SEM	KR20	N	Mean	SD	SEM	KR20
Language	30	1380	20.2	5.0	2.25	0.79	1214	18.3	5.4	2.36	0.81

* In terms of raw scores

Table C–16. KR20 Reliability Coefficients, Standard Errors of Measurement, and Related Summary Data for Primary 2, Spring, Grade 2, Abbreviated*

Subtest	Number of Items	Form A					Form B				
		N	Mean	SD	SEM	KR20	N	Mean	SD	SEM	KR20
Total Reading	70	3443	48.7	12.5	3.25	0.93	2972	49.5	12.7	3.20	0.94
Word Study Skills	20	3542	14.2	3.3	1.58	0.77	3042	14.4	3.1	1.60	0.74
Reading Vocabulary	20	3527	15.1	4.5	1.62	0.87	3031	15.2	4.3	1.61	0.86
Reading Comprehension	30	3558	19.1	6.3	2.26	0.87	3057	19.9	6.6	2.20	0.89
Total Mathematics	50	3546	34.7	8.7	2.86	0.89	3060	35.7	8.6	2.77	0.90
Mathematics Problem Solving	30	3568	20.4	5.4	2.22	0.83	3075	21.2	5.4	2.16	0.84
Mathematics Procedures	20	3579	14.3	4.1	1.76	0.81	3082	14.5	4.0	1.71	0.81
Language	30	2768	22.8	5.3	2.07	0.85	2075	21.9	5.6	2.17	0.85
Spelling	30	3595	23.6	5.6	1.99	0.87	3069	24.1	5.6	1.90	0.89
Environment	30	3564	22.2	3.7	2.19	0.65	3008	21.1	4.0	2.21	0.69
Listening	30	3555	18.5	5.2	2.34	0.80	2994	20.8	4.9	2.25	0.79

Subtest	Number of Items	Form D					Form E				
		N	Mean	SD	SEM	KR20	N	Mean	SD	SEM	KR20
Language	30	1102	21.0	4.6	2.26	0.76	1304	21.4	5.2	2.23	0.81

* In terms of raw scores

Table C–17. KR20 Reliability Coefficients, Standard Errors of Measurement, and Related Summary Data for Primary 3, Spring, Grade 3, Abbreviated*

Subtest	Number of Items	Form A					Form B				
		N	Mean	SD	SEM	KR20	N	Mean	SD	SEM	KR20
Total Reading	70	2110	47.7	12.8	3.36	0.93	2395	44.8	12.9	3.49	0.93
Word Study Skills	20	2147	13.8	3.6	1.72	0.78	2424	13.3	4.0	1.80	0.79
Reading Vocabulary	20	2148	13.8	4.2	1.81	0.82	2433	14.0	4.3	1.77	0.83
Reading Comprehension	30	2160	20.0	6.4	2.20	0.88	2455	17.4	6.5	2.33	0.87
Total Mathematics	50	2161	32.3	10.1	2.95	0.92	2454	32.5	9.5	2.94	0.90
Mathematics Problem Solving	30	2170	20.1	6.2	2.26	0.87	2463	20.5	5.7	2.21	0.85
Mathematics Procedures	20	2177	12.2	4.8	1.84	0.85	2473	11.9	4.6	1.89	0.83
Language	30	1805	19.2	6.1	2.31	0.86	2062	19.0	5.9	2.30	0.85
Spelling	30	2159	16.9	6.2	2.37	0.86	2458	17.4	6.1	2.41	0.84
Science	30	2158	18.7	5.3	2.33	0.81	2445	19.0	5.5	2.34	0.82
Social Science	30	2147	17.5	6.4	2.31	0.87	2426	20.1	5.8	2.23	0.85
Listening	30	2141	20.4	5.1	2.29	0.80	2417	18.9	4.7	2.34	0.76

Subtest	Number of Items	Form D					Form E				
		N	Mean	SD	SEM	KR20	N	Mean	SD	SEM	KR20
Language	30	669	16.9	6.0	2.43	0.83	638	16.2	5.6	2.44	0.81
Prewriting	9	673	5.7	2.2	1.26	0.68	646	5.0	2.0	1.32	0.56
Composing	10	670	4.7	2.3	1.44	0.61	644	5.1	2.2	1.43	0.58
Editing	11	671	6.5	2.5	1.47	0.65	640	6.0	2.4	1.47	0.63

* In terms of raw scores

Table C–18. KR20 Reliability Coefficients, Standard Errors of Measurement, and Related Summary Data for Intermediate 1, Spring, Grade 4, Abbreviated*

Subtest	Number of Items	Form A					Form B				
		N	Mean	SD	SEM	KR20	N	Mean	SD	SEM	KR20
Total Reading	70	2876	44.3	13.3	3.51	0.93	3518	45.0	14.3	3.49	0.94
Word Study Skills	20	2939	12.9	4.4	1.77	0.84	3575	12.8	4.5	1.83	0.84
Reading Vocabulary	20	2922	13.6	4.5	1.83	0.83	3564	14.1	4.9	1.72	0.88
Reading Comprehension	30	2963	17.5	6.2	2.35	0.86	3593	17.9	6.7	2.33	0.88
Total Mathematics	50	2947	29.7	9.5	2.99	0.90	3567	30.9	10.5	2.96	0.92
Mathematics Problem Solving	30	2971	17.0	5.6	2.30	0.83	3591	18.4	6.2	2.30	0.86
Mathematics Procedures	20	2961	12.6	4.6	1.88	0.84	3585	12.4	5.0	1.83	0.86
Language	30	2488	19.0	6.2	2.34	0.86	2228	18.3	6.4	2.36	0.87
Language Mechanics	15	2488	9.5	3.2	1.65	0.74	2228	9.4	3.4	1.63	0.76
Language Expression	15	2488	9.5	3.5	1.64	0.79	2228	8.9	3.7	1.67	0.80
Spelling	30	2911	17.7	6.1	2.37	0.85	3527	18.2	6.3	2.29	0.87
Science	30	2888	19.0	5.7	2.36	0.83	3506	18.6	6.2	2.31	0.86
Social Science	30	2879	17.4	6.3	2.37	0.86	3506	18.4	6.7	2.32	0.88
Listening	30	2875	17.8	4.8	2.41	0.75	3486	18.8	5.6	2.30	0.83

Subtest	Number of Items	Form D					Form E				
		N	Mean	SD	SEM	KR20	N	Mean	SD	SEM	KR20
Language	30	334	17.3	5.7	2.44	0.82	1218	19.0	5.9	2.35	0.84
Prewriting	8	336	5.0	1.8	1.24	0.53	1222	5.5	2.0	1.14	0.66
Composing	11	336	6.2	2.7	1.45	0.71	1219	6.5	2.4	1.50	0.60
Editing	11	336	6.1	2.4	1.49	0.62	1223	7.0	2.5	1.40	0.68

* In terms of raw scores

Table C–19. KR20 Reliability Coefficients, Standard Errors of Measurement, and Related Summary Data for Intermediate 2, Spring, Grade 5, Abbreviated*

Subtest	Number of Items	Form A					Form B				
		N	Mean	SD	SEM	KR20	N	Mean	SD	SEM	KR20
Total Reading	50	3055	31.7	9.3	3.00	0.90	2704	31.9	9.7	2.99	0.91
Reading Vocabulary	20	3087	14.3	4.2	1.78	0.82	2707	13.8	4.2	1.83	0.81
Reading Comprehension	30	3132	17.3	5.9	2.38	0.84	2751	18.0	6.3	2.34	0.86
Total Mathematics	50	3129	28.7	9.9	3.08	0.90	2718	27.2	9.4	3.09	0.89
Mathematics Problem Solving	30	3141	17.5	6.1	2.38	0.85	2737	16.0	5.6	2.42	0.81
Mathematics Procedures	20	3153	11.1	4.6	1.91	0.83	2728	11.2	4.5	1.89	0.83
Language	30	2332	19.6	5.8	2.28	0.85	2324	18.1	6.0	2.36	0.84
Language Mechanics	15	2332	10.0	2.9	1.58	0.70	2324	9.6	3.2	1.58	0.75
Language Expression	15	2332	9.6	3.5	1.62	0.78	2324	8.5	3.4	1.73	0.74
Spelling	30	3005	18.9	6.1	2.34	0.85	2692	19.8	5.7	2.30	0.84
Science	30	2992	16.2	5.4	2.48	0.79	2678	16.7	5.8	2.44	0.83
Social Science	30	2975	17.8	6.2	2.34	0.86	2671	16.0	5.8	2.41	0.83
Listening	30	2977	18.8	4.8	2.35	0.76	2676	18.6	5.0	2.38	0.77

Subtest	Number of Items	Form D					Form E				
		N	Mean	SD	SEM	KR20	N	Mean	SD	SEM	KR20
Language	30	611	15.0	6.0	2.46	0.83	342	18.9	5.2	2.34	0.80
Prewriting	8	613	4.4	1.9	1.28	0.53	343	6.3	1.6	1.04	0.60
Composing	11	614	4.8	2.4	1.49	0.61	343	6.3	2.1	1.47	0.50
Editing	11	615	5.8	2.6	1.48	0.68	342	6.3	2.4	1.50	0.61

* In terms of raw scores

Table C–20. KR20 Reliability Coefficients, Standard Errors of Measurement, and Related Summary Data for Intermediate 3, Spring, Grade 6, Abbreviated*

Subtest	Number of Items	Form A					Form B				
		N	Mean	SD	SEM	KR20	N	Mean	SD	SEM	KR20
Total Reading	50	3020	35.0	8.7	2.87	0.89	3472	31.0	9.2	3.01	0.89
Reading Vocabulary	20	3048	14.7	3.6	1.78	0.76	3491	13.9	4.0	1.81	0.80
Reading Comprehension	30	3053	20.3	5.8	2.24	0.85	3500	17.1	5.8	2.39	0.83
Total Mathematics	50	3046	27.9	10.4	3.10	0.91	3484	27.2	9.9	3.12	0.90
Mathematics Problem Solving	30	3072	16.9	6.2	2.42	0.85	3507	16.3	6.1	2.45	0.84
Mathematics Procedures	20	3053	10.9	4.8	1.91	0.84	3504	10.9	4.5	1.91	0.82
Language	30	2620	20.8	5.8	2.20	0.85	2468	20.1	6.1	2.28	0.86
Language Mechanics	15	2620	10.8	2.8	1.51	0.70	2468	10.1	3.1	1.59	0.74
Language Expression	15	2620	10.0	3.5	1.58	0.80	2468	9.9	3.5	1.62	0.79
Spelling	30	3059	19.8	5.8	2.30	0.84	3487	18.9	5.9	2.31	0.85
Science	30	3056	18.9	4.8	2.36	0.76	3482	17.1	5.3	2.44	0.79
Social Science	30	3046	17.8	5.9	2.40	0.84	3483	16.4	6.0	2.42	0.84
Listening	30	3022	19.0	5.0	2.37	0.78	3468	18.8	5.4	2.31	0.82

Subtest	Number of Items	Form D					Form E				
		N	Mean	SD	SEM	KR20	N	Mean	SD	SEM	KR20
Language	30	338	19.0	5.5	2.38	0.81	973	15.6	5.9	2.45	0.83
Prewriting	8	340	5.6	1.8	1.17	0.57	974	4.7	1.9	1.23	0.59
Composing	11	342	6.2	2.4	1.51	0.62	975	5.1	2.4	1.52	0.61
Editing	11	340	7.2	2.3	1.40	0.63	975	5.7	2.5	1.46	0.66

* In terms of raw scores

Table C–21.   KR20 Reliability Coefficients, Standard Errors of Measurement, and Related Summary Data for Advanced 1, Spring, Grade 7, Abbreviated*

Subtest	Number of Items	Form A					Form B				
		N	Mean	SD	SEM	KR20	N	Mean	SD	SEM	KR20
Total Reading	50	2307	31.4	10.5	2.95	0.92	2697	31.1	10.0	3.00	0.91
Reading Vocabulary	20	2316	13.2	4.7	1.81	0.85	2710	13.1	4.3	1.85	0.82
Reading Comprehension	30	2345	18.1	6.4	2.31	0.87	2760	17.9	6.5	2.34	0.87
Total Mathematics	50	2320	23.8	10.7	3.09	0.92	2696	25.0	10.3	3.10	0.91
Mathematics Problem Solving	30	2348	13.9	6.3	2.42	0.85	2751	15.3	6.3	2.41	0.85
Mathematics Procedures	20	2331	9.9	5.1	1.89	0.87	2724	9.6	4.9	1.91	0.85
Language	30	1640	19.7	6.4	2.26	0.88	1884	18.4	6.7	2.34	0.88
Language Mechanics	15	1640	10.3	3.1	1.56	0.76	1883	9.4	3.4	1.63	0.77
Language Expression	15	1640	9.5	3.8	1.61	0.82	1884	9.0	3.8	1.67	0.81
Spelling	30	2332	18.4	5.6	2.35	0.83	2753	17.9	5.4	2.36	0.81
Science	30	2337	16.8	5.6	2.42	0.81	2648	17.0	5.3	2.39	0.80
Social Science	30	2310	15.1	5.6	2.48	0.81	2646	15.3	5.7	2.47	0.81
Listening	30	2208	18.4	5.2	2.39	0.79	2623	18.5	5.5	2.35	0.82

Subtest	Number of Items	Form D					Form E				
		N	Mean	SD	SEM	KR20	N	Mean	SD	SEM	KR20
Language	30	582	14.5	6.4	2.47	0.85	696	16.5	5.9	2.43	0.83
Prewriting	8	582	3.8	2.0	1.28	0.61	698	4.2	1.6	1.26	0.41
Composing	11	585	5.2	2.5	1.50	0.64	707	6.3	2.6	1.47	0.68
Editing	11	585	5.4	2.8	1.48	0.71	707	5.9	2.7	1.44	0.71

* In terms of raw scores

Table C–22.　KR20 Reliability Coefficients, Standard Errors of Measurement, and Related Summary Data for Advanced 2, Spring, Grade 8, Abbreviated*

Subtest	Number of Items	Form A					Form B				
		N	Mean	SD	SEM	KR20	N	Mean	SD	SEM	KR20
Total Reading	50	1924	32.4	9.5	2.97	0.90	1752	31.3	9.5	3.00	0.90
Reading Vocabulary	20	1935	13.9	4.1	1.82	0.80	1771	13.2	4.1	1.87	0.79
Reading Comprehension	30	1937	18.5	6.1	2.33	0.85	1904	18.3	5.9	2.33	0.84
Total Mathematics	50	1824	23.3	10.1	3.18	0.90	1848	23.6	10.6	3.12	0.91
Mathematics Problem Solving	30	1855	14.1	6.1	2.47	0.84	1905	13.8	6.6	2.41	0.87
Mathematics Procedures	20	1849	9.1	4.7	1.97	0.82	1868	9.6	4.8	1.95	0.83
Language	30	1626	18.8	6.5	2.33	0.87	1047	19.8	6.4	2.25	0.88
Language Mechanics	15	1626	9.7	3.2	1.63	0.74	1047	10.5	3.4	1.55	0.79
Language Expression	15	1626	9.0	3.9	1.64	0.83	1047	9.3	3.6	1.62	0.80
Spelling	30	1927	18.9	5.4	2.28	0.82	1750	19.2	5.8	2.26	0.85
Science	30	1949	16.0	5.4	2.49	0.79	1880	15.1	5.3	2.47	0.78
Social Science	30	1872	14.3	5.8	2.49	0.82	1864	13.4	5.7	2.49	0.81
Listening	30	1816	18.0	5.0	2.37	0.78	1698	17.0	5.8	2.43	0.82

Subtest	Number of Items	Form D					Form E				
		N	Mean	SD	SEM	KR20	N	Mean	SD	SEM	KR20
Language	30	276	21.0	5.8	2.16	0.86	675	16.3	6.4	2.42	0.86
Prewriting	8	276	6.5	1.7	0.97	0.68	675	5.1	2.0	1.20	0.62
Composing	11	276	7.3	2.6	1.33	0.74	675	5.0	2.4	1.52	0.61
Editing	11	276	7.2	2.4	1.37	0.66	675	6.1	2.9	1.43	0.76

* In terms of raw scores

Table C–23. KR20 Reliability Coefficients, Standard Errors of Measurement, and Related Summary Data for Advanced 2, Spring, Grade 9, Abbreviated*

Subtest	Number of Items	Form A					Form B				
		N	Mean	SD	SEM	KR20	N	Mean	SD	SEM	KR20
Total Reading	50	1498	35.0	8.8	2.86	0.89	1826	33.5	9.1	2.90	0.90
Reading Vocabulary	20	1498	14.5	4.2	1.75	0.82	1854	14.0	4.3	1.77	0.83
Reading Comprehension	30	1515	20.4	5.7	2.22	0.85	1853	19.3	5.8	2.27	0.85
Total Mathematics	50	1470	27.4	10.3	3.13	0.91	1828	24.8	11.0	3.11	0.92
Mathematics Problem Solving	30	1495	16.6	6.5	2.41	0.87	1861	14.8	6.9	2.40	0.88
Mathematics Procedures	20	1481	10.7	4.7	1.95	0.83	1848	9.9	5.0	1.93	0.85
Language	30	881	20.8	6.4	2.20	0.88	1088	19.8	6.8	2.22	0.89
Language Mechanics	15	881	10.4	3.1	1.55	0.75	1088	10.1	3.6	1.55	0.81
Language Expression	15	881	10.3	3.7	1.54	0.83	1088	9.7	3.7	1.58	0.81
Spelling	30	1515	20.1	5.2	2.20	0.82	1865	19.2	5.8	2.28	0.84
Science	30	1500	17.2	5.5	2.43	0.80	1862	16.3	5.6	2.45	0.81
Social Science	30	1499	15.5	6.4	2.44	0.85	1851	14.2	5.9	2.48	0.82
Listening	30	1418	19.5	4.9	2.31	0.78	1814	18.3	5.7	2.38	0.83

Subtest	Number of Items	Form D					Form E				
		N	Mean	SD	SEM	KR20	N	Mean	SD	SEM	KR20
Language	30	634	18.9	7.2	2.23	0.91	754	17.2	6.2	2.39	0.85
Prewriting	8	634	5.8	2.3	1.03	0.79	757	5.6	1.9	1.13	0.64
Composing	11	638	6.7	3.1	1.36	0.81	757	5.4	2.5	1.52	0.62
Editing	11	638	6.4	2.8	1.39	0.76	754	6.2	2.9	1.42	0.76

* In terms of raw scores

Table C–24. KR20 Reliability Coefficients, Standard Errors of Measurement, and Related Summary Data for TASK 1, Spring, Grade 9, Abbreviated*

Subtest	Number of Items	Form A					Form B				
		N	Mean	SD	SEM	KR20	N	Mean	SD	SEM	KR20
Total Reading	50	1715	33.0	8.9	2.92	0.89	1116	33.9	9.4	2.86	0.91
Reading Vocabulary	20	1721	14.4	4.0	1.74	0.81	1116	13.5	3.9	1.80	0.79
Reading Comprehension	30	1721	18.5	5.7	2.32	0.84	1121	20.4	6.2	2.19	0.88
Mathematics	30	1733	13.4	5.5	2.43	0.80	1092	14.2	5.7	2.42	0.82
Language	30	913	18.9	6.8	2.26	0.89	672	19.7	6.8	2.22	0.90
Language Mechanics	15	913	9.3	3.2	1.62	0.75	672	10.2	3.2	1.54	0.78
Language Expression	15	913	9.6	4.1	1.55	0.85	672	9.6	4.1	1.55	0.86
Spelling	30	1711	18.0	5.8	2.34	0.84	1087	17.4	5.7	2.38	0.83
Science	30	1689	15.6	5.2	2.42	0.78	1104	14.9	5.4	2.46	0.79
Social Science	30	1681	13.4	5.2	2.52	0.76	1120	13.4	5.5	2.47	0.80

Subtest	Number of Items	Form D					Form E				
		N	Mean	SD	SEM	KR20	N	Mean	SD	SEM	KR20
Language	30	774	19.1	6.0	2.34	0.85	419	16.8	7.0	2.37	0.89
Prewriting	8	777	5.2	2.0	1.19	0.64	419	4.9	1.9	1.25	0.59
Composing	11	779	7.3	2.6	1.37	0.73	421	6.1	3.2	1.39	0.81
Editing	11	776	6.6	2.3	1.47	0.61	421	5.8	2.8	1.44	0.73

* In terms of raw scores

Table C–25. KR20 Reliability Coefficients, Standard Errors of Measurement, and Related Summary Data for TASK 2, Spring, Grade 10, Abbreviated*

Subtest	Number of Items	Form A					Form B				
		N	Mean	SD	SEM	KR20	N	Mean	SD	SEM	KR20
Total Reading	50	2426	34.5	8.7	2.85	0.89	2729	32.3	8.8	2.94	0.89
Reading Vocabulary	20	2451	14.9	3.5	1.67	0.78	2736	14.6	3.7	1.75	0.78
Reading Comprehension	30	2443	19.6	6.0	2.27	0.86	2744	17.7	5.9	2.34	0.84
Mathematics	30	2337	13.9	5.5	2.43	0.81	2670	12.1	5.1	2.43	0.77
Language	30	1767	18.7	6.1	2.36	0.85	1830	19.3	6.1	2.30	0.86
Language Mechanics	15	1767	8.9	3.1	1.70	0.70	1830	9.7	3.0	1.62	0.72
Language Expression	15	1767	9.9	3.5	1.62	0.79	1830	9.7	3.6	1.61	0.80
Spelling	30	2457	19.2	5.9	2.27	0.85	2678	19.0	5.7	2.24	0.85
Science	30	2343	16.4	5.7	2.41	0.82	2616	15.1	5.2	2.44	0.78
Social Science	30	2339	12.7	4.9	2.51	0.73	2612	11.7	4.6	2.51	0.70

Subtest	Number of Items	Form D					Form E				
		N	Mean	SD	SEM	KR20	N	Mean	SD	SEM	KR20
Language	30	657	18.1	6.4	2.32	0.87	790	17.8	6.0	2.34	0.85
Prewriting	8	659	5.7	1.8	1.10	0.61	792	5.6	1.8	1.10	0.63
Composing	11	657	6.3	2.8	1.43	0.74	791	5.9	2.5	1.45	0.67
Editing	11	659	6.1	2.7	1.45	0.71	793	6.3	2.6	1.45	0.69

* In terms of raw scores

Table C–26. KR20 Reliability Coefficients, Standard Errors of Measurement, and Related Summary Data for TASK 3, Spring, Grade 11, Abbreviated*

Subtest	Number of Items	Form A					Form B				
		N	Mean	SD	SEM	KR20	N	Mean	SD	SEM	KR20
Total Reading	50	1992	31.8	11.0	2.93	0.93	1650	32.6	10.1	2.91	0.92
Reading Vocabulary	20	1996	14.4	4.9	1.66	0.88	1663	13.5	4.0	1.81	0.80
Reading Comprehension	30	1996	17.4	7.0	2.37	0.88	1651	19.1	6.8	2.25	0.89
Mathematics	30	1956	12.3	5.3	2.49	0.78	1626	10.6	4.7	2.43	0.73
Language	30	1384	17.3	7.0	2.34	0.89	894	15.9	6.3	2.42	0.86
Language Mechanics	15	1384	8.5	3.5	1.65	0.78	894	7.7	3.0	1.71	0.68
Language Expression	15	1381	8.7	4.0	1.64	0.83	894	8.2	3.9	1.68	0.81
Spelling	30	1939	17.8	6.7	2.27	0.89	1588	18.3	6.7	2.29	0.88
Science	30	1952	13.3	5.5	2.44	0.80	1564	13.1	5.1	2.45	0.77
Social Science	30	1962	11.2	5.0	2.47	0.76	1569	10.4	4.7	2.46	0.73

Subtest	Number of Items	Form D					Form E				
		N	Mean	SD	SEM	KR20	N	Mean	SD	SEM	KR20
Language	30	509	16.9	7.2	2.32	0.90	641	18.5	6.8	2.29	0.89
Prewriting	8	511	5.2	2.3	1.09	0.78	643	5.1	2.1	1.14	0.72
Composing	11	512	6.0	2.9	1.44	0.75	645	6.9	3.0	1.36	0.79
Editing	11	510	5.7	2.8	1.43	0.74	643	6.4	2.6	1.43	0.69

* In terms of raw scores

Table C–27. KR20 Reliability Coefficients, Standard Errors of Measurement, and Related Summary Data for TASK 3, Spring, Grade 12, Abbreviated*

Subtest	Number of Items	Form A					Form B				
		N	Mean	SD	SEM	KR20	N	Mean	SD	SEM	KR20
Total Reading	50	2029	31.6	11.1	2.91	0.93	1297	29.9	11.9	2.94	0.94
Reading Vocabulary	20	2054	14.8	4.6	1.66	0.87	1299	12.5	4.8	1.83	0.85
Reading Comprehension	30	2046	16.8	7.5	2.33	0.90	1302	17.4	7.8	2.26	0.92
Mathematics	30	2045	11.7	5.1	2.49	0.76	1282	9.8	5.0	2.39	0.77
Language	30	1460	16.6	7.4	2.34	0.90	697	15.3	7.1	2.36	0.89
Language Mechanics	15	1459	8.2	3.7	1.65	0.81	697	7.3	3.5	1.69	0.76
Language Expression	15	1460	8.3	4.2	1.63	0.85	697	7.9	4.1	1.63	0.84
Spelling	30	2052	18.4	6.7	2.24	0.89	1226	16.9	7.9	2.21	0.92
Science	30	2019	13.0	5.6	2.45	0.81	1205	11.9	5.6	2.41	0.82
Social Science	30	2002	10.9	5.0	2.46	0.75	1209	10.1	4.9	2.43	0.75

Subtest	Number of Items	Form D					Form E				
		N	Mean	SD	SEM	KR20	N	Mean	SD	SEM	KR20
Language	30	528	18.0	7.1	2.27	0.90	472	14.8	7.9	2.30	0.92
Prewriting	8	538	5.5	2.3	1.05	0.79	474	4.0	2.4	1.15	0.77
Composing	11	531	6.3	2.8	1.43	0.73	477	5.5	3.2	1.40	0.81
Editing	11	528	6.0	2.9	1.38	0.78	477	5.3	3.0	1.40	0.79

* In terms of raw scores

Table C–28. KR20 Reliability Coefficients, Standard Errors of Measurement, and Related Summary Data for SESAT 1, Fall, Grade K, Full-Length*

Subtest	Number of Items	Form A			
		N	Mean	SD	SEM	KR20
Total Reading	70	925	45.4	11.3	3.48	0.91
Sounds and Letters	40	937	30.7	6.7	2.39	0.87
Word Reading	30	1111	14.5	5.8	2.47	0.82
Mathematics	40	1096	26.7	6.8	2.55	0.86
Environment	40	1096	27.8	6.4	2.58	0.84
Listening to Words and Stories	40	1085	28.2	6.0	2.54	0.82

* In terms of raw scores

Table C–29. KR20 Reliability Coefficients, Standard Errors of Measurement, and Related Summary Data for SESAT 2, Fall, Grade 1, Full-Length*

Subtest	Number of Items	Form A			
		N	Mean	SD	SEM	KR20
Total Reading	100	1220	80.0	16.0	3.45	0.95
Sounds and Letters	40	1241	36.1	4.2	1.72	0.83
Word Reading	30	1349	21.7	6.2	2.12	0.88
Sentence Reading	30	1220	21.9	7.0	1.99	0.92
Mathematics	40	1359	32.8	5.4	2.09	0.85
Environment	40	1359	30.2	4.4	2.34	0.72
Listening to Words and Stories	40	1349	27.8	5.4	2.56	0.78

* In terms of raw scores

Table C–30. KR20 Reliability Coefficients, Standard Errors of Measurement, and Related Summary Data for Primary 1, Fall, Grade 2, Full-Length*

Subtest	Number of Items	Form A					Form B				
		N	Mean	SD	SEM	KR20	N	Mean	SD	SEM	KR20
Total Reading	130	1466	107.8	19.0	3.63	0.96	1802	108.2	20.1	3.59	0.97
Word Study Skills	30	1491	22.8	5.1	1.89	0.86	1878	22.9	4.6	1.92	0.83
Word Reading	30	1517	26.0	5.3	1.56	0.91	1886	25.8	5.3	1.62	0.91
Sentence Reading	30	1511	26.4	4.5	1.54	0.89	1853	26.9	5.0	1.40	0.92
Reading Comprehension	40	1494	32.1	7.0	2.12	0.91	1814	32.3	7.1	2.10	0.97
Total Mathematics	72	1505	52.5	11.7	3.31	0.92	1870	50.3	12.3	3.28	0.93
Mathematics Problem Solving	42	1509	32.1	6.9	2.43	0.88	1881	32.3	7.2	2.38	0.89
Mathematics Procedures	30	1517	20.4	5.8	2.20	0.86	1875	17.9	6.1	2.20	0.87
Language	40	1233	28.1	7.1	2.61	0.87	1137	28.5	7.7	2.52	0.89
Spelling	36	1513	27.9	7.6	2.08	0.93	1873	27.9	6.9	2.11	0.91
Environment	40	1505	31.4	4.5	2.33	0.73	1882	31.3	4.1	2.24	0.70
Listening	40	1485	28.8	7.3	2.53	0.88	1809	29.5	6.7	2.49	0.86

Subtest	Number of Items	Form D					Form E				
		N	Mean	SD	SEM	KR20	N	Mean	SD	SEM	KR20
Language	40	621	28.9	6.5	2.49	0.85	932	29.2	6.0	2.48	0.83

* In terms of raw scores

Table C–31. KR20 Reliability Coefficients, Standard Errors of Measurement, and Related Summary Data for Primary 2, Fall, Grade 3, Full-Length*

Subtest	Number of Items	Form A					Form B				
		N	Mean	SD	SEM	KR20	N	Mean	SD	SEM	KR20
Total Reading	100	1605	77.7	16.0	3.54	0.95	917	76.8	16.5	3.60	0.95
Word Study Skills	30	1620	23.4	4.4	1.83	0.83	920	23.1	4.8	1.86	0.85
Reading Vocabulary	30	1618	24.8	5.6	1.77	0.90	925	24.4	5.7	1.81	0.90
Reading Comprehension	40	1630	29.3	7.7	2.39	0.90	928	29.2	7.7	2.43	0.90
Total Mathematics	74	1618	55.3	12.6	3.25	0.93	923	57.2	11.6	3.12	0.93
Mathematics Problem Solving	44	1622	32.4	7.6	2.55	0.89	930	33.7	6.9	2.43	0.88
Mathematics Procedures	30	1628	22.9	5.9	1.97	0.89	923	23.5	5.6	1.91	0.88
Language	48	1030	39.0	7.4	2.39	0.90	647	39.0	8.0	2.41	0.91
Spelling	36	1628	30.3	5.9	1.93	0.89	922	30.9	6.1	1.81	0.91
Environment	40	1594	31.7	4.1	2.30	0.69	923	29.5	4.6	2.47	0.72
Listening	40	1621	28.7	6.0	2.53	0.82	910	28.8	6.4	2.53	0.85

Subtest	Number of Items	Form D					Form E				
		N	Mean	SD	SEM	KR20	N	Mean	SD	SEM	KR20
Language	40	854	30.4	5.3	2.41	0.79	412	30.7	6.2	2.43	0.85

* In terms of raw scores

Table C–32. KR20 Reliability Coefficients, Standard Errors of Measurement, and Related Summary Data for Primary 3, Fall, Grade 4, Full-Length*

Subtest	Number of Items	Form A					Form B				
		N	Mean	SD	SEM	KR20	N	Mean	SD	SEM	KR20
Total Reading	114	4225	77.6	22.7	4.27	0.97	2944	79.9	21.0	4.26	0.96
Word Study Skills	30	4271	20.3	5.9	2.14	0.87	2969	21.2	5.6	2.17	0.85
Reading Vocabulary	30	4267	21.1	6.7	2.13	0.90	2961	22.6	6.2	2.03	0.89
Reading Comprehension	54	4291	36.0	12.1	2.93	0.94	2972	35.9	11.3	2.97	0.93
Total Mathematics	76	4257	48.8	16.3	3.60	0.95	2960	51.5	15.3	3.51	0.95
Mathematics Problem Solving	46	4275	30.7	10.1	2.75	0.93	2976	33.2	9.5	2.59	0.93
Mathematics Procedures	30	4285	18.1	7.1	2.27	0.90	2967	18.2	6.8	2.32	0.89
Language	48	3371	32.6	10.3	2.80	0.93	1687	30.7	9.9	2.88	0.92
Spelling	38	4268	22.4	8.1	2.66	0.89	2969	23.3	7.9	2.63	0.89
Science	40	4262	26.3	8.0	2.61	0.89	2964	25.5	7.5	2.68	0.87
Social Science	40	4256	25.6	8.5	2.61	0.91	2978	27.8	7.8	2.51	0.90
Listening	40	4248	27.2	7.4	2.63	0.88	2969	26.9	6.7	2.63	0.85

Subtest	Number of Items	Form D					Form E				
		N	Mean	SD	SEM	KR20	N	Mean	SD	SEM	KR20
Language	45	831	22.5	9.0	2.99	0.89	1235	26.4	8.6	2.93	0.88
Prewriting	14	841	7.6	3.2	1.60	0.74	1241	8.8	3.1	1.57	0.74
Composing	16	833	7.2	3.5	1.82	0.73	1240	8.4	3.3	1.80	0.71
Editing	15	839	7.7	3.4	1.72	0.75	1241	9.1	3.4	1.68	0.76

* In terms of raw scores

Table C–33. KR20 Reliability Coefficients, Standard Errors of Measurement, and Related Summary Data for Intermediate 1, Fall, Grade 5, Full-Length*

Subtest	Number of Items	Form A					Form B				
		N	Mean	SD	SEM	KR20	N	Mean	SD	SEM	KR20
Total Reading	114	3956	76.3	22.4	4.32	0.96	3392	79.3	22.6	4.24	0.97
Word Study Skills	30	4046	20.0	6.5	2.15	0.89	3461	20.2	6.5	2.20	0.89
Reading Vocabulary	30	3989	21.9	6.6	2.09	0.90	3420	22.9	6.7	1.96	0.92
Reading Comprehension	54	4074	34.1	11.3	3.03	0.93	3464	35.8	11.5	2.97	0.93
Total Mathematics	80	4071	50.4	16.6	3.70	0.95	3463	52.7	16.7	3.63	0.95
Mathematics Problem Solving	48	4084	29.7	9.9	2.84	0.92	3469	31.9	9.8	2.77	0.92
Mathematics Procedures	32	4083	20.7	7.6	2.31	0.91	3484	20.8	7.7	2.30	0.91
Language	48	2637	31.6	10.1	2.87	0.92	2126	33.2	9.6	2.80	0.92
Language Mechanics	24	2637	15.4	5.1	2.05	0.84	2126	16.5	4.8	1.97	0.83
Language Expression	24	2637	16.3	5.6	1.98	0.88	2126	16.7	5.5	1.96	0.87
Spelling	40	4052	24.7	8.3	2.65	0.90	3473	25.5	8.3	2.62	0.90
Science	40	4006	26.4	7.8	2.65	0.88	3464	26.0	7.5	2.59	0.88
Social Science	40	4002	24.9	8.5	2.67	0.90	3459	26.9	8.5	2.58	0.91
Listening	40	4019	25.5	7.0	2.67	0.85	3431	26.3	7.0	2.62	0.86

Subtest	Number of Items	Form D					Form E				
		N	Mean	SD	SEM	KR20	N	Mean	SD	SEM	KR20
Language	48	1294	27.4	9.7	3.03	0.90	1303	29.6	9.7	2.95	0.91
Prewriting	12	1311	7.4	2.8	1.47	0.72	1303	8.1	3.0	1.38	0.78
Composing	18	1320	10.1	4.1	1.87	0.79	1310	10.5	3.9	1.85	0.78
Editing	18	1310	9.6	4.0	1.86	0.79	1310	10.9	3.9	1.81	0.78

* In terms of raw scores

Table C–34. KR20 Reliability Coefficients, Standard Errors of Measurement, and Related Summary Data for Intermediate 2 Fall, Grade 6, Full-Length*

Subtest	Number of Items	Form A					Form B				
		N	Mean	SD	SEM	KR20	N	Mean	SD	SEM	KR20
Total Reading	84	4651	56.9	15.6	3.76	0.94	4125	57.6	16.3	3.71	0.95
Reading Vocabulary	30	4657	22.5	5.7	2.11	0.86	4139	22.3	5.8	2.11	0.87
Reading Comprehension	54	4704	34.3	10.7	3.10	0.92	4176	35.2	11.4	3.03	0.93
Total Mathematics	80	4677	48.4	15.6	3.80	0.94	4182	44.9	15.8	3.84	0.94
Mathematics Problem Solving	48	4694	29.9	9.7	2.93	0.91	4193	26.9	9.7	2.99	0.91
Mathematics Procedures	32	4688	18.4	6.8	2.38	0.88	4191	18.0	7.0	2.37	0.89
Language	48	3411	32.5	9.7	2.83	0.92	2664	30.3	9.7	2.95	0.91
Language Mechanics	24	3411	16.7	4.9	1.95	0.84	2664	15.9	5.0	2.01	0.84
Language Expression	24	3411	15.8	5.4	2.03	0.86	2664	14.5	5.5	2.14	0.85
Spelling	40	4678	26.4	8.3	2.60	0.90	4188	27.0	8.1	2.58	0.90
Science	40	4678	24.0	7.0	2.79	0.84	4173	24.6	7.6	2.71	0.87
Social Science	40	4675	25.5	8.2	2.64	0.90	4150	24.0	8.1	2.72	0.89
Listening	40	4610	25.4	6.5	2.69	0.83	4116	25.6	6.6	2.69	0.83

Subtest	Number of Items	Form D					Form E				
		N	Mean	SD	SEM	KR20	N	Mean	SD	SEM	KR20
Language	48	1107	27.9	9.3	3.01	0.90	1477	30.3	9.4	2.95	0.90
Prewriting	12	1108	7.9	2.7	1.45	0.72	1477	8.7	2.6	1.36	0.73
Composing	18	1116	9.9	4.0	1.87	0.78	1479	10.7	3.8	1.85	0.76
Editing	18	1115	10.1	3.7	1.84	0.75	1479	10.9	4.0	1.84	0.79

* In terms of raw scores

Table C-35. KR20 Reliability Coefficients, Standard Errors of Measurement, and Related Summary Data for Intermediate 3, Fall, Grade 7, Full-Length*

Subtest	Number of Items	Form A					Form B				
		N	Mean	SD	SEM	KR20	N	Mean	SD	SEM	KR20
Total Reading	84	3989	55.4	17.0	3.77	0.95	4247	55.6	16.8	3.72	0.95
Reading Vocabulary	30	4000	21.4	6.4	2.16	0.89	4267	22.0	6.3	2.08	0.89
Reading Comprehension	54	4032	33.8	11.6	3.06	0.93	4340	33.3	11.5	3.06	0.93
Total Mathematics	80	4021	41.9	16.0	3.96	0.94	4333	43.1	16.7	3.91	0.95
Mathematics Problem Solving	48	4042	25.2	10.2	3.07	0.91	4344	26.0	10.5	3.05	0.92
Mathematics Procedures	32	4026	16.7	7.0	2.44	0.88	4363	17.1	7.5	2.37	0.90
Language	48	2644	30.3	9.9	2.94	0.91	2845	31.5	10.9	2.84	0.93
Language Mechanics	24	2644	15.2	4.9	2.06	0.83	2851	16.0	5.4	2.00	0.86
Language Expression	24	2644	15.1	5.6	2.08	0.86	2845	15.5	6.0	2.00	0.89
Spelling	40	4029	25.2	8.4	2.67	0.90	4349	25.4	8.4	2.65	0.90
Science	40	4011	25.0	7.2	2.72	0.86	4330	24.2	7.1	2.78	0.85
Social Science	40	4004	21.9	7.7	2.80	0.87	4313	22.6	8.2	2.76	0.89
Listening	40	4010	24.8	7.4	2.74	0.86	4305	25.5	7.6	2.67	0.88

Subtest	Number of Items	Form D					Form E				
		N	Mean	SD	SEM	KR20	N	Mean	SD	SEM	KR20
Language	48	1224	29.7	9.4	2.98	0.90	1345	28.6	9.9	2.93	0.91
Prewriting	12	1226	7.9	2.5	1.44	0.68	1346	7.9	2.8	1.41	0.75
Composing	18	1232	10.3	4.0	1.87	0.78	1348	10.2	3.8	1.85	0.77
Editing	18	1230	11.3	4.0	1.77	0.81	1349	10.5	4.2	1.77	0.82

* In terms of raw scores

Table C–36. KR20 Reliability Coefficients, Standard Errors of Measurement, and Related Summary Data for Advanced 1, Fall, Grade 8, Full-Length*

Subtest	Number of Items	Form A					Form B				
		N	Mean	SD	SEM	KR20	N	Mean	SD	SEM	KR20
Total Reading	84	5076	54.7	17.1	3.77	0.95	3759	56.7	17.4	3.67	0.96
Reading Vocabulary	30	5085	21.0	6.7	2.14	0.90	3771	21.2	6.8	2.09	0.91
Reading Comprehension	54	5108	33.7	11.2	3.07	0.93	3830	35.3	11.5	2.98	0.93
Total Mathematics	80	4937	40.8	17.1	3.89	0.95	3755	43.4	18.0	3.85	0.95
Mathematics Problem Solving	48	5008	24.4	10.1	3.03	0.91	3824	26.6	11.1	2.96	0.93
Mathematics Procedures	32	4958	16.3	7.9	2.40	0.91	3799	16.6	7.9	2.41	0.91
Language	48	3506	30.6	10.9	2.88	0.93	2565	31.3	10.9	2.85	0.93
Language Mechanics	24	3506	15.6	5.3	2.02	0.85	2565	16.1	5.4	1.98	0.86
Language Expression	24	3507	15.1	6.2	2.02	0.89	2565	15.2	6.0	2.03	0.89
Spelling	40	5039	25.2	7.7	2.67	0.88	3811	24.9	7.6	2.67	0.88
Science	40	4956	23.9	8.0	2.72	0.89	3751	23.6	7.5	2.72	0.87
Social Science	40	4944	21.2	7.7	2.84	0.87	3716	20.8	7.9	2.82	0.87
Listening	40	4899	26.1	7.1	2.70	0.86	3764	26.0	7.5	2.66	0.87

Subtest	Number of Items	Form D					Form E				
		N	Mean	SD	SEM	KR20	N	Mean	SD	SEM	KR20
Language	48	1365	30.5	9.5	2.91	0.91	1152	28.8	9.8	2.98	0.91
Prewriting	12	1374	8.1	2.6	1.42	0.71	1156	6.8	2.5	1.51	0.63
Composing	18	1372	11.1	4.0	1.82	0.79	1155	10.9	4.2	1.83	0.81
Editing	18	1373	11.2	3.9	1.76	0.79	1155	11.1	4.1	1.80	0.81

* In terms of raw scores

Table C–37. KR20 Reliability Coefficients, Standard Errors of Measurement, and Related Summary Data for Advanced 2, Fall, Grade 9, Full-Length*

Subtest	Number of Items	Form A					Form B				
		N	Mean	SD	SEM	KR20	N	Mean	SD	SEM	KR20
Total Reading	84	1293	54.8	15.3	3.90	0.94	1169	55.9	15.3	3.78	0.94
Reading Vocabulary	30	1309	20.9	6.3	2.20	0.88	1183	21.8	5.7	2.12	0.86
Reading Comprehension	54	1311	33.7	10.6	3.15	0.91	1181	33.9	10.7	3.09	0.92
Total Mathematics	80	1261	37.7	15.6	3.99	0.94	1004	41.8	16.9	3.90	0.95
Mathematics Problem Solving	48	1299	23.0	9.8	3.07	0.90	1152	23.7	10.4	3.03	0.92
Mathematics Procedures	32	1267	14.6	7.0	2.49	0.87	1021	16.9	8.1	2.36	0.92
Language	48	795	30.2	10.0	2.94	0.91	707	30.1	9.8	2.92	0.91
Language Mechanics	24	795	15.6	4.7	2.04	0.81	707	15.4	4.9	2.05	0.82
Language Expression	24	795	14.6	6.0	2.07	0.88	707	14.7	5.5	2.07	0.86
Spelling	40	1284	25.7	7.1	2.61	0.86	1137	26.2	7.4	2.58	0.88
Science	40	1257	20.5	6.9	2.84	0.83	1113	21.0	7.3	2.84	0.85
Social Science	40	1260	18.9	8.5	2.82	0.89	1061	18.5	7.7	2.85	0.86
Listening	40	1139	24.9	6.7	2.71	0.83	1011	24.2	7.3	2.77	0.86

Subtest	Number of Items	Form D					Form E				
		N	Mean	SD	SEM	KR20	N	Mean	SD	SEM	KR20
Language	48	446	32.4	9.4	2.84	0.91	371	32.6	9.1	2.83	0.90
Prewriting	12	446	8.9	2.7	1.33	0.76	374	9.0	2.5	1.31	0.72
Composing	18	450	11.6	3.9	1.78	0.80	376	10.7	3.7	1.84	0.75
Editing	18	450	11.8	3.8	1.75	0.79	373	12.6	4.2	1.66	0.84

* In terms of raw scores

Table C–38. KR20 Reliability Coefficients, Standard Errors of Measurement, and Related Summary Data for TASK 1, Fall, Grade 9, Full-Length*

Subtest	Number of Items	Form A					Form B				
		N	Mean	SD	SEM	KR20	N	Mean	SD	SEM	KR20
Total Reading	84	1647	53.1	15.6	3.85	0.94	1253	55.9	14.6	3.74	0.94
Reading Vocabulary	30	1662	20.9	6.1	2.15	0.88	1265	20.4	5.1	2.22	0.81
Reading Comprehension	54	1667	32.0	10.5	3.16	0.91	1274	35.2	10.8	2.97	0.92
Mathematics	50	1644	20.5	8.2	3.11	0.86	1276	21.6	8.1	3.11	0.85
Language	48	942	29.0	9.9	2.95	0.91	644	28.9	10.4	2.93	0.92
Language Mechanics	24	942	14.5	4.7	2.08	0.81	644	15.1	4.8	2.04	0.82
Language Expression	24	942	14.5	5.9	2.06	0.88	644	13.9	6.3	2.07	0.89
Spelling	40	1617	24.4	7.2	2.68	0.86	1280	24.1	7.1	2.73	0.85
Science	40	1629	20.4	6.6	2.80	0.82	1247	20.5	7.1	2.81	0.84
Social Science	40	1623	17.9	7.0	2.89	0.83	1247	17.5	6.9	2.86	0.83

Subtest	Number of Items	Form D					Form E				
		N	Mean	SD	SEM	KR20	N	Mean	SD	SEM	KR20
Language	48	647	32.1	9.1	2.87	0.90	577	30.0	9.2	2.92	0.90
Prewriting	12	648	8.0	2.6	1.46	0.69	577	7.3	2.4	1.43	0.65
Composing	18	651	12.5	4.1	1.69	0.83	579	11.3	3.9	1.81	0.79
Editing	18	650	11.5	3.5	1.78	0.74	577	11.3	3.9	1.77	0.80

* In terms of raw scores

Table C–39. KR20 Reliability Coefficients, Standard Errors of Measurement, and Related Summary Data for TASK 2, Fall, Grade 10, Full-Length*

Subtest	Number of Items	Form A					Form B				
		N	Mean	SD	SEM	KR20	N	Mean	SD	SEM	KR20
Total Reading	84	1505	49.4	17.9	3.82	0.95	1284	46.6	18.5	3.86	0.96
Reading Vocabulary	30	1610	18.7	6.8	2.20	0.90	1290	18.5	6.9	2.23	0.90
Reading Comprehension	54	1512	30.7	11.7	3.11	0.93	1339	27.4	12.6	3.13	0.94
Mathematics	50	1591	18.9	7.6	3.15	0.83	1316	18.5	7.4	3.16	0.82
Language	48	1412	26.1	10.5	3.03	0.92	736	28.8	10.6	2.92	0.92
Language Mechanics	24	1412	12.4	5.1	2.14	0.83	736	14.4	5.0	2.06	0.83
Language Expression	24	1412	13.6	6.0	2.12	0.87	736	14.5	6.2	2.03	0.89
Spelling	40	1611	21.1	8.9	2.67	0.91	1342	21.7	9.5	2.62	0.93
Science	40	1578	19.7	7.1	2.84	0.84	1325	19.2	7.2	2.84	0.84
Social Science	40	1574	14.5	5.9	2.87	0.77	1326	14.2	5.9	2.86	0.77

Subtest	Number of Items	Form D					Form E				
		N	Mean	SD	SEM	KR20	N	Mean	SD	SEM	KR20
Language	48	158	28.3	8.7	3.01	0.88	568	27.6	9.2	2.93	0.90
Prewriting	12	158	7.5	2.5	1.49	0.63	572	7.9	2.6	1.38	0.71
Composing	18	158	10.2	3.6	1.85	0.74	575	9.7	3.7	1.86	0.75
Editing	18	158	10.6	3.7	1.84	0.76	573	9.9	4.1	1.78	0.81

* In terms of raw scores

Table C–40. KR20 Reliability Coefficients, Standard Errors of Measurement, and Related Summary Data for TASK 3, Fall, Grade 11, Full-Length*

Subtest	Number of Items	Form A					Form B				
		N	Mean	SD	SEM	KR20	N	Mean	SD	SEM	KR20
Total Reading	84	1330	47.3	18.9	3.85	0.96	1338	47.8	18.2	3.88	0.96
Reading Vocabulary	30	1339	18.4	7.4	2.20	0.91	1340	18.6	6.4	2.27	0.88
Reading Comprehension	54	1344	28.8	12.2	3.14	0.93	1350	29.0	12.6	3.12	0.94
Mathematics	50	1330	17.7	7.2	3.18	0.81	1348	16.3	6.5	3.12	0.77
Language	48	949	23.5	10.7	3.00	0.92	1164	27.4	9.4	3.04	0.89
Language Mechanics	24	949	11.4	5.1	2.13	0.83	1164	13.0	4.5	2.17	0.77
Language Expression	24	949	12.1	6.2	2.09	0.89	1164	14.4	5.5	2.11	0.85
Spelling	40	1311	21.6	8.5	2.67	0.90	1343	22.3	8.6	2.71	0.90
Science	40	1368	16.0	6.1	2.85	0.78	1365	17.5	6.7	2.84	0.82
Social Science	40	1359	12.6	5.2	2.80	0.71	1353	13.5	5.2	2.84	0.70

Subtest	Number of Items	Form D					Form E				
		N	Mean	SD	SEM	KR20	N	Mean	SD	SEM	KR20
Language	48	396	25.7	9.7	3.00	0.90	195	16.0	10.1	2.67	0.93
Prewriting	12	400	7.0	2.8	1.50	0.72	198	4.5	2.5	1.35	0.70
Composing	18	406	9.2	3.8	1.85	0.77	199	6.2	4.1	1.68	0.83
Editing	18	402	9.3	4.1	1.80	0.80	198	5.2	4.1	1.56	0.86

* In terms of raw scores

Table C–41. KR20 Reliability Coefficients, Standard Errors of Measurement, and Related Summary Data for TASK 3, Fall, Grade 12, Full-Length*

Subtest	Number of Items	Form A					Form B				
		N	Mean	SD	SEM	KR20	N	Mean	SD	SEM	KR20
Total Reading	84	869	51.4	17.6	3.80	0.95	1119	51.3	16.5	3.85	0.95
Reading Vocabulary	30	874	21.2	6.6	2.07	0.90	1125	20.5	5.6	2.21	0.84
Reading Comprehension	54	878	30.0	12.1	3.14	0.93	1127	30.8	12.0	3.11	0.93
Mathematics	50	818	19.6	8.1	3.20	0.84	1119	17.0	7.6	3.09	0.84
Language	48	592	25.9	11.4	2.98	0.93	761	27.4	9.9	3.01	0.91
Language Mechanics	24	592	13.0	5.3	2.14	0.84	760	13.3	4.5	2.16	0.78
Language Expression	24	592	12.9	6.7	2.04	0.91	760	14.1	6.0	2.07	0.88
Spelling	40	807	24.1	8.5	2.62	0.91	1066	24.4	8.1	2.68	0.89
Science	40	813	17.2	6.9	2.86	0.83	1044	17.2	7.0	2.82	0.84
Social Science	40	806	13.5	5.8	2.84	0.76	1052	13.8	5.6	2.85	0.74

Subtest	Number of Items	Form D					Form E				
		N	Mean	SD	SEM	KR20	N	Mean	SD	SEM	KR20
Language	48	197	29.6	10.0	2.90	0.92	259	24.3	10.9	3.01	0.92
Prewriting	12	199	8.0	2.7	1.42	0.73	260	6.2	2.9	1.49	0.73
Composing	18	199	10.8	4.2	1.79	0.82	264	9.3	4.4	1.87	0.82
Editing	18	197	10.8	4.1	1.76	0.82	265	8.6	4.5	1.83	0.83

* In terms of raw scores

Table C–42. KR20 Reliability Coefficients, Standard Errors of Measurement, and Related Summary Data for Primary 1, Fall, Grade 2, Abbreviated*

Subtest	Number of Items	Form A					Form B				
		N	Mean	SD	SEM	KR20	N	Mean	SD	SEM	KR20
Total Reading	90	1466	74.8	13.6	3.00	0.95	1802	74.5	14.3	3.04	0.96
Word Study Skills	20	1491	15.2	3.6	1.53	0.82	1878	15.4	3.2	1.60	0.74
Word Reading	20	1517	17.4	3.7	1.23	0.89	1886	17.1	3.8	1.34	0.87
Sentence Reading	20	1511	17.7	3.0	1.25	0.83	1853	17.8	3.5	1.18	0.89
Reading Comprehension	30	1494	41.8	7.8	2.24	0.92	1814	41.9	8.4	2.19	0.93
Total Mathematics	50	1505	36.5	7.9	2.75	0.88	1870	35.8	8.4	2.69	0.90
Mathematics Problem Solving	30	1509	23.0	4.9	2.05	0.83	1881	23.2	5.1	2.01	0.85
Mathematics Procedures	20	1517	13.4	3.9	1.80	0.78	1875	12.6	4.1	1.75	0.82
Language	30	1233	21.0	5.5	2.26	0.83	1137	21.3	5.8	2.18	0.86
Spelling	30	1513	23.6	6.3	1.86	0.91	1873	23.5	6.0	1.91	0.90
Environment	30	1505	23.5	3.4	2.06	0.64	1882	23.5	3.1	1.92	0.62
Listening	30	1485	21.6	5.6	2.17	0.85	1809	22.1	5.0	2.16	0.81

Subtest	Number of Items	Form D					Form E				
		N	Mean	SD	SEM	KR20	N	Mean	SD	SEM	KR20
Language	30	621	21.3	4.9	2.17	0.81	932	21.7	4.7	2.16	0.79

* In terms of raw scores

Table C–43. KR20 Reliability Coefficients, Standard Errors of Measurement, and Related Summary Data for Primary 2, Fall, Grade 3, Abbreviated*

Subtest	Number of Items	Form A					Form B				
		N	Mean	SD	SEM	KR20	N	Mean	SD	SEM	KR20
Total Reading	70	1605	53.8	11.2	3.00	0.93	917	53.7	11.6	3.02	0.93
Word Study Skills	20	1620	15.6	3.1	1.49	0.77	920	15.4	3.2	1.53	0.78
Reading Vocabulary	20	1618	16.7	3.7	1.43	0.85	925	16.4	3.9	1.45	0.86
Reading Comprehension	30	1630	21.4	5.8	2.13	0.87	928	21.9	5.8	2.11	0.87
Total Mathematics	50	1618	37.2	8.6	2.69	0.90	923	38.4	7.9	2.57	0.89
Mathematics Problem Solving	30	1622	21.9	5.3	2.12	0.84	930	22.9	4.8	2.01	0.83
Mathematics Procedures	20	1628	15.4	4.0	1.62	0.84	923	15.5	3.8	1.57	0.83
Language	30	1030	24.4	4.7	1.89	0.84	647	24.1	5.2	1.93	0.86
Spelling	30	1628	25.0	5.0	1.79	0.87	922	25.7	5.1	1.67	0.90
Environment	30	1594	23.7	3.2	2.03	0.59	923	22.2	3.7	2.16	0.65
Listening	30	1621	20.9	4.7	2.24	0.77	910	21.8	4.8	2.18	0.79

Subtest	Number of Items	Form D					Form E				
		N	Mean	SD	SEM	KR20	N	Mean	SD	SEM	KR20
Language	30	854	22.7	4.1	2.10	0.73	412	22.8	5.0	2.12	0.82

* In terms of raw scores

Table C–44. KR20 Reliability Coefficients, Standard Errors of Measurement, and Related Summary Data for Primary 3, Fall, Grade 4, Abbreviated*

Subtest	Number of Items	Form A					Form B				
		N	Mean	SD	SEM	KR20	N	Mean	SD	SEM	KR20
Total Reading	70	4225	48.5	13.8	3.28	0.94	2944	47.4	13.2	3.39	0.93
Word Study Skills	20	4271	13.6	3.9	1.72	0.80	2969	13.9	4.0	1.77	0.80
Reading Vocabulary	20	4267	14.3	4.5	1.73	0.85	2961	14.9	4.3	1.66	0.85
Reading Comprehension	30	4291	20.5	6.9	2.13	0.91	2972	18.5	6.6	2.29	0.88
Total Mathematics	50	4257	32.1	10.6	2.94	0.92	2960	33.5	10.1	2.88	0.92
Mathematics Problem Solving	30	4275	20.0	6.5	2.24	0.88	2976	21.1	6.2	2.15	0.88
Mathematics Procedures	20	4285	12.1	4.9	1.85	0.86	2967	12.5	4.6	1.87	0.84
Language	30	3371	20.1	6.4	2.24	0.88	1687	18.8	6.4	2.29	0.87
Spelling	30	4268	17.8	6.5	2.35	0.87	2969	18.2	6.4	2.35	0.87
Science	30	4262	19.8	6.0	2.25	0.86	2964	19.3	5.8	2.31	0.84
Social Science	30	4256	18.9	6.7	2.24	0.89	2978	20.6	5.8	2.18	0.86
Listening	30	4248	20.4	5.5	2.27	0.83	2969	19.5	5.0	2.31	0.79

Subtest	Number of Items	Form D					Form E				
		N	Mean	SD	SEM	KR20	N	Mean	SD	SEM	KR20
Language	30	831	15.8	6.4	2.43	0.86	1235	18.3	5.8	2.38	0.83
Prewriting	9	841	5.5	2.4	1.24	0.73	1241	5.8	2.0	1.25	0.63
Composing	10	833	4.4	2.3	1.45	0.62	1240	5.7	2.3	1.40	0.63
Editing	11	839	5.9	2.8	1.47	0.72	1241	6.8	2.5	1.44	0.68

* In terms of raw scores

Table C–45. KR20 Reliability Coefficients, Standard Errors of Measurement, and Related Summary Data for Intermediate 1, Fall, Grade 5, Abbreviated*

Subtest	Number of Items	Form A					Form B				
		N	Mean	SD	SEM	KR20	N	Mean	SD	SEM	KR20
Total Reading	70	3956	46.8	13.7	3.38	0.94	3392	48.7	14.1	3.33	0.94
Word Study Skills	20	4046	13.4	4.4	1.72	0.85	3461	13.4	4.5	1.78	0.84
Reading Vocabulary	20	3989	14.5	4.5	1.73	0.85	3420	15.3	4.6	1.60	0.88
Reading Comprehension	30	4074	18.8	6.4	2.28	0.87	3464	19.7	6.6	2.25	0.88
Total Mathematics	50	4071	31.2	10.2	2.91	0.92	3463	32.6	10.3	2.88	0.92
Mathematics Problem Solving	30	4084	18.1	6.0	2.24	0.86	3469	19.5	6.0	2.22	0.86
Mathematics Procedures	20	4083	13.1	4.9	1.82	0.86	3484	13.0	4.9	1.80	0.86
Language	30	2637	20.0	6.3	2.26	0.87	2126	20.5	6.1	2.24	0.87
Language Mechanics	15	2637	9.8	3.2	1.62	0.75	2126	10.1	3.1	1.59	0.74
Language Expression	15	2637	10.2	3.6	1.55	0.82	2126	10.4	3.5	1.55	0.81
Spelling	30	4052	18.6	6.3	2.30	0.87	3473	19.4	6.3	2.24	0.87
Science	30	4006	20.2	6.0	2.26	0.86	3464	20.2	5.8	2.20	0.86
Social Science	30	4002	18.6	6.6	2.29	0.88	3459	20.0	6.5	2.24	0.88
Listening	30	4019	18.8	5.1	2.33	0.79	3431	20.4	5.5	2.20	0.84

Subtest	Number of Items	Form D					Form E				
		N	Mean	SD	SEM	KR20	N	Mean	SD	SEM	KR20
Language	30	1294	18.1	6.2	2.38	0.85	1303	19.6	6.1	2.29	0.86
Prewriting	8	1311	5.3	2.0	1.18	0.64	1303	5.5	2.0	1.13	0.68
Composing	11	1320	6.6	2.6	1.44	0.70	1310	6.8	2.5	1.43	0.67
Editing	11	1310	6.0	2.6	1.47	0.68	1310	7.3	2.5	1.37	0.71

* In terms of raw scores

Table C–46. KR20 Reliability Coefficients, Standard Errors of Measurement, and Related Summary Data for Intermediate 2, Fall, Grade 6, Abbreviated*

Subtest	Number of Items	Form A					Form B				
		N	Mean	SD	SEM	KR20	N	Mean	SD	SEM	KR20
Total Reading	50	4651	33.8	9.3	2.91	0.90	4125	34.0	9.8	2.88	0.91
Reading Vocabulary	20	4657	14.9	4.0	1.72	0.82	4139	14.6	4.1	1.76	0.81
Reading Comprehension	30	4704	18.8	5.9	2.34	0.84	4176	19.3	6.4	2.26	0.88
Total Mathematics	50	4677	29.9	10.0	2.99	0.91	4182	28.2	9.9	3.03	0.91
Mathematics Problem Solving	30	4694	18.7	6.2	2.32	0.86	4193	17.1	6.0	2.36	0.85
Mathematics Procedures	20	4688	11.2	4.5	1.87	0.83	4191	11.1	4.6	1.88	0.83
Language	30	3411	20.1	6.1	2.23	0.87	2664	18.8	6.1	2.32	0.86
Language Mechanics	15	3411	10.2	3.1	1.56	0.75	2664	10.0	3.1	1.55	0.76
Language Expression	15	3411	9.9	3.5	1.58	0.80	2664	8.8	3.5	1.70	0.77
Spelling	30	4678	19.6	6.3	2.28	0.87	4188	20.5	6.1	2.22	0.87
Science	30	4678	18.1	5.5	2.42	0.81	4173	18.3	5.9	2.37	0.84
Social Science	30	4675	19.6	6.3	2.23	0.88	4150	17.6	6.0	2.36	0.85
Listening	30	4610	19.2	5.0	2.30	0.79	4116	19.8	5.1	2.31	0.80

Subtest	Number of Items	Form D					Form E				
		N	Mean	SD	SEM	KR20	N	Mean	SD	SEM	KR20
Language	30	1107	18.2	6.0	2.39	0.84	1477	18.6	5.8	2.35	0.84
Prewriting	8	1108	5.2	2.0	1.19	0.63	1477	6.0	1.8	1.07	0.66
Composing	11	1116	5.9	2.5	1.48	0.64	1479	6.3	2.3	1.45	0.61
Editing	11	1115	7.0	2.5	1.43	0.68	1479	6.3	2.6	1.49	0.66

* In terms of raw scores

Table C–47.   KR20 Reliability Coefficients, Standard Errors of Measurement, and Related Summary Data for Intermediate 3, Fall, Grade 7, Abbreviated*

Subtest	Number of Items	Form A					Form B				
		N	Mean	SD	SEM	KR20	N	Mean	SD	SEM	KR20
Total Reading	50	3989	33.9	9.9	2.89	0.92	4247	31.7	10.0	2.94	0.91
Reading Vocabulary	20	4000	14.1	4.2	1.79	0.82	4267	14.2	4.5	1.74	0.85
Reading Comprehension	30	4032	19.7	6.4	2.26	0.88	4340	17.4	6.2	2.35	0.86
Total Mathematics	50	4021	26.7	10.2	3.12	0.91	4333	26.9	10.7	3.08	0.92
Mathematics Problem Solving	30	4042	16.4	6.5	2.42	0.86	4344	16.4	6.7	2.41	0.87
Mathematics Procedures	20	4026	10.2	4.6	1.92	0.83	4363	10.5	4.8	1.87	0.85
Language	30	2644	19.3	6.3	2.28	0.87	2845	19.5	7.0	2.24	0.90
Language Mechanics	15	2644	9.9	3.1	1.59	0.75	2851	9.8	3.5	1.58	0.80
Language Expression	15	2644	9.4	3.7	1.63	0.80	2854	9.7	3.9	1.57	0.84
Spelling	30	4029	19.0	6.4	2.30	0.87	4349	19.1	6.3	2.29	0.87
Science	30	4011	18.8	5.4	2.35	0.81	4330	17.8	5.6	2.40	0.82
Social Science	30	4004	17.0	6.1	2.42	0.84	4313	16.9	6.1	2.39	0.85
Listening	30	4010	18.5	5.4	2.40	0.80	4305	19.1	5.9	2.29	0.85

Subtest	Number of Items	Form D					Form E				
		N	Mean	SD	SEM	KR20	N	Mean	SD	SEM	KR20
Language	30	1224	19.3	5.7	2.33	0.83	1345	17.6	6.0	2.35	0.85
Prewriting	8	1226	5.8	1.8	1.14	0.60	1346	5.3	1.9	1.17	0.62
Composing	11	1232	6.2	2.5	1.48	0.66	1348	6.0	2.5	1.46	0.65
Editing	11	1230	7.3	2.4	1.37	0.68	1349	6.3	2.5	1.41	0.69

* In terms of raw scores

Table C–48. KR20 Reliability Coefficients, Standard Errors of Measurement, and Related Summary Data for Advanced 1, Fall, Grade 8, Abbreviated*

Subtest	Number of Items	Form A					Form B				
		N	Mean	SD	SEM	KR20	N	Mean	SD	SEM	KR20
Total Reading	50	5076	33.2	10.4	2.88	0.92	3759	32.8	10.7	2.90	0.93
Reading Vocabulary	20	5085	13.9	4.6	1.75	0.86	3771	13.7	4.6	1.77	0.85
Reading Comprehension	30	5108	19.2	6.4	2.26	0.87	3830	18.9	6.8	2.26	0.89
Total Mathematics	50	4937	25.2	11.0	3.08	0.92	3755	26.5	11.2	3.05	0.93
Mathematics Problem Solving	30	5008	15.0	6.6	2.40	0.87	3824	16.4	6.9	2.35	0.89
Mathematics Procedures	20	4958	10.1	5.1	1.89	0.87	3799	10.0	5.0	1.90	0.86
Language	30	3506	19.5	6.9	2.23	0.90	2565	19.2	6.9	2.27	0.89
Language Mechanics	15	3506	10.1	3.4	1.55	0.79	2565	9.7	3.5	1.58	0.80
Language Expression	15	3507	9.4	4.1	1.57	0.85	2565	9.5	3.9	1.61	0.83
Spelling	30	5039	18.9	5.8	2.31	0.84	3811	18.6	5.8	2.32	0.84
Science	30	4956	17.4	6.0	2.38	0.84	3751	17.6	5.6	2.35	0.82
Social Science	30	4944	16.1	5.9	2.46	0.82	3716	16.2	6.2	2.43	0.84
Listening	30	4899	19.3	5.2	2.37	0.79	3764	19.5	5.6	2.29	0.84

Subtest	Number of Items	Form D					Form E				
		N	Mean	SD	SEM	KR20	N	Mean	SD	SEM	KR20
Language	30	1365	18.8	6.3	2.34	0.86	1152	17.7	6.4	2.36	0.86
Prewriting	8	1374	5.1	1.9	1.23	0.59	1156	4.4	1.8	1.22	0.55
Composing	11	1372	6.9	2.5	1.42	0.69	1155	6.8	2.8	1.40	0.75
Editing	11	1373	6.8	2.8	1.38	0.76	1155	6.5	2.6	1.43	0.71

* In terms of raw scores

Table C–49.   KR20 Reliability Coefficients, Standard Errors of Measurement, and Related Summary Data for Advanced 2, Fall, Grade 9, Abbreviated*

Subtest	Number of Items	Form A					Form B				
		N	Mean	SD	SEM	KR20	N	Mean	SD	SEM	KR20
Total Reading	50	1293	32.6	9.1	2.99	0.89	1169	33.0	8.8	2.94	0.89
Reading Vocabulary	20	1309	13.9	4.3	1.81	0.82	1183	13.9	4.0	1.80	0.80
Reading Comprehension	30	1311	18.6	6.1	2.32	0.85	1181	19.0	5.6	2.31	0.83
Total Mathematics	50	1261	23.6	9.7	3.18	0.89	1004	25.8	10.6	3.09	0.92
Mathematics Problem Solving	30	1299	14.5	6.3	2.44	0.85	1152	14.7	6.7	2.40	0.87
Mathematics Procedures	20	1267	9.1	4.4	1.99	0.79	1021	10.4	5.0	1.88	0.86
Language	30	795	18.7	6.3	2.34	0.86	707	19.1	6.4	2.30	0.87
Language Mechanics	15	795	9.7	3.1	1.62	0.72	707	9.9	3.3	1.61	0.76
Language Expression	15	795	8.9	3.8	1.66	0.81	707	9.2	3.6	1.63	0.79
Spelling	30	1284	19.4	5.3	2.24	0.82	1137	19.9	5.9	2.20	0.86
Science	30	1257	15.7	5.3	2.47	0.79	1113	15.8	5.6	2.46	0.81
Social Science	30	1260	14.2	6.2	2.46	0.84	1061	13.6	5.8	2.47	0.82
Listening	30	1139	18.6	5.2	2.34	0.80	1011	18.2	5.6	2.39	0.82

Subtest	Number of Items	Form D					Form E				
		N	Mean	SD	SEM	KR20	N	Mean	SD	SEM	KR20
Language	30	446	20.2	6.5	2.22	0.88	371	19.1	6.3	2.30	0.87
Prewriting	8	446	6.1	2.0	1.02	0.75	374	5.8	1.9	1.09	0.69
Composing	11	450	7.2	2.9	1.34	0.78	376	6.0	2.6	1.48	0.67
Editing	11	450	6.8	2.6	1.39	0.71	373	7.3	2.8	1.34	0.77

* In terms of raw scores

Table C–50. KR20 Reliability Coefficients, Standard Errors of Measurement, and Related Summary Data for TASK 1, Fall, Grade 9, Abbreviated*

Subtest	Number of Items	Form A					Form B				
		N	Mean	SD	SEM	KR20	N	Mean	SD	SEM	KR20
Total Reading	50	1647	32.5	9.2	2.94	0.90	1253	33.6	8.8	2.89	0.89
Reading Vocabulary	20	1662	14.0	4.1	1.77	0.81	1265	13.4	3.6	1.84	0.74
Reading Comprehension	30	1667	18.4	5.9	2.33	0.84	1274	20.0	6.1	2.21	0.87
Mathematics	30	1644	12.6	5.3	2.43	0.79	1276	13.6	5.4	2.41	0.80
Language	30	942	18.1	6.5	2.33	0.87	644	18.0	6.3	2.34	0.86
Language Mechanics	15	942	8.9	3.1	1.64	0.73	644	9.3	3.0	1.63	0.70
Language Expression	15	942	9.2	3.9	1.62	0.83	644	8.7	3.9	1.65	0.82
Spelling	30	1617	17.7	5.6	2.35	0.82	1280	17.5	5.4	2.37	0.81
Science	30	1629	15.3	5.2	2.41	0.79	1247	14.9	5.2	2.44	0.78
Social Science	30	1623	12.9	5.4	2.51	0.79	1247	12.8	5.2	2.48	0.77

Subtest	Number of Items	Form D					Form E				
		N	Mean	SD	SEM	KR20	N	Mean	SD	SEM	KR20
Language	30	647	19.3	6.3	2.30	0.87	577	19.2	6.5	2.26	0.88
Prewriting	8	648	5.3	2.0	1.18	0.64	577	5.1	2.0	1.15	0.66
Composing	11	651	7.3	2.8	1.34	0.77	579	7.1	2.8	1.38	0.75
Editing	11	650	6.7	2.4	1.45	0.63	577	7.0	2.6	1.36	0.74

* In terms of raw scores

Table C–51. KR20 Reliability Coefficients, Standard Errors of Measurement, and Related Summary Data for TASK 2, Fall, Grade 10, Abbreviated*

Subtest	Number of Items	Form A					Form B				
		N	Mean	SD	SEM	KR20	N	Mean	SD	SEM	KR20
Total Reading	50	1505	29.3	11.1	2.94	0.93	1284	27.4	11.5	2.96	0.93
Reading Vocabulary	20	1610	12.4	4.8	1.78	0.87	1290	12.1	5.4	1.79	0.89
Reading Comprehension	30	1512	16.8	6.8	2.32	0.88	1339	15.0	6.9	2.33	0.89
Mathematics	30	1591	11.5	5.0	2.42	0.77	1316	10.5	4.6	2.41	0.73
Language	30	1412	16.0	6.5	2.43	0.86	736	18.1	6.6	2.32	0.88
Language Mechanics	15	1412	7.7	3.2	1.73	0.71	736	9.2	3.2	1.63	0.74
Language Expression	15	1412	8.3	3.9	1.68	0.81	736	8.9	3.9	1.62	0.83
Spelling	30	1611	15.5	6.9	2.30	0.89	1342	16.1	7.2	2.24	0.90
Science	30	1578	14.8	5.4	2.46	0.79	1325	14.5	5.5	2.44	0.81
Social Science	30	1574	10.6	4.5	2.48	0.70	1326	10.3	4.4	2.47	0.69

Subtest	Number of Items	Form D					Form E				
		N	Mean	SD	SEM	KR20	N	Mean	SD	SEM	KR20
Language	30	158	16.2	6.0	2.41	0.84	568	17.5	5.7	2.34	0.83
Prewriting	8	158	5.1	1.8	1.16	0.61	572	5.6	1.7	1.10	0.56
Composing	11	158	5.5	2.7	1.49	0.69	575	5.6	2.2	1.49	0.55
Editing	11	158	5.7	2.5	1.47	0.66	573	6.1	2.8	1.41	0.74

* In terms of raw scores

141

Table C–52.   KR20 Reliability Coefficients, Standard Errors of Measurement, and Related Summary Data for TASK 3, Fall, Grade 11, Abbreviated*

Subtest	Number of Items	Form A					Form B				
		N	Mean	SD	SEM	KR20	N	Mean	SD	SEM	KR20
Total Reading	50	1330	28.0	11.7	2.98	0.94	1338	29.2	10.8	3.00	0.92
Reading Vocabulary	20	1339	12.4	5.4	1.77	0.89	1340	11.9	4.3	1.90	0.80
Reading Comprehension	30	1344	15.5	7.0	2.36	0.89	1350	17.3	7.1	2.30	0.90
Mathematics	30	1330	10.7	4.5	2.48	0.70	1348	9.5	4.1	2.40	0.67
Language	30	949	15.2	6.5	2.41	0.86	1164	16.8	5.9	2.40	0.83
Language Mechanics	15	949	7.4	3.3	1.70	0.74	1164	8.1	2.9	1.71	0.64
Language Expression	15	949	7.8	3.8	1.69	0.80	1164	8.8	3.6	1.67	0.79
Spelling	30	1311	16.0	6.3	2.32	0.87	1343	16.3	7.0	2.29	0.89
Science	30	1368	11.3	5.1	2.43	0.77	1365	12.9	4.9	2.46	0.74
Social Science	30	1359	9.2	4.1	2.40	0.65	1353	10.0	4.1	2.45	0.63

Subtest	Number of Items	Form D					Form E				
		N	Mean	SD	SEM	KR20	N	Mean	SD	SEM	KR20
Language	30	396	15.1	6.7	2.36	0.88	195	8.7	6.9	2.07	0.91
Prewriting	8	400	4.7	2.2	1.17	0.73	198	2.1	2.2	0.98	0.81
Composing	11	406	5.2	2.7	1.45	0.71	199	3.7	2.7	1.33	0.75
Editing	11	402	5.1	2.8	1.42	0.74	198	2.8	2.6	1.24	0.77

* In terms of raw scores

Table C–53. KR20 Reliability Coefficients, Standard Errors of Measurement, and Related Summary Data for TASK 3, Fall, Grade 12, Abbreviated*

Subtest	Number of Items	Form A					Form B				
		N	Mean	SD	SEM	KR20	N	Mean	SD	SEM	KR20
Total Reading	50	869	30.6	10.8	2.94	0.93	1119	31.5	9.6	2.97	0.90
Reading Vocabulary	20	874	14.4	4.5	1.70	0.86	1125	13.1	3.7	1.86	0.75
Reading Comprehension	30	878	16.1	7.2	2.35	0.89	1127	18.3	6.6	2.28	0.88
Mathematics	30	818	11.7	4.9	2.50	0.74	1119	9.9	4.9	2.38	0.76
Language	30	592	16.1	7.3	2.35	0.90	761	16.9	6.2	2.38	0.85
Language Mechanics	15	592	8.2	3.6	1.68	0.78	760	8.2	2.9	1.70	0.67
Language Expression	15	592	7.9	4.2	1.63	0.85	760	8.7	3.9	1.64	0.83
Spelling	30	807	17.8	6.4	2.27	0.88	1066	17.9	6.6	2.29	0.88
Science	30	813	12.7	5.4	2.45	0.80	1044	12.5	5.2	2.44	0.78
Social Science	30	806	10.0	4.5	2.45	0.71	1052	10.2	4.3	2.47	0.68

Subtest	Number of Items	Form D					Form E				
		N	Mean	SD	SEM	KR20	N	Mean	SD	SEM	KR20
Language	30	197	17.9	7.1	2.28	0.90	259	14.3	7.5	2.34	0.90
Prewriting	8	199	5.3	2.2	1.09	0.75	260	3.7	2.3	1.17	0.75
Composing	11	199	6.3	2.9	1.40	0.77	264	5.5	3.1	1.42	0.79
Editing	11	197	6.2	2.9	1.38	0.77	265	4.9	2.8	1.43	0.75

* In terms of raw scores

Appendix D:

Conditional Standard Errors of Measurement

Appendix D

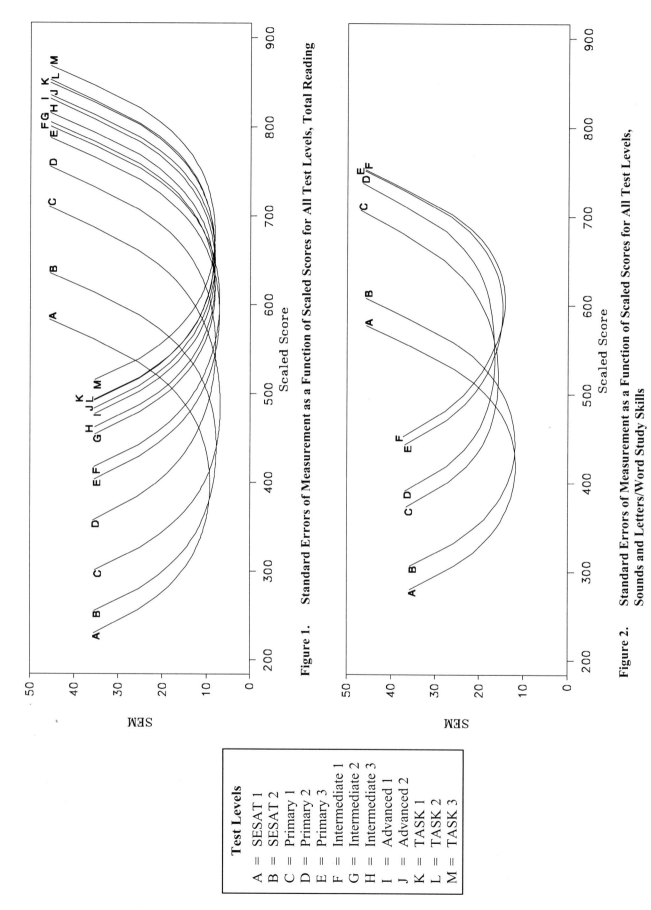

Figure 1. Standard Errors of Measurement as a Function of Scaled Scores for All Test Levels, Total Reading

Figure 2. Standard Errors of Measurement as a Function of Scaled Scores for All Test Levels, Sounds and Letters/Word Study Skills

Test Levels

A = SESAT 1
B = SESAT 2
C = Primary 1
D = Primary 2
E = Primary 3
F = Intermediate 1
G = Intermediate 2
H = Intermediate 3
I = Advanced 1
J = Advanced 2
K = TASK 1
L = TASK 2
M = TASK 3

146

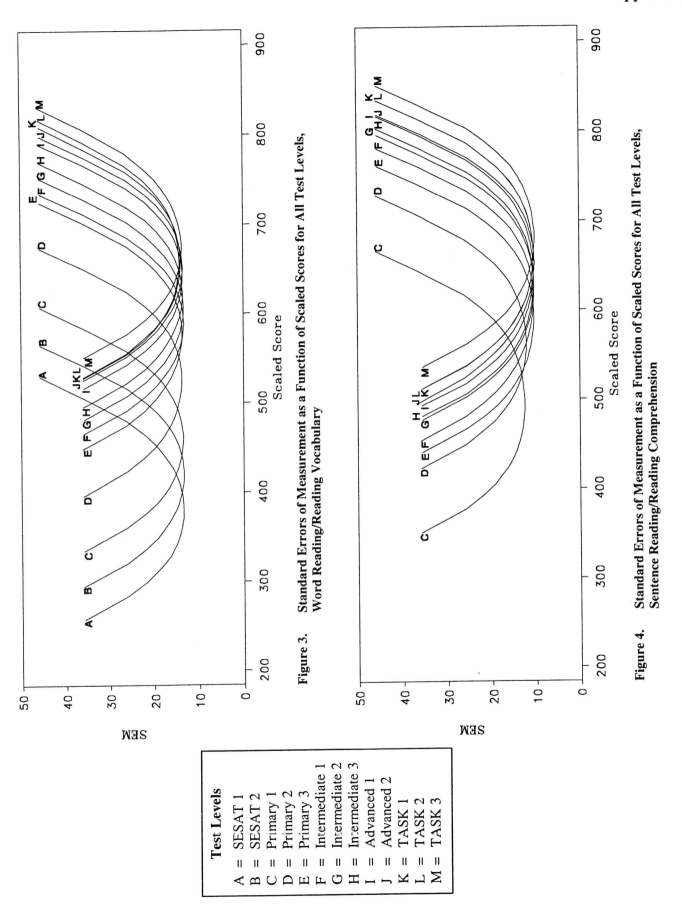

Figure 3.   Standard Errors of Measurement as a Function of Scaled Scores for All Test Levels, Word Reading/Reading Vocabulary

Figure 4.   Standard Errors of Measurement as a Function of Scaled Scores for All Test Levels, Sentence Reading/Reading Comprehension

Test Levels

A = SESAT 1
B = SESAT 2
C = Primary 1
D = Primary 2
E = Primary 3
F = Intermediate 1
G = Intermediate 2
H = Intermediate 3
I = Advanced 1
J = Advanced 2
K = TASK 1
L = TASK 2
M = TASK 3

Appendix D

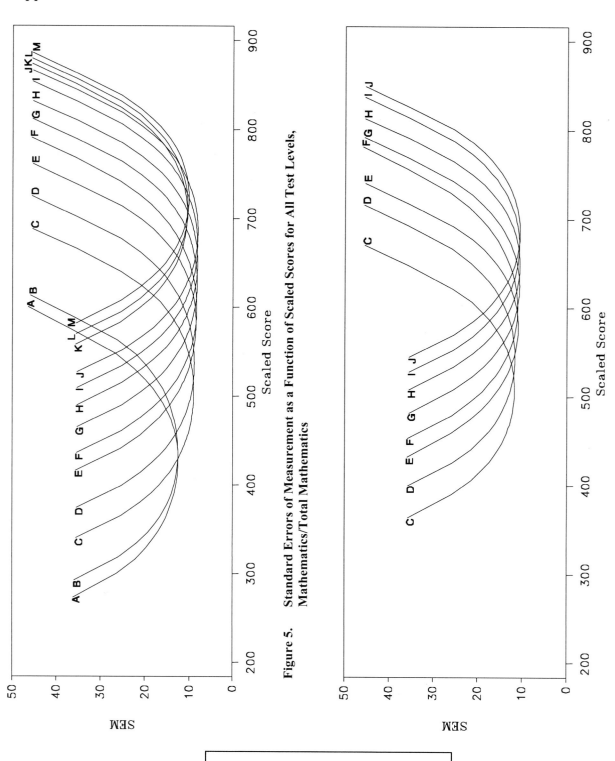

Figure 5. Standard Errors of Measurement as a Function of Scaled Scores for All Test Levels, Mathematics/Total Mathematics

Figure 6. Standard Errors of Measurement as a Function of Scaled Scores for All Test Levels, Mathematics Problem Solving

Test Levels

A = SESAT 1
B = SESAT 2
C = Primary 1
D = Primary 2
E = Primary 3
F = Intermediate 1
G = Intermediate 2
H = Intermediate 3
I = Advanced 1
J = Advanced 2
K = TASK 1
L = TASK 2
M = TASK 3

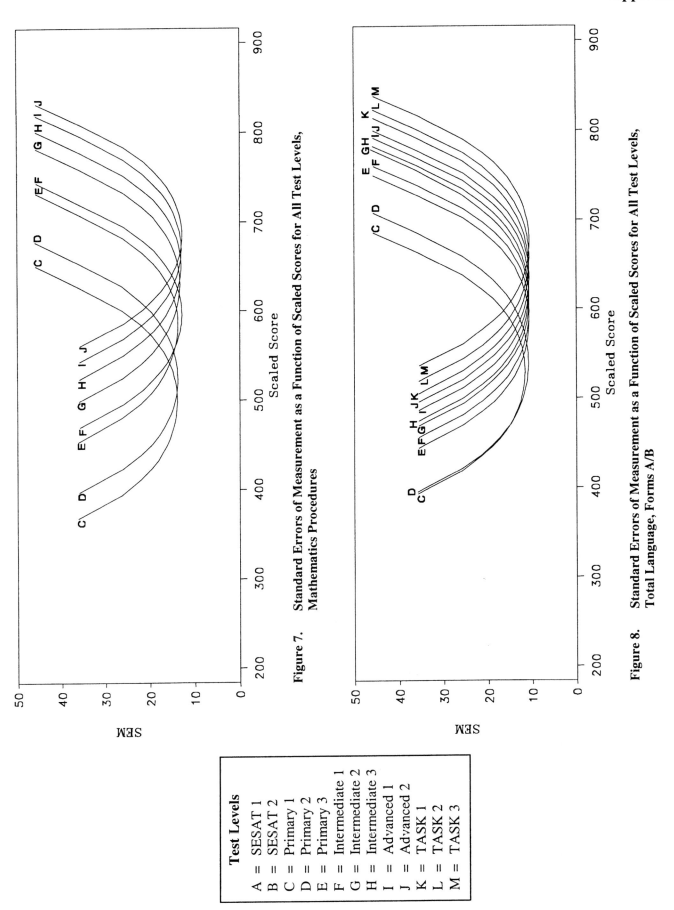

Figure 7. Standard Errors of Measurement as a Function of Scaled Scores for All Test Levels, Mathematics Procedures

Figure 8. Standard Errors of Measurement as a Function of Scaled Scores for All Test Levels, Total Language, Forms A/B

Test Levels

A = SESAT 1
B = SESAT 2
C = Primary 1
D = Primary 2
E = Primary 3
F = Intermediate 1
G = Intermediate 2
H = Intermediate 3
I = Advanced 1
J = Advanced 2
K = TASK 1
L = TASK 2
M = TASK 3

149

Appendix D

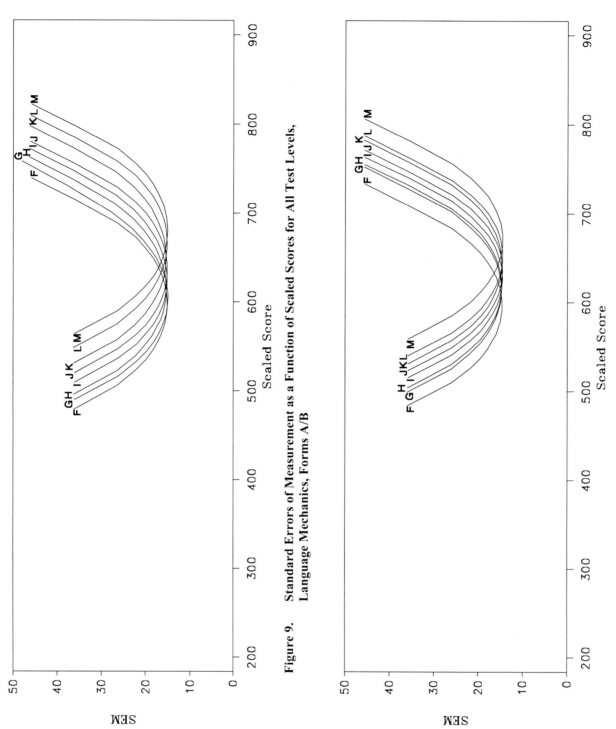

Figure 9.   Standard Errors of Measurement as a Function of Scaled Scores for All Test Levels, Language Mechanics, Forms A/B

Figure 10.   Standard Errors of Measurement as a Function of Scaled Scores for All Test Levels, Language Expression, Forms A/B

Test Levels

A = SESAT 1
B = SESAT 2
C = Primary 1
D = Primary 2
E = Primary 3
F = Intermediate 1
G = Intermediate 2
H = Intermediate 3
I = Advanced 1
J = Advanced 2
K = TASK 1
L = TASK 2
M = TASK 3

150

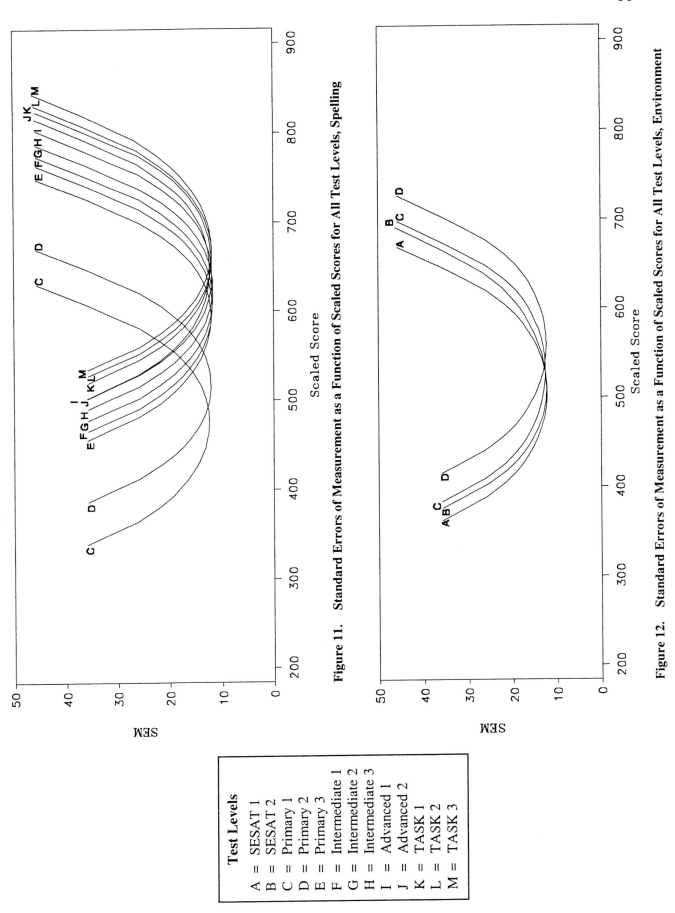

Figure 11. Standard Errors of Measurement as a Function of Scaled Scores for All Test Levels, Spelling

Figure 12. Standard Errors of Measurement as a Function of Scaled Scores for All Test Levels, Environment

Test Levels

A = SESAT 1
B = SESAT 2
C = Primary 1
D = Primary 2
E = Primary 3
F = Intermediate 1
G = Intermediate 2
H = Intermediate 3
I = Advanced 1
J = Advanced 2
K = TASK 1
L = TASK 2
M = TASK 3

Appendix D

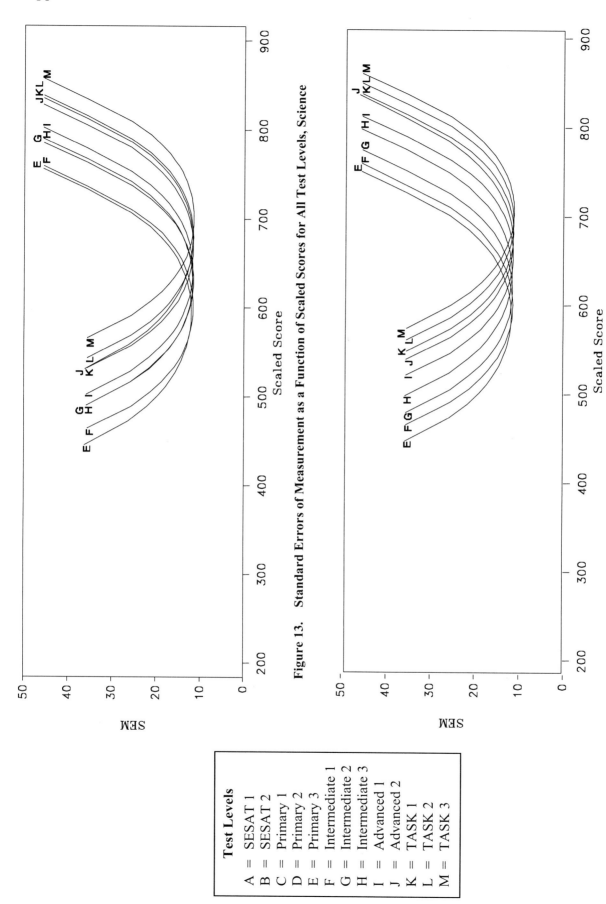

Figure 13. Standard Errors of Measurement as a Function of Scaled Scores for All Test Levels, Science

Figure 14. Standard Errors of Measurement as a Function of Scaled Scores for All Test Levels, Social Science

Test Levels

A = SESAT 1
B = SESAT 2
C = Primary 1
D = Primary 2
E = Primary 3
F = Intermediate 1
G = Intermediate 2
H = Intermediate 3
I = Advanced 1
J = Advanced 2
K = TASK 1
L = TASK 2
M = TASK 3

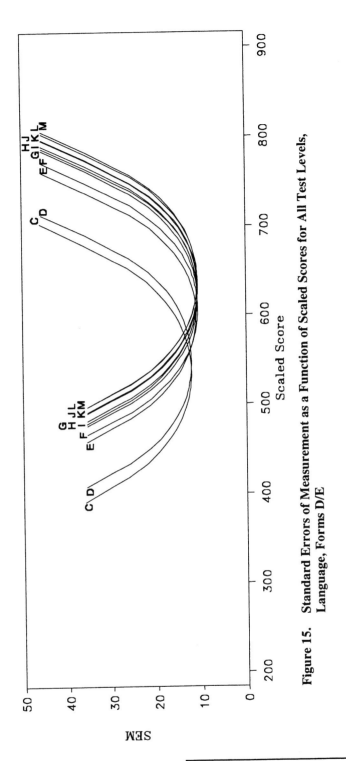

Figure 15. Standard Errors of Measurement as a Function of Scaled Scores for All Test Levels, Language, Forms D/E

Test Levels

A = SESAT 1
B = SESAT 2
C = Primary 1
D = Primary 2
E = Primary 3
F = Intermediate 1
G = Intermediate 2
H = Intermediate 3
I = Advanced 1
J = Advanced 2
K = TASK 1
L = TASK 2
M = TASK 3

153

Appendix E:

Alternate-Forms Reliability Coefficients (r), Standard Errors of Measurement, and Related Summary Data for the Equating of Forms Sample

Appendix E

Table E–1. Alternate-Forms Reliability Coefficients (r), Standard Errors of Measurement, and Related Summary Data for the Primary 1 Equating of Forms Sample*

Subtest/Total	Number of Items	N	Form A			Form B			r
			Mean	SD	SEM	Mean	SD	SEM	
Total Reading	130	730	101.1	21.4	5.6	101.8	22.3	5.8	0.93
Word Study Skills	30	730	21.3	5.0	2.4	21.9	4.8	2.3	0.76
Word Reading	30	730	24.4	6.0	2.4	24.3	6.3	2.5	0.85
Sentence Reading	30	800	25.3	5.0	2.2	25.4	5.7	2.5	0.80
Reading Comprehension	40	849	30.1	7.6	2.8	30.2	7.7	2.8	0.87
Total Mathematics	72	533	50.8	12.4	3.9	51.6	12.8	4.0	0.90
Mathematics Problem Solving	42	533	30.9	7.3	2.8	31.5	7.1	2.8	0.85
Mathematics Procedures	30	533	20.0	6.3	2.6	20.2	6.5	2.7	0.82
Language	40	589	26.2	7.2	2.8	27.0	7.7	3.0	0.85
Spelling	36	589	26.0	7.3	2.2	26.6	7.1	2.1	0.91
Environment	40	532	30.7	4.5	2.6	30.5	4.4	2.6	0.66
Listening	40	460	28.5	6.8	2.9	28.3	6.7	2.8	0.82

Subtest/Total	Number of Items	N	Form D			Form E			r
			Mean	SD	SEM	Mean	SD	SEM	
Language	40	120	29.4	6.5	2.4	29.2	7.1	2.6	0.86

* In terms of raw scores

Table E–2. Alternate-Forms Reliability Coefficients (r), Standard Errors of Measurement, and Related Summary Data for the Primary 2 Equating of Forms Sample*

Subtest/Total	Number of Items	N	Form A			Form B			r
			Mean	SD	SEM	Mean	SD	SEM	
Total Reading	100	862	71.3	17.0	4.7	70.7	17.9	4.9	0.92
Word Study Skills	30	862	22.2	4.5	2.0	22.3	4.8	2.1	0.81
Reading Vocabulary	30	862	22.8	6.1	2.4	22.1	6.3	2.5	0.85
Reading Comprehension	40	906	26.2	8.0	3.1	26.2	8.5	3.3	0.85
Total Mathematics	74	663	53.6	12.1	3.9	55.5	11.9	3.9	0.90
Mathematics Problem Solving	44	663	30.8	7.6	3.1	32.3	7.4	3.0	0.84
Mathematics Procedures	30	663	22.8	5.5	2.3	23.1	5.5	2.3	0.83
Language	48	632	36.9	7.4	3.0	37.6	6.9	2.8	0.84
Spelling	36	632	29.4	6.2	3.0	30.2	6.1	3.0	0.76
Environment	40	665	29.7	4.4	2.4	27.8	5.0	2.7	0.71
Listening	40	593	27.1	6.2	2.8	28.5	6.4	2.9	0.80

Subtest/Total	Number of Items	N	Form D			Form E			r
			Mean	SD	SEM	Mean	SD	SEM	
Language	40	108	30.1	5.5	2.7	30.5	6.3	3.1	0.76

* In terms of raw scores

Table E–3. Alternate-Forms Reliability Coefficients (r), Standard Errors of Measurement, and Related Summary Data for the Primary 3 Equating of Forms Sample*

Subtest/Total	Number of Items	N	Form A			Form B			r
			Mean	SD	SEM	Mean	SD	SEM	
Total Reading	114	788	75.2	20.8	6.7	74.3	21.6	7.0	0.90
Word Study Skills	30	788	20.3	5.6	2.4	20.2	5.7	2.4	0.82
Reading Vocabulary	30	788	20.2	6.4	2.8	20.7	6.3	2.7	0.82
Reading Comprehension	54	915	34.7	10.9	4.5	33.3	11.3	4.6	0.83
Total Mathematics	76	517	52.4	14.6	4.6	53.4	14.0	4.4	0.90
Mathematics Problem Solving	46	517	32.6	8.8	3.4	33.6	8.3	3.2	0.85
Mathematics Procedures	30	517	19.8	6.8	2.8	19.9	6.8	2.8	0.83
Language	48	536	31.5	9.4	3.5	30.3	9.8	3.7	0.86
Spelling	38	536	21.7	7.4	3.3	22.3	7.6	3.4	0.80
Science	40	524	27.4	6.7	3.0	25.9	6.7	3.0	0.80
Social Science	40	522	26.6	7.5	3.4	27.9	7.5	3.4	0.79
Listening	40	493	29.1	5.7	2.8	27.4	6.0	3.0	0.75

Subtest/Total	Number of Items	N	Form D			Form E			r
			Mean	SD	SEM	Mean	SD	SEM	
Language	45	144	25.5	8.5	3.3	26.1	8.6	3.4	0.85
Prewriting	14	144	8.3	2.6	1.4	8.5	2.9	1.7	0.68
Composing	16	144	8.1	3.5	2.0	8.3	3.5	2.0	0.68
Editing	15	144	9.1	3.4	1.6	9.3	3.3	1.6	0.78

* In terms of raw scores

Appendix E

Table E–4. **Alternate-Forms Reliability Coefficients (r), Standard Errors of Measurement, and Related Summary Data for the Intermediate 1 Equating of Forms Sample***

Subtest/Total	Number of Items	N	Form A			Form B			r
			Mean	SD	SEM	Mean	SD	SEM	
Total Reading	114	1004	75.7	22.0	6.1	74.3	23.2	6.4	0.92
Word Study Skills	30	1004	20.4	6.3	2.5	20.2	6.4	2.5	0.85
Reading Vocabulary	30	1004	21.6	6.7	2.7	21.4	7.2	2.9	0.83
Reading Comprehension	54	1113	32.9	11.2	4.3	32.0	11.9	4.6	0.85
Total Mathematics	80	600	54.0	15.0	7.0	55.5	15.3	7.1	0.78
Mathematics Problem Solving	48	600	31.2	8.9	4.5	32.6	9.0	4.6	0.74
Mathematics Procedures	32	600	22.8	7.2	3.7	22.9	7.1	3.7	0.73
Language	48	525	30.2	10.0	3.6	29.7	10.4	3.8	0.87
Language Mechanics	24	524	15.2	5.1	2.3	15.1	5.3	2.4	0.80
Language Expression	24	525	15.0	5.6	2.5	14.6	5.7	2.6	0.80
Spelling	40	525	23.9	7.9	3.3	23.5	8.2	3.4	0.83
Science	40	580	28.4	6.6	3.7	27.0	6.9	3.9	0.68
Social Science	40	577	27.9	7.5	4.2	28.4	7.8	4.3	0.69
Listening	40	534	27.0	5.7	2.9	27.2	6.1	3.1	0.74
Subtest/Total	Number of Items	N	Form D			Form E			r
			Mean	SD	SEM	Mean	SD	SEM	
Language	48	443	27.3	9.0	3.5	28.6	9.3	3.6	0.85
Prewriting	12	441	7.3	2.6	1.5	7.8	2.9	1.6	0.69
Composing	18	443	10.0	4.0	2.1	10.0	3.8	2.0	0.72
Editing	18	442	10.0	3.5	2.0	10.9	3.7	2.1	0.68

* In terms of raw scores

Table E–5. Alternate-Forms Reliability Coefficients (r), Standard Errors of Measurement, and Related Summary Data for the Intermediate 2 Equating of Forms Sample*

Subtest/Total	Number of Items	N	Form A			Form B			r
			Mean	SD	SEM	Mean	SD	SEM	
Total Reading	84	1399	54.3	15.7	5.9	54.7	16.5	6.2	0.86
Reading Vocabulary	30	1399	22.0	6.0	2.9	21.6	6.0	2.9	0.77
Reading Comprehension	54	1457	32.0	10.7	4.5	32.8	11.6	4.9	0.83
Total Mathematics	80	911	50.2	15.5	5.8	48.2	15.5	5.8	0.86
Mathematics Problem Solving	48	911	29.8	9.6	4.1	27.9	9.3	3.9	0.82
Mathematics Procedures	32	911	20.5	6.9	3.2	20.3	7.1	3.3	0.78
Language	48	741	31.1	9.9	4.0	30.5	9.9	4.0	0.83
Language Mechanics	24	741	16.1	5.0	2.5	15.9	4.9	2.4	0.76
Language Expression	24	741	15.0	5.5	2.7	14.6	5.7	2.7	0.77
Spelling	40	741	25.4	8.1	3.5	26.5	7.5	3.3	0.81
Science	40	890	24.3	7.0	3.4	24.9	7.3	3.6	0.76
Social Science	40	893	25.4	8.1	3.7	23.8	7.6	3.5	0.79
Listening	40	919	26.0	6.1	3.3	26.3	5.7	3.0	0.72

Subtest/Total	Number of Items	N	Form D			Form E			r
			Mean	SD	SEM	Mean	SD	SEM	
Language	48	604	26.2	9.0	3.7	28.0	9.6	3.9	0.84
Prewriting	12	604	7.5	2.6	1.6	8.1	2.7	1.7	0.62
Composing	18	603	9.1	3.9	2.0	9.7	3.8	2.0	0.74
Editing	18	604	9.7	3.5	1.9	10.2	4.1	2.2	0.72

* In terms of raw scores

Appendix E

Table E–6. Alternate-Forms Reliability Coefficients (r), Standard Errors of Measurement, and Related Summary Data for the Intermediate 3 Equating of Forms Sample*

Subtest/Total	Number of Items	N	Form A			Form B			r
			Mean	SD	SEM	Mean	SD	SEM	
Total Reading	84	1194	56.6	15.5	5.5	55.2	14.9	5.2	0.88
Reading Vocabulary	30	1194	22.3	5.5	2.5	22.1	5.7	2.5	0.80
Reading Comprehension	54	1222	34.1	10.9	4.5	32.9	10.2	4.2	0.83
Total Mathematics	80	671	48.7	16.0	5.3	48.4	15.2	5.1	0.89
Mathematics Problem Solving	48	671	28.5	9.7	3.8	28.2	9.2	3.6	0.85
Mathematics Procedures	32	671	20.2	7.3	3.2	20.2	6.9	3.0	0.80
Language	48	583	30.8	9.7	3.8	30.8	9.6	3.8	0.84
Language Mechanics	24	583	15.7	4.6	2.3	15.7	4.7	2.4	0.75
Language Expression	24	583	15.2	5.6	2.5	15.1	5.6	2.5	0.80
Spelling	40	583	25.1	8.1	3.1	25.4	8.1	3.1	0.85
Science	40	691	26.4	6.8	3.4	25.4	6.8	3.4	0.75
Social Science	40	686	23.9	7.4	3.1	24.3	7.6	3.2	0.83
Listening	40	654	27.1	6.6	2.8	27.5	6.1	2.6	0.82

Subtest/Total	Number of Items	N	Form D			Form E			r
			Mean	SD	SEM	Mean	SD	SEM	
Language	48	384	29.2	9.3	3.7	30.3	9.3	3.7	0.84
Prewriting	12	383	7.7	2.4	1.5	8.3	2.6	1.5	0.64
Composing	18	383	10.1	3.9	2.0	10.5	3.9	2.0	0.74
Editing	18	383	11.4	3.9	2.0	11.6	3.7	1.9	0.73

* In terms of raw scores

Table E–7. Alternate-Forms Reliability Coefficients (r), Standard Errors of Measurement, and Related Summary Data for the Advanced 1 Equating of Forms Sample*

Subtest/Total	Number of Items	N	Form A			Form B			r
			Mean	SD	SEM	Mean	SD	SEM	
Total Reading	84	592	55.7	15.1	5.8	55.5	15.7	6.0	0.85
Reading Vocabulary	30	592	21.8	5.9	2.6	21.1	6.2	2.8	0.80
Reading Comprehension	54	601	33.7	10.2	4.7	34.3	10.3	4.7	0.79
Total Mathematics	80	339	45.0	15.9	4.8	47.1	16.8	5.1	0.91
Mathematics Problem Solving	48	339	25.9	9.1	3.5	28.1	9.9	3.7	0.86
Mathematics Procedures	32	339	19.0	7.8	3.1	19.0	7.8	3.1	0.85
Language	48	298	31.5	10.0	3.6	29.8	10.9	3.9	0.87
Language Mechanics	24	298	15.9	4.9	2.3	15.5	5.6	2.5	0.79
Language Expression	24	298	15.5	5.6	2.5	14.3	6.0	2.6	0.81
Spelling	40	298	25.1	7.7	3.1	24.5	7.5	3.1	0.83
Science	40	349	25.2	6.6	3.5	24.0	6.7	3.5	0.72
Social Science	40	350	22.5	6.7	3.2	21.5	7.4	3.6	0.77
Listening	40	334	28.0	5.8	2.9	27.8	6.3	3.2	0.74
Subtest/Total	Number of Items	N	Form D			Form E			r
			Mean	SD	SEM	Mean	SD	SEM	
Language	48	204	28.4	8.8	3.3	28.5	8.8	3.4	0.86
Prewriting	12	202	7.5	2.5	1.6	6.5	2.3	1.5	0.61
Composing	18	204	10.2	3.7	2.1	10.9	3.8	2.1	0.69
Editing	18	204	10.7	3.6	1.7	11.1	3.9	1.9	0.77

* In terms of raw scores

Appendix E

Table E–8. Alternate-Forms Reliability Coefficients (r), Standard Errors of Measurement, and Related Summary Data for the Advanced 2 Equating of Forms Sample*

Subtest/Total	Number of Items	N	Form A			Form B			r
			Mean	SD	SEM	Mean	SD	SEM	
Total Reading	84	504	57.8	15.1	5.3	57.0	15.1	5.3	0.88
Reading Vocabulary	30	504	22.4	5.9	2.3	22.4	5.9	2.3	0.85
Reading Comprehension	54	513	35.2	10.0	4.4	34.4	10.2	4.4	0.81
Total Mathematics	80	333	46.0	15.6	5.7	46.3	16.0	5.9	0.87
Mathematics Problem Solving	48	333	26.4	8.9	3.5	26.5	9.5	3.7	0.85
Mathematics Procedures	32	333	19.7	7.6	3.5	19.8	7.8	3.6	0.79
Language	48	253	31.2	9.4	3.8	30.3	9.7	4.0	0.84
Language Mechanics	24	253	15.7	4.5	2.4	15.6	4.9	2.6	0.72
Language Expression	24	253	15.5	5.5	2.5	14.7	5.4	2.5	0.79
Spelling	40	253	25.5	6.5	2.8	25.4	6.8	2.9	0.82
Science	40	337	21.3	6.1	3.1	22.5	6.9	3.6	0.74
Social Science	40	334	21.0	7.8	3.5	20.9	7.8	3.5	0.80
Listening	40	272	27.2	6.2	3.0	26.7	6.5	3.1	0.77

Subtest/Total	Number of Items	N	Form D			Form E			r
			Mean	SD	SEM	Mean	SD	SEM	
Language	48	229	33.8	8.2	2.7	33.9	8.8	2.9	0.89
Prewriting	12	229	9.1	2.5	1.4	9.3	2.3	1.3	0.69
Composing	18	229	12.2	3.4	1.8	11.0	3.8	2.0	0.73
Editing	18	229	12.5	3.5	1.6	13.5	3.7	1.7	0.78

* In terms of raw scores

Table E–9. Alternate-Forms Reliability Coefficients (r), Standard Errors of Measurement, and Related Summary Data for the TASK 1 Equating of Forms Sample*

Subtest/Total	Number of Items	N	Form A			Form B			r
			Mean	SD	SEM	Mean	SD	SEM	
Total Reading	84	551	56.4	13.2	5.3	59.1	14.8	5.9	0.84
Reading Vocabulary	30	551	22.4	5.2	2.6	21.8	5.4	2.7	0.75
Reading Comprehension	54	611	34.1	9.0	4.4	37.0	10.3	5.0	0.77
Mathematics	50	202	25.5	7.8	3.4	25.4	8.1	3.6	0.81
Language	48	184	31.5	10.6	3.4	31.3	12.0	3.9	0.90
Language Mechanics	24	184	15.8	5.1	1.9	15.5	5.8	2.2	0.85
Language Expression	24	184	15.7	6.0	2.3	15.8	6.7	2.6	0.85
Spelling	40	184	24.9	8.5	3.0	25.0	8.2	2.9	0.87
Science	40	206	22.2	5.3	3.2	21.6	6.3	3.8	0.63
Social Science	40	198	18.2	6.9	3.4	18.6	7.5	3.6	0.77
Subtest/Total	Number of Items	N	Form D			Form E			r
			Mean	SD	SEM	Mean	SD	SEM	
Language	48	358	32.3	8.9	4.5	30.6	9.2	4.7	0.74
Prewriting	12	357	8.3	2.4	1.6	7.7	2.5	1.6	0.56
Composing	18	358	12.5	3.7	2.3	11.7	3.9	2.4	0.63
Editing	18	358	11.6	3.6	2.2	11.2	3.7	2.2	0.65

* In terms of raw scores

Appendix E

Table E–10. Alternate-Forms Reliability Coefficients (r), Standard Errors of Measurement, and Related Summary Data for the TASK 2 Equating of Forms Sample*

Subtest/Total	Number of Items	N	Form A			Form B			r
			Mean	SD	SEM	Mean	SD	SEM	
Total Reading	84	403	61.0	12.9	4.9	57.6	15.0	5.7	0.85
Reading Vocabulary	30	403	23.3	5.1	2.4	22.7	5.1	2.4	0.78
Reading Comprehension	54	501	37.5	8.6	4.0	34.2	10.8	5.1	0.78
Mathematics	50	257	23.2	8.9	3.3	22.7	9.2	3.4	0.87
Language	48	165	32.9	9.4	2.9	32.4	9.4	2.9	0.91
Language Mechanics	24	165	15.7	4.9	2.0	15.9	4.7	1.9	0.84
Language Expression	24	165	17.2	5.2	2.2	16.4	5.3	2.3	0.82
Spelling	40	165	27.4	7.5	2.9	26.4	7.9	3.1	0.84
Science	40	279	22.3	7.3	3.6	20.3	6.5	3.2	0.76
Social Science	40	256	17.6	6.3	3.0	16.4	6.4	3.0	0.78

Subtest/Total	Number of Items	N	Form D			Form E			r
			Mean	SD	SEM	Mean	SD	SEM	
Language	48	252	32.9	9.0	4.0	30.6	10.1	4.5	0.80
Prewriting	12	247	8.7	2.5	1.7	8.5	2.7	1.8	0.54
Composing	18	251	12.0	3.6	2.0	10.2	3.7	2.1	0.70
Editing	18	248	12.5	3.5	1.8	12.2	4.1	2.2	0.72

* In terms of raw scores

Table E–11. Alternate-Forms Reliability Coefficients (r), Standard Errors of Measurement, and Related Summary Data for the TASK 3 Equating of Forms Sample*

Subtest/Total	Number of Items	N	Form A			Form B			r
			Mean	SD	SEM	Mean	SD	SEM	
Total Reading	84	481	58.6	14.0	6.1	58.4	14.5	6.3	0.81
Reading Vocabulary	30	481	23.5	5.0	2.5	22.4	4.8	2.4	0.75
Reading Comprehension	54	522	34.9	9.9	5.0	35.6	10.9	5.5	0.74
Mathematics	50	307	21.3	8.6	3.9	19.6	7.7	3.5	0.80
Language	48	186	31.2	9.7	3.3	31.1	9.5	3.2	0.88
Language Mechanics	24	186	15.1	4.6	2.1	15.1	4.7	2.2	0.79
Language Expression	24	186	16.1	5.5	2.1	16.0	5.5	2.1	0.85
Spelling	40	186	25.9	7.9	2.7	26.2	8.1	2.7	0.89
Science	40	307	19.6	6.0	3.5	19.4	6.5	3.8	0.66
Social Science	40	300	15.9	5.8	3.4	14.8	5.7	3.3	0.66

Subtest/Total	Number of Items	N	Form D			Form E			r
			Mean	SD	SEM	Mean	SD	SEM	
Language	48	294	30.6	8.6	4.2	29.5	9.2	4.5	0.76
Prewriting	12	294	8.4	2.3	1.6	7.7	2.7	1.9	0.53
Composing	18	292	11.3	3.5	2.1	11.3	3.8	2.2	0.66
Editing	18	294	11.0	3.8	2.2	10.5	3.7	2.2	0.66

* In terms of raw scores

Appendix F:

Stanford 10 Student Completion Rates

Appendix F

Table F–1. Student Completion Rates for SESAT 2, Spring, Grade K, Full-Length

Subtest	Number of Items	Form A
Sounds and Letters	40	99
Word Reading	30	98
Sentence Reading	30	92
Mathematics	40	94
Environment	40	96
Listening to Words and Stories	40	97

Table F–2. Student Completion Rates for Primary 1, Spring, Grade 1, Full-Length

Subtest	Number of Items	Form A	Form B
Word Study Skills	30	98	98
Word Reading	30	96	95
Sentence Reading	30	98	98
Reading Comprehension	40	96	96
Mathematics Problem Solving	42	98	98
Mathematics Procedures	30	96	95
Language	40	99	99
Spelling	36	98	99
Environment	40	99	99
Listening	40	99	99
Subtest	**Number of Items**	**Form D**	**Form E**
Language	40	98	98

168

Table F–3. Student Completion Rates for Primary 2, Spring, Grade 2, Full-Length

Subtest	Number of Items	Form A	Form B
Word Study Skills	30	99	100
Reading Vocabulary	30	99	98
Reading Comprehension	40	94	94
Mathematics Problem Solving	44	99	99
Mathematics Procedures	30	97	97
Language	48	99	100
Spelling	36	99	99
Environment	40	100	99
Listening	40	99	100
Subtest	**Number of Items**	**Form D**	**Form E**
Language	40	98	99

Table F–4. Student Completion Rates for Primary 3, Spring, Grade 3, Full-Length

Subtest	Number of Items	Form A	Form B
Word Study Skills	30	99	98
Reading Vocabulary	30	98	98
Reading Comprehension	54	93	92
Mathematics Problem Solving	46	96	95
Mathematics Procedures	30	93	90
Language	48	96	95
Spelling	38	99	99
Science	40	96	95
Social Science	40	96	96
Listening	40	99	100
Subtest	**Number of Items**	**Form D**	**Form E**
Language	45	95	96

Appendix F

Table F–5. Student Completion Rates for Intermediate 1, Spring, Grade 4, Full-Length

Subtest	Number of Items	Form A	Form B
Word Study Skills	30	99	99
Reading Vocabulary	30	99	98
Reading Comprehension	54	93	91
Mathematics Problem Solving	48	95	94
Mathematics Procedures	32	93	93
Language	48	98	97
Spelling	40	99	99
Science	40	95	95
Social Science	40	93	94
Listening	40	99	99
Subtest	**Number of Items**	**Form D**	**Form E**
Language	48	95	99

Table F–6. Student Completion Rates for Intermediate 2, Spring, Grade 5, Full-Length

Subtest	Number of Items	Form A	Form B
Reading Vocabulary	30	98	98
Reading Comprehension	54	91	90
Mathematics Problem Solving	48	93	92
Mathematics Procedures	32	90	91
Language	48	99	99
Spelling	40	99	99
Science	40	95	94
Social Science	40	95	94
Listening	40	99	100
Subtest	**Number of Items**	**Form D**	**Form E**
Language	48	95	96

Table F–7. Student Completion Rates for Intermediate 3, Spring, Grade 6, Full-Length

Subtest	Number of Items	Form A	Form B
Reading Vocabulary	30	99	99
Reading Comprehension	54	92	94
Mathematics Problem Solving	48	94	93
Mathematics Procedures	32	91	90
Language	48	99	100
Spelling	40	100	100
Science	40	96	97
Social Science	40	96	95
Listening	40	99	99
Subtest	Number of Items	Form D	Form E
Language	48	98	99

Table F–8. Student Completion Rates for Advanced 1, Spring, Grade 7, Full-Length

Subtest	Number of Items	Form A	Form B
Reading Vocabulary	30	98	99
Reading Comprehension	54	95	94
Mathematics Problem Solving	48	93	95
Mathematics Procedures	32	90	91
Language	48	99	99
Spelling	40	100	100
Science	40	95	94
Social Science	40	95	97
Listening	40	99	99
Subtest	Number of Items	Form D	Form E
Language	48	98	98

Table F–9. Student Completion Rates for Advanced 2, Spring, Grade 8, Full-Length

Subtest	Number of Items	Form A	Form B
Reading Vocabulary	30	98	99
Reading Comprehension	54	94	94
Mathematics Problem Solving	48	93	94
Mathematics Procedures	32	92	94
Language	48	99	99
Spelling	40	99	100
Science	40	95	94
Social Science	40	95	95
Listening	40	99	99
Subtest	**Number of Items**	**Form D**	**Form E**
Language	48	95	100

Table F–10. Student Completion Rates for Advanced 2, Spring, Grade 9, Full-Length

Subtest	Number of Items	Form A	Form B
Reading Vocabulary	30	99	98
Reading Comprehension	54	95	94
Mathematics Problem Solving	48	91	93
Mathematics Procedures	32	89	92
Language	48	100	99
Spelling	40	99	100
Science	40	92	91
Social Science	40	93	95
Listening	40	100	99
Subtest	**Number of Items**	**Form D**	**Form E**
Language	48	98	99

Table F–11. Student Completion Rates for TASK 1, Spring, Grade 9, Full-Length

Subtest	Number of Items	Form A	Form B
Reading Vocabulary	30	100	99
Reading Comprehension	54	86	93
Mathematics	50	67	80
Language	48	99	99
Spelling	40	100	100
Science	40	91	90
Social Science	40	87	88
Subtest	Number of Items	Form D	Form E
Language	48	98	98

Table F–12. Student Completion Rates for TASK 2, Spring, Grade 10, Full-Length

Subtest	Number of Items	Form A	Form B
Reading Vocabulary	30	100	99
Reading Comprehension	54	91	91
Mathematics	50	80	79
Language	48	99	99
Spelling	40	100	100
Science	40	91	85
Social Science	40	92	81
Subtest	Number of Items	Form D	Form E
Language	48	93	99

Appendix F

Table F–13. Student Completion Rates for TASK 3, Spring, Grade 11, Full-Length

Subtest	Number of Items	Form A	Form B
Reading Vocabulary	30	99	100
Reading Comprehension	54	94	94
Mathematics	50	86	85
Language	48	100	97
Spelling	40	100	100
Science	40	88	89
Social Science	40	90	90
Subtest	**Number of Items**	**Form D**	**Form E**
Language	48	98	98

Table F–14. Student Completion Rates for TASK 3, Spring, Grade 12, Full-Length

Subtest	Number of Items	Form A	Form B
Reading Vocabulary	30	99	99
Reading Comprehension	54	94	95
Mathematics	50	83	81
Language	48	99	99
Spelling	40	100	99
Science	40	90	88
Social Science	40	93	91
Subtest	**Number of Items**	**Form D**	**Form E**
Language	48	90	99

174

Table F–15. Student Completion Rates for Primary 1, Spring, Grade 1, Abbreviated

Subtest	Number of Items	Form A	Form B
Word Study Skills	20	98	99
Word Reading	20	96	96
Sentence Reading	20	98	99
Reading Comprehension	30	96	96
Mathematics Problem Solving	30	99	98
Mathematics Procedures	20	96	96
Language	30	99	99
Spelling	30	99	99
Environment	30	99	99
Listening	30	99	98
Subtest	**Number of Items**	**Form D**	**Form E**
Language	30	99	98

Table F–16. Student Completion Rates for Primary 2, Spring, Grade 2, Abbreviated

Subtest	Number of Items	Form A	Form B
Word Study Skills	20	99	100
Reading Vocabulary	20	100	98
Reading Comprehension	30	95	96
Mathematics Problem Solving	30	99	99
Mathematics Procedures	20	97	97
Language	30	99	100
Spelling	30	99	99
Environment	30	100	99
Listening	30	99	99
Subtest	**Number of Items**	**Form D**	**Form E**
Language	30	98	99

Appendix F

Table F–17. Student Completion Rates for Primary 3, Spring, Grade 3, Abbreviated

Subtest	Number of Items	Form A	Form B
Word Study Skills	20	99	98
Reading Vocabulary	20	98	98
Reading Comprehension	30	94	92
Mathematics Problem Solving	30	96	95
Mathematics Procedures	20	93	90
Language	30	96	95
Spelling	30	99	99
Science	30	96	95
Social Science	30	96	96
Listening	30	99	100
Subtest	Number of Items	Form D	Form E
Language	30	96	98

Table F–18. Student Completion Rates for Intermediate 1, Spring, Grade 4, Abbreviated

Subtest	Number of Items	Form A	Form B
Word Study Skills	20	99	99
Reading Vocabulary	20	99	98
Reading Comprehension	30	93	91
Mathematics Problem Solving	30	95	94
Mathematics Procedures	20	93	95
Language	30	98	99
Spelling	30	99	99
Science	30	95	95
Social Science	30	93	94
Listening	30	100	99
Subtest	Number of Items	Form D	Form E
Language	30	99	99

Table F–19. Student Completion Rates for Intermediate 2, Spring, Grade 5, Abbreviated

Subtest	Number of Items	Form A	Form B
Reading Vocabulary	20	98	98
Reading Comprehension	30	91	90
Mathematics Problem Solving	30	93	92
Mathematics Procedures	20	90	91
Language	30	99	99
Spelling	30	99	99
Science	30	95	94
Social Science	30	95	94
Listening	30	100	99
Subtest	Number of Items	Form D	Form E
Language	30	98	99

Table F–20. Student Completion Rates for Intermediate 3, Spring, Grade 6, Abbreviated

Subtest	Number of Items	Form A	Form B
Reading Vocabulary	20	99	99
Reading Comprehension	30	92	94
Mathematics Problem Solving	30	94	93
Mathematics Procedures	20	91	90
Language	30	99	100
Spelling	30	100	100
Science	30	96	97
Social Science	30	96	95
Listening	30	99	99
Subtest	Number of Items	Form D	Form E
Language	30	100	98

Appendix F

Table F-21. Student Completion Rates for Advanced 1, Spring, Grade 7, Abbreviated

Subtest	Number of Items	Form A	Form B
Reading Vocabulary	20	98	99
Reading Comprehension	30	96	96
Mathematics Problem Solving	30	94	95
Mathematics Procedures	20	91	91
Language	30	99	99
Spelling	30	100	100
Science	30	95	94
Social Science	30	95	97
Listening	30	100	99
Subtest	**Number of Items**	**Form D**	**Form E**
Language	30	98	98

Table F-22. Student Completion Rates for Advanced 2, Spring, Grade 8, Abbreviated

Subtest	Number of Items	Form A	Form B
Reading Vocabulary	20	98	99
Reading Comprehension	30	94	94
Mathematics Problem Solving	30	93	94
Mathematics Procedures	20	92	94
Language	30	99	99
Spelling	30	99	100
Science	30	94	95
Social Science	30	96	95
Listening	30	99	99
Subtest	**Number of Items**	**Form D**	**Form E**
Language	30	95	100

Table F–23. Student Completion Rates for Advanced 2, Spring, Grade 9, Abbreviated

Subtest	Number of Items	Form A	Form B
Reading Vocabulary	20	99	98
Reading Comprehension	30	95	94
Mathematics Problem Solving	30	91	93
Mathematics Procedures	20	89	92
Language	30	100	99
Spelling	30	99	99
Science	30	92	91
Social Science	30	93	95
Listening	30	100	99
Subtest	Number of Items	Form D	Form E
Language	30	98	99

Table F–24. Student Completion Rates for TASK 1, Spring, Grade 9, Abbreviated

Subtest	Number of Items	Form A	Form B
Reading Vocabulary	20	100	99
Reading Comprehension	30	86	93
Mathematics	30	67	80
Language	30	99	99
Spelling	30	100	100
Science	30	91	90
Social Science	30	87	88
Subtest	Number of Items	Form D	Form E
Language	30	98	98

Appendix F

Table F–25. Student Completion Rates for TASK 2, Spring, Grade 10, Abbreviated

Subtest	Number of Items	Form A	Form B
Reading Vocabulary	20	100	100
Reading Comprehension	30	91	92
Mathematics	30	81	79
Language	30	99	99
Spelling	30	100	100
Science	30	91	85
Social Science	30	92	81
Subtest	**Number of Items**	**Form D**	**Form E**
Language	30	93	99

Table F–26. Student Completion Rates for TASK 3, Spring, Grade 11, Abbreviated

Subtest	Number of Items	Form A	Form B
Reading Vocabulary	20	99	100
Reading Comprehension	30	94	96
Mathematics	30	87	85
Language	30	100	97
Spelling	30	100	100
Science	30	88	88
Social Science	30	90	90
Subtest	**Number of Items**	**Form D**	**Form E**
Language	30	98	98

Table F–27. Student Completion Rates for TASK 3, Spring, Grade 12, Abbreviated

Subtest	Number of Items	Form A	Form B
Reading Vocabulary	20	99	99
Reading Comprehension	30	94	97
Mathematics	30	85	81
Language	30	99	99
Spelling	30	100	99
Science	30	90	88
Social Science	30	93	91
Subtest	**Number of Items**	**Form D**	**Form E**
Language	30	90	99

Appendix G:

Item *p*-Values for the Stanford 10 Standardization Samples

Appendix G

Table G–1. Item *p*–Values for SESAT 1, Forms A/D, Spring, Grade K

Item Number	Sounds and Letters	Word Reading	Mathematics	Environment	Listening to Words and Stories
1	.94	.92	.99	.95	.91
2	.90	.85	.98	.86	.96
3	.89	.81	.97	.75	.77
4	.85	.86	.97	.63	.93
5	.91	.85	.82	.83	.95
6	.89	.89	.82	.38	.73
7	.93	.83	.69	.76	.94
8	.90	.78	.87	.93	.82
9	.94	.85	.96	.54	.78
10	.86	.89	.67	.70	.96
11	.90	.82	.85	.78	.84
12	.83	.83	.65	.70	.42
13	.88	.92	.86	.87	.77
14	.79	.80	.93	.80	.75
15	.94	.87	.23	.95	.82
16	.88	.88	.92	.80	.57
17	.77	.83	.85	.91	.91
18	.76	.69	.91	.86	.85
19	.97	.80	.32	.62	.86
20	.97	.88	.87	.90	.83
21	.87	.78	.63	.96	.91
22	.99	.53	.62	.74	.90
23	.97	.72	.74	.97	.58
24	.96	.72	.89	.90	.73
25	.99	.65	.82	.83	.60
26	.98	.80	.69	.70	.70
27	.98	.79	.40	.75	.68
28	.98	.77	.82	.78	.77
29	.97	.67	.66	.59	.92
30	.98	.72	.78	.91	.66
31	.96		.96	.66	.90
32	.96		.81	.88	.88
33	.97		.85	.63	.36
34	.94		.83	.53	.56
35	.93		.83	.86	.71
36	.96		.98	.63	.61
37	.96		.97	.60	.76
38	.95		.78	.87	.79
39	.95		.92	.41	.82
40	.95		.75	.72	.89

Table G–2. Item *p*–Values for SESAT 2, Forms A/D, Spring, Grade K

Item Number	Sounds and Letters	Word Reading	Sentence Reading	Mathematics	Environment	Listening to Words and Stories
1	.90	.87	.73	.90	.92	.87
2	.85	.77	.80	.93	.79	.84
3	.85	.58	.79	.81	.82	.62
4	.89	.64	.74	.65	.76	.92
5	.82	.59	.71	.71	.87	.72
6	.87	.70	.63	.80	.88	.85
7	.86	.70	.64	.80	.72	.70
8	.90	.63	.69	.93	.78	.68
9	.95	.61	.61	.81	.80	.84
10	.83	.56	.33	.82	.76	.84
11	.82	.43	.39	.92	.41	.89
12	.84	.71	.65	.94	.86	.53
13	.91	.54	.49	.72	.79	.56
14	.86	.59	.68	.46	.93	.55
15	.86	.59	.57	.69	.74	.82
16	.71	.59	.38	.57	.29	.59
17	.75	.60	.68	.68	.83	.66
18	.73	.42	.52	.62	.86	.58
19	.93	.54	.54	.72	.59	.33
20	.97	.48	.51	.81	.83	.48
21	.86	.48	.56	.77	.85	.53
22	.95	.53	.38	.66	.83	.40
23	.96	.73	.50	.82	.81	.51
24	.96	.67	.43	.36	.78	.35
25	.89	.63	.62	.79	.71	.59
26	.87	.59	.40	.49	.14	.38
27	.77	.49	.27	.41	.69	.36
28	.83	.59	.44	.88	.42	.66
29	.83	.45	.33	.88	.82	.25
30	.78	.69	.60	.63	.57	.54
31	.77			.93	.52	.60
32	.50			.82	.92	.57
33	.94			.88	.78	.48
34	.95			.84	.35	.73
35	.67			.42	.50	.36
36	.92			.80	.58	.58
37	.92			.34	.67	.41
38	.65			.93	.38	.49
39	.70			.81	.85	.83
40	.54			.46	.74	.71

Appendix G

Table G–3. Item *p*–Values for SESAT 2, Forms A/D, Spring, Grade 1

Item Number	Sounds and Letters	Word Reading	Sentence Reading	Mathematics	Environment	Listening to Words and Stories
1	.96	.99	.93	.95	.96	.95
2	.93	.97	.92	.97	.83	.91
3	.94	.90	.96	.95	.91	.68
4	.95	.95	.95	.87	.83	.97
5	.92	.93	.96	.86	.95	.84
6	.92	.94	.97	.94	.92	.91
7	.92	.92	.95	.93	.78	.83
8	.96	.95	.95	.98	.92	.81
9	.97	.87	.94	.93	.91	.94
10	.95	.89	.77	.95	.86	.92
11	.90	.81	.82	.97	.51	.95
12	.93	.96	.95	.99	.96	.54
13	.96	.94	.91	.91	.89	.64
14	.94	.89	.97	.62	.97	.66
15	.91	.91	.92	.90	.78	.93
16	.77	.90	.88	.78	.31	.64
17	.87	.92	.96	.98	.90	.67
18	.82	.83	.93	.84	.89	.73
19	.98	.89	.91	.91	.74	.54
20	.98	.86	.93	.95	.93	.71
21	.98	.74	.87	.95	.96	.73
22	.98	.89	.89	.89	.89	.44
23	.99	.96	.92	.95	.86	.67
24	.98	.94	.91	.87	.85	.44
25	.96	.91	.94	.93	.86	.73
26	.98	.80	.92	.79	.20	.67
27	.93	.92	.78	.42	.83	.39
28	.86	.92	.89	.90	.44	.78
29	.98	.87	.89	.98	.86	.40
30	.96	.94	.88	.89	.61	.64
31	.96			.99	.65	.77
32	.94			.93	.96	.81
33	.99			.97	.88	.69
34	.99			.93	.43	.83
35	.90			.76	.58	.44
36	.98			.91	.69	.70
37	.99			.50	.71	.42
38	.92			.97	.48	.59
39	.93			.99	.94	.95
40	.89			.79	.90	.85

Table G–4. Item *p*–Values for Primary 1, Forms A/D, Spring, Grade 1

Item Number	Word Study Skills	Word Reading	Sentence Reading	Reading Comprehension	Mathematics Problem Solving	Mathematics Procedures	Language A	Spelling	Environment	Listening	Language D
1	.84*	.91*	.95*	.94	.61	.76*	.61*	.93*	.95*	.77*	.73*
2	.78*	.90*	.91*	.93*	.58	.88*	.90*	.85*	.76	.68*	.82*
3	.72*	.76*	.90	.70*	.89*	.55*	.74	.67*	.62*	.70*	.81
4	.75	.86	.93	.79	.71	.71	.53*	.73*	.84*	.44*	.96*
5	.92	.81	.79*	.94*	.66*	.75*	.67*	.75*	.56*	.46*	.92*
6	.95*	.82	.92*	.80*	.75*	.52	.57*	.66*	.78*	.78*	.94
7	.88*	.88	.75*	.92*	.82*	.52*	.79*	.91*	.67*	.50*	.73
8	.91	.79	.94	.80*	.59	.64	.45	.70*	.92*	.60*	.66
9	.94*	.76	.94*	.88*	.83	.85	.78*	.69*	.81*	.77*	.83*
10	.97	.73	.90*	.85	.47*	.72*	.34*	.80*	.56*	.47*	.82*
11	.93*	.69	.86	.92*	.86*	.78	.58*	.55	.73*	.78*	.67*
12	.95*	.65	.90*	.79*	.46*	.50*	.63	.58	.67	.58*	.78*
13	.92	.88*	.84*	.62*	.70*	.43	.66*	.86*	.89	.90*	.70*
14	.72*	.76*	.95*	.82	.86*	.52*	.61*	.51*	.87*	.84*	.46*
15	.78*	.91*	.85	.86	.68*	.79*	.70*	.64*	.48	.76*	.52*
16	.80*	.67*	.86*	.89	.69*	.48*	.55*	.81*	.72*	.69*	.81*
17	.49*	.82*	.63	.86*	.78*	.60*	.77	.90*	.92	.80	.69*
18	.72*	.80*	.56*	.77*	.70*	.58*	.53	.80	.39*	.68	.55*
19	.44	.87*	.90*	.80*	.83*	.25*	.71*	.65*	.86*	.67	.81*
20	.76	.74*	.72*	.84*	.46	.54	.38*	.41*	.92*	.50*	.41*
21	.81*	.76*	.81*	.89*	.65*	.81*	.78	.71*	.94	.57*	.69*
22	.28	.87*	.65*	.85*	.83	.76*	.62*	.43*	.92*	.45*	.34*
23	.54	.85*	.83	.85*	.43*	.57*	.63*	.66*	.75*	.56	.87*
24	.27*	.60*	.78	.82*	.76*	.57	.65*	.44*	.79*	.55	.25*
25	.68*	.75*	.83*	.65*	.76*	.54	.75*	.80*	.80*	.67	.62*
26	.65*	.88*	.88	.78*	.61*	.45*	.66	.73*	.52	.79*	.70*
27	.49*	.87*	.88*	.68*	.63	.49	.63*	.75	.96*	.70*	.74*
28	.49	.74*	.81*	.68*	.71*	.30*	.59	.71*	.87*	.52*	.73
29	.52*	.68*	.83*	.51*	.70*	.34*	.58	.56*	.48	.77*	.47*
30	.18*	.77	.83	.58*	.86*	.44*	.59*	.57*	.62*	.73*	.81*
31				.46*	.69		.58	.72*	.78*	.55*	.77*
32				.53	.59*		.84*	.46	.49*	.56*	.73*
33				.52	.93		.63*	.72*	.93	.67*	.55*
34				.63	.91*		.71*	.64*	.84*	.49*	.44*
35				.73*	.50*		.62*	.67*	.65*	.73*	.68
36				.51*	.88*		.67*	.56	.75*	.70*	.63
37				.47*	.55*		.50*		.72*	.65	.75
38				.61*	.42*		.62*		.92	.65	.63
39				.46*	.74*		.57*		.52*	.59	.68
40				.45	.85*		.46*		.93*	.50	.48
41					.70						
42					.54						

* Item is included in the Abbreviated Battery.

Appendix G

Table G–5. Item p–Values for Primary 1, Forms A/D, Spring, Grade 2

Item Number	Word Study Skills	Word Reading	Sentence Reading	Reading Comprehension	Mathematics Problem Solving	Mathematics Procedures	Language A	Spelling	Environment	Listening	Language D
1	.96*	.99*	.99*	.99	.81	.93*	.86*	.98*	.96*	.90*	.81*
2	.95*	.98*	.98*	.99*	.73	.95*	.98*	.98*	.92	.75*	.87*
3	.95*	.95*	.98	.91*	.92*	.87*	.88	.91*	.81*	.83*	.87
4	.94	.97	.99	.95	.88	.92	.78*	.95*	.91*	.62*	.98*
5	.98	.95	.93*	.98*	.78*	.84*	.92*	.96*	.61*	.55*	.93*
6	.99*	.96	.95*	.95*	.89*	.73	.59	.93*	.91*	.89*	.95
7	.97*	.97	.90*	.98*	.84*	.78*	.92*	.99*	.71*	.68*	.87
8	.98	.95	.99	.93*	.79	.83	.77	.91*	.95*	.84*	.76
9	.97*	.95	.97*	.96*	.93	.92	.91*	.95*	.89*	.87*	.90*
10	.99	.92	.97*	.97	.67*	.90*	.75*	.95*	.69*	.68*	.90*
11	.98*	.87	.96	.98*	.92*	.92	.88*	.89	.82*	.91*	.76*
12	.98*	.86	.98*	.95*	.58*	.76*	.91	.87	.79	.75*	.88*
13	.97	.96*	.97*	.84*	.86*	.75	.88*	.98*	.96	.95*	.82*
14	.90*	.95*	.99*	.97	.91*	.79*	.84*	.86*	.93*	.87*	.57*
15	.91*	.96*	.95	.94	.88*	.94*	.92*	.92*	.60	.87*	.77*
16	.85*	.89*	.96*	.99	.84*	.82*	.74*	.94*	.85*	.82*	.92*
17	.72*	.97*	.80	.96*	.88*	.90*	.93	.95*	.97	.81	.76*
18	.84*	.95*	.77*	.92*	.88*	.88*	.79	.96	.57*	.77	.72*
19	.73	.97*	.98*	.96*	.93*	.64*	.81*	.91*	.87*	.83	.93*
20	.82	.91*	.82*	.93*	.67	.86	.49*	.67*	.96*	.60*	.59*
21	.85*	.91*	.95*	.98*	.85*	.94*	.90	.96*	.98	.66*	.87*
22	.50	.99*	.80*	.96*	.94	.90*	.87*	.70*	.94*	.66*	.61*
23	.75	.97*	.94	.97*	.69*	.81*	.87*	.94*	.84*	.67	.91*
24	.48*	.85*	.88	.97*	.83*	.82	.87*	.88*	.88*	.71	.33*
25	.81*	.94*	.95*	.81*	.91*	.83	.87*	.95*	.88*	.80	.78*
26	.77*	.97*	.98	.94*	.78*	.77*	.88	.92*	.63	.89*	.83*
27	.70*	.97*	.98*	.87*	.82	.82	.83*	.97	.98*	.85*	.87*
28	.59	.95*	.96*	.88*	.79*	.74*	.78	.89*	.92*	.72*	.91*
29	.68*	.94*	.92*	.69*	.90*	.75*	.74	.89*	.56	.86*	.51*
30	.37*	.95	.87	.72*	.97*	.77*	.72*	.87*	.70*	.89*	.92*
31				.78*	.84		.73	.85*	.80*	.72*	.88*
32				.80	.72*		.93*	.81	.65*	.71*	.83*
33				.75	.97		.78*	.96*	.96	.81*	.78*
34				.88	.95*		.87*	.96*	.86*	.64*	.70*
35				.94*	.69*		.77*	.94*	.82*	.79*	.81
36				.83*	.94*		.82*	.91	.87*	.78*	.77
37				.77*	.68*		.77*		.74*	.77	.82
38				.88*	.79*		.76*		.96	.76	.71
39				.66*	.87*		.61*		.66*	.72	.71
40				.74	.92*		.58*		.95*	.56	.80
41					.88						
42					.80						

* Item is included in the Abbreviated Battery.

Table G–6. Item *p*–Values for Primary 2, Forms A/D, Spring, Grade 2

Item Number	Word Study Skills	Reading Vocabulary	Reading Comprehension	Mathematics Problem Solving	Mathematics Procedures	Language A	Spelling	Environment	Listening	Language D
1	.93*	.90	.86	.84	.92*	.87*	.89	.92*	.89*	.88*
2	.91*	.88*	.90	.83*	.90	.97*	.80*	.84*	.55*	.79*
3	.87*	.83*	.67	.78*	.87*	.81	.84*	.77	.29*	.83
4	.91	.79*	.58*	.85*	.86*	.86	.72*	.76*	.40*	.85
5	.98*	.90*	.71*	.67	.68*	.83*	.84*	.87	.54*	.59
6	.95	.91*	.79*	.85*	.72	.70*	.92*	.72	.72*	.92*
7	.93*	.72	.69*	.88*	.67*	.76*	.89*	.60*	.47*	.93
8	.98	.65*	.59*	.70*	.75	.85	.77*	.84	.58*	.77*
9	.98*	.48	.79*	.77	.87	.82*	.90*	.77*	.46*	.88*
10	.96	.62*	.75*	.61*	.85*	.76*	.75*	.81*	.54*	.71*
11	.97*	.85*	.68*	.63*	.73*	.74*	.82*	.50*	.96*	.77*
12	.90*	.90	.56*	.59*	.65*	.85*	.77*	.73	.78*	.53*
13	.84*	.77*	.45*	.74*	.66	.82	.82*	.95*	.95*	.57*
14	.70*	.69*	.80*	.46	.79*	.77*	.80*	.93*	.69*	.81*
15	.58*	.63*	.77*	.54	.78*	.50*	.90*	.82*	.79*	.82*
16	.67*	.71	.86*	.71	.51	.77	.84*	.52*	.70*	.77*
17	.80*	.62*	.44*	.86	.79*	.86	.85*	.87*	.63*	.52*
18	.61	.88	.87*	.87*	.78*	.65	.73*	.43*	.64*	.41
19	.34*	.78*	.80	.38*	.79	.81*	.71*	.57*	.78*	.63*
20	.53*	.68*	.85	.45	.74*	.62*	.79*	.75*	.82	.45*
21	.31	.69*	.74	.92*	.72*	.86	.62*	.96	.87	.84*
22	.50*	.73*	.30*	.55*	.46*	.75*	.61*	.74*	.84	.78
23	.79	.64	.71*	.49*	.74*	.87*	.90	.74*	.56*	.60*
24	.85	.72	.69*	.47*	.81	.40	.83	.73*	.55*	.79
25	.55	.86*	.47*	.70*	.37	.90	.60*	.67*	.50*	.79*
26	.39	.80*	.76*	.85*	.55*	.77*	.82	.84*	.56*	.63*
27	.23*	.77*	.63*	.50*	.72*	.80*	.88*	.82*	.63*	.43*
28	.82*	.65*	.73*	.70	.58*	.84	.82*	.49	.52*	.80*
29	.54*	.75	.59*	.75	.66	.86*	.79*	.72*	.59*	.66*
30	.24*	.66	.39*	.52*	.45*	.82	.70*	.86*	.40*	.73*
31			.60*	.59*		.74*	.81	.65*	.31*	.77*
32			.49*	.79*		.89*	.78*	.48*	.88*	.44*
33			.33*	.69		.69	.77*	.73*	.58*	.84
34			.77*	.65*		.63*	.76*	.88	.50	.86*
35			.68*	.87*		.62	.80*	.84	.69	.75*
36			.65*	.93		.81*	.70	.94*	.51	.49
37			.58	.56*		.82		.44*	.72	.40*
38			.57	.65*		.61		.86*	.57	.62*
39			.60	.67		.53*		.35	.64	.58
40			.44	.39		.86*		.91*	.80	.84*
41				.67*		.68*				
42				.85*		.72				
43				.28*		.69*				
44				.74*		.82*				
45						.77				
46						.88*				
47						.74*				
48						.53*				

* Item is included in the Abbreviated Battery.

Appendix G

Table G–7. Item *p*–Values for Primary 2, Forms A/D, Spring, Grade 3

Item Number	Word Study Skills	Reading Vocabulary	Reading Compre-hension	Mathematics Problem Solving	Mathematics Procedures	Language A	Spelling	Environment	Listening	Language D
1	.97*	.96	.93	.87	.96*	.93*	.93	.95*	.95*	.91*
2	.95*	.95*	.95	.91*	.95	.99*	.92*	.95*	.68*	.85*
3	.95*	.95*	.74	.89*	.93*	.87	.86*	.86	.35*	.89
4	.97	.90*	.68*	.91*	.91*	.94	.76*	.78*	.54*	.87
5	.99*	.97*	.79*	.76	.76*	.93*	.90*	.92	.64*	.62
6	.98	.97*	.88*	.88*	.78	.81*	.97*	.80	.84*	.94*
7	.98*	.83	.76*	.91*	.75*	.82*	.93*	.69*	.59*	.92
8	.99	.82*	.75*	.87*	.83	.95	.89*	.85	.74*	.87*
9	.98*	.72	.87*	.87	.92	.90*	.96*	.86*	.62*	.89*
10	.98	.74*	.83*	.72*	.91*	.86*	.84*	.90*	.68*	.73*
11	.99*	.93*	.79*	.79*	.81*	.86*	.92*	.59*	.97*	.85*
12	.93*	.94	.69*	.69*	.79*	.92*	.87*	.84	.89*	.61*
13	.84*	.90*	.50*	.82*	.80	.92	.89*	.97*	.97*	.68*
14	.72*	.84*	.89*	.61	.87*	.86*	.89*	.97*	.71*	.88*
15	.72*	.71*	.87*	.74	.89*	.60*	.97*	.84*	.84*	.86*
16	.72*	.81	.92*	.82	.77	.90	.92*	.57*	.74*	.83*
17	.88*	.76*	.48*	.92	.88*	.91	.92*	.93*	.73*	.66*
18	.67	.96	.95*	.94*	.86*	.86	.86*	.44*	.65*	.45
19	.48*	.88*	.89	.50*	.88	.87*	.83*	.63*	.84*	.66*
20	.60*	.84*	.87	.54	.83*	.77*	.91*	.75*	.87	.60*
21	.41	.84*	.81	.94*	.84*	.92	.74*	.98	.91	.89*
22	.66*	.85*	.43*	.77*	.59*	.91*	.65*	.78*	.89	.85
23	.82	.79	.76*	.63*	.86*	.94*	.96	.75*	.65*	.75*
24	.88	.83	.78*	.62*	.88	.56	.94	.75*	.55*	.89
25	.67	.95*	.55*	.82*	.50	.94	.72*	.78*	.52*	.84*
26	.55	.90*	.84*	.91*	.76*	.90*	.92	.89*	.72*	.65*
27	.32*	.88*	.70*	.68*	.84*	.91*	.95*	.83*	.74*	.44*
28	.90*	.79*	.80*	.88	.87*	.93	.91*	.60	.64*	.85*
29	.69*	.85	.73*	.84	.89	.92*	.88*	.82*	.67*	.71*
30	.39*	.80	.48*	.68*	.76*	.91	.82*	.87*	.45*	.85*
31			.67*	.77*		.82*	.88	.79*	.43*	.80*
32			.59*	.88*		.96*	.89*	.52*	.95*	.45*
33			.40*	.88		.75	.89*	.78*	.70*	.90
34			.85*	.80*		.80*	.89*	.91	.62	.95*
35			.77*	.87*		.76	.86*	.87	.78	.83*
36			.74*	.96		.92*	.84	.95*	.60	.55
37			.71	.63*		.92		.53*	.81	.38*
38			.68	.71*		.71		.90*	.67	.69*
39			.71	.76		.71*		.37	.75	.65
40			.62	.48		.90*		.95*	.85	.89*
41				.67*		.82*				
42				.89*		.82				
43				.47*		.82*				
44				.80*		.87*				
45						.89				
46						.95*				
47						.89*				
48						.69*				

* Item is included in the Abbreviated Battery.

Table G–8. Item *p*–Values for Primary 3, Forms A/D, Spring, Grade 3

Item Number	Word Study Skills	Reading Vocabulary	Reading Comprehension	Mathematics Problem Solving	Mathematics Procedures	Language A	Spelling	Science	Social Science	Listening	Language D
1	.96*	.88	.71*	.65	.81*	.90*	.82*	.94*	.83	.86*	.62*
2	.96	.78*	.79	.50*	.89*	.87*	.65	.92*	.87*	.73*	.82*
3	.92*	.80*	.85*	.88	.62	.39*	.49*	.67	.83*	.38*	.70*
4	.93*	.66	.71*	.49*	.73*	.63*	.62*	.62	.52*	.91*	.65*
5	.84*	.66*	.87*	.82*	.84	.68*	.45*	.70	.71	.35*	.48*
6	.84*	.61	.84*	.52	.63*	.71	.61*	.65*	.62	.73*	.41*
7	.74	.86*	.71*	.75*	.62*	.51	.85*	.66*	.83*	.54*	.52*
8	.74	.79*	.79*	.50	.81	.69*	.76*	.51*	.45*	.55*	.46*
9	.60*	.63*	.59	.86	.47*	.72	.46*	.76*	.63*	.71*	.47*
10	.41*	.48*	.85*	.47*	.54*	.79	.46	.66*	.67*	.62*	.73*
11	.56*	.91*	.60*	.79*	.53	.58	.57*	.91*	.67	.88*	.53*
12	.60	.42	.57*	.58*	.86*	.49*	.29	.41*	.43*	.70*	.54*
13	.81*	.61*	.83	.70*	.76*	.78	.68	.47*	.52	.83*	.35*
14	.77*	.70	.75	.58*	.46	.72*	.36*	.74*	.31*	.59*	.50*
15	.59*	.50*	.63	.55*	.76*	.59	.55*	.55*	.42	.88*	.71*
16	.53*	.63	.88	.66*	.67	.74*	.80*	.71*	.55*	.58*	.72*
17	.78	.70*	.68	.62	.67	.72*	.72*	.64	.61	.67*	.71*
18	.37	.62*	.60	.72*	.55	.82*	.42*	.35	.56*	.79*	.57*
19	.79*	.78*	.40*	.80*	.56	.57*	.53*	.84	.76*	.63	.38*
20	.69	.72*	.74*	.55	.49*	.61*	.33*	.42*	.61*	.62	.52*
21	.95*	.57	.92*	.61*	.46*	.69	.75*	.67*	.66*	.73	.31*
22	.81*	.82*	.46	.75*	.52*	.73	.40*	.55	.38*	.83	.69*
23	.55	.68	.69*	.64	.54*	.69*	.43*	.80*	.77*	.67	.44*
24	.37*	.73*	.63*	.44*	.50*	.38*	.65*	.71*	.61*	.74	.42*
25	.74*	.74	.67	.68	.54*	.70	.52*	.53*	.64*	.72*	.50*
26	.63*	.54*	.45	.96*	.42	.81*	.60*	.69*	.71*	.71*	.49*
27	.60	.71*	.59	.82*	.59*	.56*	.70*	.58*	.44*	.58*	.67*
28	.39*	.61*	.70	.53	.47*	.67	.56*	.49*	.57*	.74*	.58*
29	.63	.54*	.65	.61*	.61*	.70	.68	.39	.77*	.61*	.62*
30	.33*	.55	.69	.80	.35*	.55*	.46*	.51*	.46*	.74*	.77*
31			.46*	.65*		.68*	.40*	.59*	.52	.58*	.61
32			.71*	.72*		.79*	.40*	.38*	.56*	.73*	.31
33			.81*	.67*		.60*	.75*	.74*	.68	.40*	.29
34			.76*	.69		.58*	.52	.43*	.25*	.79*	.45
35			.81*	.59*		.40*	.49	.81*	.61*	.83*	.68
36			.80	.53		.70*	.65	.29*	.59*	.66*	.43
37			.53	.78		.71	.48*	.71	.34*	.61	.60
38			.52	.65*		.66	.46*	.35*	.64	.66	.65
39			.68	.65*		.75		.67	.41*	.84	.50
40			.58	.74*		.70*		.81*	.69*	.65	.44
41			.64	.66*		.72*					.47
42			.50*	.57		.48*					.78
43			.55*	.60*		.74					.46
44			.54*	.70		.49					.31
45			.63*	.85*		.52*					.45
46			.47	.69*		.70*					
47			.32*			.66					
48			.57*			.53*					
49			.63*								
50			.50								
51			.62*								
52			.65*								
53			.54*								
54			.45								

* Item is included in the Abbreviated Battery.

Appendix G

Table G–9. Item *p*–Values for Primary 3, Forms A/D, Spring, Grade 4

Item Number	Word Study Skills	Reading Vocabulary	Reading Comprehension	Mathematics Problem Solving	Mathematics Procedures	Language A	Spelling	Science	Social Science	Listening	Language D
1	.96*	.90	.75*	.76	.85*	.93*	.90*	.92*	.93	.89*	.57*
2	.95	.83*	.82	.61*	.86*	.84*	.77	.91*	.93*	.85*	.86*
3	.94*	.86*	.88*	.87	.74	.61*	.65*	.72	.88*	.51*	.74*
4	.94*	.76	.77*	.60*	.77*	.77*	.67*	.69	.64*	.93*	.65*
5	.85*	.76*	.91*	.83*	.85	.69*	.57*	.73	.82	.40*	.48*
6	.90*	.77	.83*	.62	.73*	.78	.73*	.72*	.77	.79*	.40*
7	.81	.91*	.77*	.85*	.69*	.73	.90*	.73*	.89*	.56*	.43*
8	.79	.90*	.84*	.64	.82	.81*	.78*	.57*	.62*	.64*	.48*
9	.74*	.73*	.59	.90	.62*	.77	.58*	.78*	.75*	.79*	.54*
10	.49*	.57*	.85*	.54*	.67*	.82	.54	.74*	.72*	.82*	.65*
11	.66*	.92*	.70*	.85*	.67	.71	.67*	.88*	.69	.86*	.49*
12	.73	.58	.64*	.70*	.90*	.57*	.45	.52*	.61*	.72*	.56*
13	.85*	.70*	.84	.81*	.82*	.80	.75	.58*	.66	.84*	.41*
14	.79*	.81	.81	.66*	.60	.87*	.47*	.80*	.49*	.68*	.46*
15	.65*	.62*	.71	.59*	.83*	.65	.68*	.66*	.54	.90*	.66*
16	.56*	.70	.86	.78*	.77	.85*	.86*	.73*	.70*	.64*	.70*
17	.83	.72*	.75	.74	.74	.75*	.83*	.69	.76	.71*	.68*
18	.44	.71*	.62	.82*	.67	.87*	.47*	.56	.65*	.84*	.63*
19	.81*	.83*	.51*	.84*	.71	.75*	.68*	.87	.84*	.65	.52*
20	.74	.77*	.83*	.69	.63*	.78*	.52*	.58*	.70*	.70	.62*
21	.91*	.69	.91*	.70*	.62*	.84	.77*	.73*	.77*	.77	.38*
22	.78*	.86*	.59	.77*	.69*	.83	.53*	.64	.51*	.84	.63*
23	.58	.78	.77*	.75	.60*	.77*	.55*	.80*	.85*	.72	.46*
24	.39*	.80*	.73*	.59*	.66*	.59*	.71*	.79*	.77*	.75	.39*
25	.75*	.81	.79	.74	.68*	.81	.63*	.57*	.70*	.76*	.55*
26	.65*	.68*	.52	.94*	.59	.84*	.73*	.76*	.79*	.74*	.52*
27	.64	.75*	.69	.85*	.68*	.69*	.73*	.61*	.55*	.61*	.57*
28	.42*	.74*	.74	.69	.63*	.75	.66*	.56*	.69*	.79*	.56*
29	.63	.68*	.76	.72*	.70*	.80	.75	.49	.83*	.68*	.61*
30	.35*	.71	.80	.83	.53*	.70*	.65*	.63*	.58*	.75*	.69*
31			.56*	.73*		.74*	.57*	.64*	.64	.66*	.63
32			.77*	.80*		.86*	.56*	.49*	.69*	.78*	.34
33			.84*	.76*		.76*	.75*	.80*	.73	.52*	.32
34			.82*	.76		.75*	.55	.66*	.50*	.80*	.34
35			.86*	.74*		.58*	.61	.84*	.73*	.83*	.68
36			.82	.58		.76*	.68	.37*	.71*	.71*	.38
37			.66	.81		.80	.60*	.74	.49*	.67	.58
38			.69	.67*		.80	.58*	.45*	.77	.66	.63
39			.80	.66*		.87		.79	.55*	.83	.42
40			.71	.72*		.77*		.82*	.72*	.67	.41
41			.70	.70*		.84*					.49
42			.58*	.67		.64*					.72
43			.67*	.64*		.84					.56
44			.61*	.69		.61					.31
45			.70*	.85*		.73*					.42
46			.52	.75*		.81*					
47			.33*			.80					
48			.65*			.62*					
49			.69*								
50			.59								
51			.69*								
52			.73*								
53			.62*								
54			.54								

* Item is included in the Abbreviated Battery.

Table G–10. Item *p*–Values for Intermediate 1, Forms A/D, Spring, Grade 4

Item Number	Word Study Skills	Reading Vocab- ulary	Reading Compre- hension	Mathematics Problem Solving	Mathematics Procedures	Language A	Spelling	Science	Social Science	Listening	Language D
1	.98*	.86	.94	.67	.75*	.83*	.79*	.80*	.84*	.57*	.61*
2	.87*	.64*	.54	.72*	.80*	.64	.70	.74*	.80*	.91*	.74*
3	.93*	.47*	.69	.51	.78	.84*	.68*	.60	.74*	.49*	.83*
4	.94	.79*	.88	.61	.63*	.65*	.43*	.64	.47*	.67*	.65*
5	.80	.68*	.53	.63*	.71*	.50*	.71*	.79*	.58*	.28*	.51*
6	.65	.58	.55	.16*	.62	.54*	.47*	.49*	.63	.61*	.56
7	.84*	.67*	.70*	.73*	.67	.80	.76*	.71*	.54*	.39*	.60*
8	.72*	.64	.64*	.69	.67*	.50*	.80*	.86*	.52*	.63*	.80*
9	.67*	.52*	.77*	.71	.69*	.61	.68*	.44	.68*	.52*	.75*
10	.54	.62*	.58	.83*	.73	.75*	.56*	.54	.58*	.64*	.43*
11	.59*	.71*	.71*	.40*	.81*	.40	.49*	.41*	.65*	.85*	.45*
12	.62*	.52	.38*	.74*	.57*	.59*	.81	.76*	.52	.49*	.56*
13	.71*	.86*	.87*	.67*	.45	.66	.57	.66	.41*	.70*	.61*
14	.86*	.77*	.77	.57	.76*	.69*	.56*	.71*	.47*	.46	.72*
15	.57	.77*	.73*	.93*	.48*	.70	.60	.51*	.66	.64	.75*
16	.50*	.85	.66*	.51	.53	.72*	.80*	.62*	.45*	.50	.51*
17	.41*	.74*	.46*	.62*	.74*	.70	.50	.75*	.64*	.67	.64*
18	.55*	.68	.74*	.67	.67*	.73*	.68*	.83*	.67*	.44*	.48*
19	.54	.60*	.89*	.81*	.49*	.52*	.84*	.54*	.51	.36*	.39*
20	.62*	.76	.56*	.14*	.63	.40*	.57*	.71*	.62*	.64*	.67*
21	.70*	.66*	.44*	.47*	.53*	.71*	.41*	.66	.40*	.77	.36*
22	.54	.76	.36*	.48*	.50*	.43	.45*	.64*	.53	.77	.60*
23	.42*	.76*	.85*	.44	.64*	.53*	.67*	.47	.68*	.88	.54*
24	.50	.74*	.49	.38	.48*	.44	.46*	.48*	.69*	.62*	.53*
25	.54*	.62*	.60*	.83	.58	.61*	.43*	.35	.45	.81*	.49*
26	.62*	.65*	.49*	.80*	.58*	.64*	.32*	.75*	.41*	.25*	.58*
27	.45*	.74	.41*	.69*	.50	.75*	.59*	.74	.56*	.62*	.34*
28	.66	.65*	.56*	.57*	.60	.52	.37	.43*	.65*	.61*	.36*
29	.55	.59	.57	.73*	.54	.58	.65*	.49*	.60	.71*	.60*
30	.34*	.71*	.43*	.76	.55*	.53*	.72*	.51*	.48	.27*	.70
31			.72	.58	.43	.65	.76*	.76*	.66*	.66*	.58*
32			.65	.49*	.57*	.52*	.32	.67*	.69	.44*	.61*
33			.54	.34*		.52*	.49	.69	.68*	.84*	.67
34			.78	.59*		.60	.37*	.48*	.41*	.69*	.60
35			.64	.47		.64	.80	.48*	.52*	.62*	.31
36			.61	.63*		.72*	.48*	.69*	.65*	.73*	.50
37			.54*	.70		.68	.53*	.55*	.55*	.79*	.43
38			.70*	.56*		.79*	.44*	.44*	.65	.58	.38
39			.50*	.56		.52*	.57*	.78*	.21*	.66	.47
40			.55*	.29*		.67	.58	.59*	.69*	.57	.52
41			.57	.34*		.70					.41
42			.49*	.40*		.64					.47
43			.56	.43		.76*					.28
44			.71	.74*		.67*					.71
45			.48	.51		.66*					.29
46			.29	.24*		.72*					.61
47			.39	.57*		.56*					.50
48			.42	.73*		.56*					.46
49			.43*								
50			.60*								
51			.62*								
52			.30*								
53			.52*								
54			.43								

* Item is included in the Abbreviated Battery.

Appendix G

Table G–11. Item *p*–Values for Intermediate 1, Forms A/D, Spring, Grade 5

Item Number	Word Study Skills	Reading Vocab-ulary	Reading Compre-hension	Mathematics Problem Solving	Mathematics Procedures	Language A	Spelling	Science	Social Science	Listening	Language D
1	.98*	.91	.96	.79	.81*	.91*	.85*	.85*	.88*	.67*	.63*
2	.92*	.73*	.68	.82*	.84*	.75	.72	.81*	.86*	.94*	.79*
3	.96*	.61*	.78	.62	.85	.90*	.80*	.63	.82*	.59*	.81*
4	.96	.89*	.92	.72	.76*	.78*	.44*	.76	.62*	.78*	.68*
5	.85	.79*	.61	.70*	.79*	.61*	.80*	.82*	.70*	.34*	.55*
6	.80	.69	.62	.28*	.72	.67*	.60*	.58*	.72	.72*	.61
7	.92*	.76*	.81*	.80*	.78	.87	.82*	.79*	.56*	.46*	.69*
8	.80*	.81	.74*	.81	.77*	.57*	.86*	.89*	.58*	.69*	.80*
9	.79*	.66*	.86*	.80	.83*	.75	.75*	.50	.78*	.54*	.78*
10	.62	.78*	.67	.86*	.86	.83*	.67*	.65	.72*	.84*	.41*
11	.74*	.77*	.79*	.56*	.88*	.59	.57*	.52*	.74*	.89*	.47*
12	.72*	.65	.48*	.82*	.68*	.64*	.88	.85*	.64	.56*	.55*
13	.78*	.89*	.92*	.79*	.69	.76	.70	.75	.54*	.79*	.63*
14	.89*	.86*	.85	.68	.84*	.81*	.57*	.79*	.60*	.54	.68*
15	.59	.81*	.81*	.95*	.62*	.77	.73	.59*	.74	.69	.68*
16	.55*	.87	.71*	.67	.68	.78*	.81*	.74*	.47*	.60	.53*
17	.42*	.78*	.57*	.76*	.81*	.78	.64	.78*	.72*	.73	.66*
18	.63*	.73	.83*	.78	.77*	.81*	.77*	.86*	.78*	.52*	.55*
19	.56	.74*	.91*	.87*	.62*	.61*	.87*	.55*	.56	.45*	.48*
20	.71*	.84	.65*	.18*	.75	.54*	.64*	.79*	.70*	.68*	.75*
21	.76*	.77*	.51*	.57*	.69*	.78*	.50*	.75	.41*	.82	.43*
22	.60	.86	.47*	.57*	.65*	.45	.53*	.73*	.63	.86	.59*
23	.45*	.82*	.89*	.59	.77*	.62*	.77*	.56	.76*	.91	.63*
24	.60	.81*	.58	.53	.61*	.57	.52*	.60*	.75*	.70*	.50*
25	.63*	.71*	.70*	.88	.69	.71*	.59*	.44	.54	.89*	.57*
26	.71*	.77*	.59*	.80*	.75*	.73*	.45*	.80*	.46*	.27*	.60*
27	.55*	.86	.49*	.79*	.57	.80*	.68*	.81	.68*	.64*	.35*
28	.75	.78*	.67*	.67*	.74	.66	.45	.55*	.76*	.61*	.39*
29	.63	.73	.70	.85*	.67	.67	.74*	.50*	.73	.78*	.63*
30	.37*	.82*	.51*	.84	.71*	.63*	.84*	.60*	.59	.33*	.67
31			.81	.70	.58	.71	.83*	.82*	.75*	.71*	.61*
32			.73	.61*	.70*	.65*	.39	.75*	.74	.47*	.58*
33			.62	.49*		.59*	.64	.74	.75*	.89*	.69
34			.85	.71*		.65	.46*	.54*	.54*	.72*	.67
35			.77	.60		.74	.86	.54*	.60*	.69*	.34
36			.69	.71*		.78*	.55*	.71*	.74*	.87*	.56
37			.60*	.79		.72	.62*	.63*	.64*	.84*	.41
38			.72*	.66*		.82*	.55*	.48*	.74	.68	.43
39			.68*	.67		.66*	.69*	.82*	.24*	.66	.50
40			.64*	.40*		.77	.62	.70*	.76*	.62	.62
41			.69	.54*		.78					.47
42			.54*	.53*		.70					.56
43			.65	.51		.84*					.35
44			.79	.80*		.70*					.69
45			.59	.61		.73*					.30
46			.37	.32*		.80*					.65
47			.45	.69*		.68*					.51
48			.51	.79*		.64*					.44
49			.52*								
50			.68*								
51			.66*								
52			.38*								
53			.61*								
54			.53								

* Item is included in the Abbreviated Battery.

Table G–12. Item *p*–Values for Intermediate 2, Forms A/D, Spring, Grade 5

Item Number	Reading Vocabulary	Reading Comprehension	Mathematics Problem Solving	Mathematics Procedures	Language A	Spelling	Science	Social Science	Listening	Language D
1	.87*	.71*	.75	.85	.89*	.86	.89	.83*	.91*	.66*
2	.77*	.86*	.73*	.83*	.74*	.71	.61*	.64	.61*	.43*
3	.66	.55*	.25	.73	.86	.73*	.55*	.63*	.84*	.41*
4	.65*	.77	.73*	.62*	.72*	.73*	.59*	.53*	.22*	.76*
5	.83	.80*	.59*	.73*	.71*	.51*	.57*	.79*	.59*	.45*
6	.57*	.68*	.19*	.65*	.73*	.77*	.52*	.74*	.30*	.39*
7	.59*	.70*	.69*	.85	.66	.76	.82*	.70*	.59*	.49*
8	.51*	.72	.64*	.42*	.84*	.86*	.49	.60*	.78*	.68
9	.62*	.69*	.64	.65*	.65*	.64*	.37*	.62*	.38*	.52*
10	.72	.53*	.47*	.38*	.71	.51*	.38*	.81*	.73*	.60*
11	.72*	.60*	.52	.50	.71	.58*	.79	.46*	.81	.51*
12	.53	.81*	.47	.74	.56*	.57*	.47*	.66	.65	.61*
13	.86*	.58	.67*	.30*	.83*	.73*	.50*	.45	.37	.48*
14	.83	.60	.64	.65	.62*	.53*	.58*	.58	.75*	.68*
15	.77*	.81	.84*	.62*	.68	.68	.51	.72*	.83*	.37*
16	.80	.68	.59*	.55*	.14*	.52*	.39*	.67	.33*	.43*
17	.77*	.68	.52*	.73*	.81	.51*	.73*	.64*	.63*	.59*
18	.65*	.60	.51	.69*	.64	.80	.55*	.45*	.74*	.50*
19	.73*	.61*	.52	.59	.80	.73*	.46*	.56*	.54*	.53*
20	.79*	.33*	.43*	.50	.60*	.55*	.47*	.74*	.44*	.56*
21	.57	.54	.49*	.52	.70*	.72*	.53	.44*	.48*	.41*
22	.78	.31*	.53*	.64*	.73	.55*	.43	.50	.58*	.29*
23	.80*	.61*	.62	.47	.60*	.36	.63*	.79*	.77*	.39*
24	.74*	.57*	.66	.35	.72*	.75*	.42*	.53*	.63*	.39*
25	.74	.84	.79*	.69*	.77*	.33*	.48*	.45	.48*	.22*
26	.75*	.49	.49*	.32*	.64	.73*	.49*	.72*	.52*	.65*
27	.82*	.42	.62	.52*	.76*	.59*	.42*	.52*	.70*	.67*
28	.70*	.66	.68	.41*	.67*	.76*	.52*	.44	.51*	.61*
29	.64*	.32	.64*	.41*	.63*	.66*	.64*	.59	.73*	.37*
30	.70	.43	.63*	.47*	.67	.76*	.39	.71*	.76*	.50*
31		.82*	.85*	.44	.42*	.64*	.40	.50	.86*	.56*
32		.57*	.54*	.49*	.61	.54	.75*	.48*	.75*	.24
33		.46*	.69*		.51	.43*	.41	.77*	.82*	.81
34		.59*	.52		.79*	.72*	.68*	.37*	.38	.37
35		.51*	.57		.67*	.48	.46*	.51*	.60	.59
36		.60	.50*		.66	.75	.66	.51*	.72	.62
37		.42	.61*		.57	.46	.31*	.42*	.49	.37
38		.55	.44		.75	.79*	.32*	.60*	.66	.43
39		.66	.53*		.85*	.53*	.76*	.28*	.60	.47
40		.65	.62*		.66*	.50*	.71*	.37*	.68	.31
41		.35	.47		.67*					.32
42		.53	.54*		.55*					.50
43		.64	.50*		.60*					.39
44		.62*	.29*		.69					.43
45		.37*	.58*		.56					.42
46		.60*	.57*		.59*					.33
47		.52*	.74		.48*					.62
48		.42*	.79		.44*					.15
49		.27*								
50		58								
51		.60*								
52		.38*								
53		.66*								
54		.57*								

* Item is included in the Abbreviated Battery.

Table G–13. Item *p*–Values for Intermediate 2, Forms A/D, Spring, Grade 6

Item Number	Reading Vocabulary	Reading Comprehension	Mathematics Problem Solving	Mathematics Procedures	Language A	Spelling	Science	Social Science	Listening	Language D
1	.94*	.72*	.76	.88	.88*	.90	.90	.87*	.93*	.73*
2	.89*	.89*	.78*	.87*	.78*	.75	.70*	.70	.68*	.54*
3	.76	.63*	.41	.73	.86	.77*	.59*	.71*	.91*	.46*
4	.75*	.79	.78*	.70*	.82*	.77*	.65*	.59*	.21*	.85*
5	.89	.82*	.70*	.80*	.81*	.58*	.66*	.85*	.62*	.54*
6	.67*	.73*	.28*	.69*	.70*	.82*	.63*	.83*	.37*	.41*
7	.69*	.75*	.74*	.89	.72	.80	.89*	.77*	.56*	.65*
8	.62*	.80	.74*	.50*	.85*	.87*	.59	.66*	.85*	.80
9	.68*	.72*	.73	.73*	.73*	.73*	.44*	.69*	.55*	.63*
10	.79	.59*	.61*	.49*	.68	.55*	.43*	.84*	.75*	.71*
11	.77*	.68*	.60	.62	.76	.60*	.83	.54*	.84	.46*
12	.60	.87*	.52	.75	.59*	.62*	.63*	.67	.69	.72*
13	.91*	.68	.77*	.46*	.86*	.77*	.56*	.50	.44	.49*
14	.89	.72	.70	.65	.70*	.61*	.70*	.68	.78*	.81*
15	.86*	.84	.87*	.64*	.66	.70	.64	.77*	.83*	.51*
16	.85	.77	.65*	.64*	.24*	.59*	.45*	.72	.31*	.57*
17	.83*	.73	.60*	.78*	.84	.60*	.80*	.70*	.67*	.62*
18	.70*	.73	.60	.73*	.67	.87	.63*	.50*	.77*	.56*
19	.80*	.65*	.60	.71	.82	.82*	.56*	.61*	.58*	.69*
20	.82*	.42*	.55*	.57	.69*	.62*	.60*	.79*	.54*	.69*
21	.65	.63	.60*	.60	.77*	.79*	.60	.51*	.50*	.61*
22	.85	.39*	.60*	.72*	.82	.63*	.51	.61	.63*	.41*
23	.86*	.71*	.74	.55	.68*	.43	.70*	.82*	.83*	.51*
24	.78*	.66*	.70	.45	.76*	.82*	.54*	.61*	.54*	.39*
25	.79	.88	.83*	.77*	.80*	.44*	.59*	.53	.49*	.32*
26	.82*	.57	.63*	.45*	.73	.78*	.60*	.78*	.57*	.70*
27	.85*	.52	.74	.65*	.79*	.66*	.48*	.67*	.72*	.68*
28	.75*	.66	.76	.51*	.76*	.79*	.66*	.45	.55*	.70*
29	.70*	.38	.69*	.54*	.67*	.74*	.72*	.63	.79*	.50*
30	.77	.49	.73*	.61*	.70	.81*	.41	.76*	.80*	.62*
31		.86*	.91*	.54	.52*	.68*	.48	.59	.89*	.76*
32		.56*	.68*	.58*	.69	.63	.76*	.56*	.83*	.27
33		.43*	.73*		.59	.48*	.56	.83*	.86*	.84
34		.62*	.64		.79*	.77*	.72*	.47*	.47	.49
35		.57*	.67		.71*	.60	.55*	.65*	.63	.68
36		.65	.58*		.69	.80	.74	.59*	.76	.70
37		.48	.69*		.66	.55	.47*	.58*	.55	.52
38		.57	.53		.79	.83*	.44*	.72*	.71	.42
39		.70	.60*		.82*	.64*	.83*	.41*	.59	.64
40		.70	.67*		.72*	.58*	.81*	.43*	.74	.45
41		.40	.57		.70*					.43
42		.59	.58*		.65*					.66
43		.73	.62*		.65*					.47
44		.69*	.42*		.73					.54
45		.48*	.68*		.60					.53
46		.69*	.72*		.65*					.45
47		.63*	.81		.64*					.72
48		.47*	.87		.56*					.13
49		.41*								
50		.65								
51		.70*								
52		.50*								
53		.76*								
54		.66*								

* Item is included in the Abbreviated Battery.

Table G–14. Item *p*–Values for Intermediate 3, Forms A/D, Spring, Grade 6

Item Number	Reading Vocabulary	Reading Comprehension	Mathematics Problem Solving	Mathematics Procedures	Language A	Spelling	Science	Social Science	Listening	Language D
1	.83*	.90*	.59*	.76	.68*	.86*	.93*	.88*	.73*	.72*
2	.74*	.74*	.51*	.67*	.88*	.82*	.76*	.82*	.87*	.67*
3	.65	.70*	.40	.66	.78*	.65*	.59	.60*	.40*	.65*
4	.83*	.71*	.59	.53*	.82*	.62*	.76*	.67*	.37*	.56*
5	.84·	.65*	.64*	.82	.90*	.65*	.74*	.71*	.67*	.61*
6	.81*	.84	.59*	.75*	.76	.79*	.58	.87*	.58*	.57*
7	.60*	.92*	.35	.55*	.52	.82*	.89*	.29	.52*	.55*
8	.64	.86*	.70*	.73*	.85*	.78*	.70	.52*	.94*	.86
9	.53*	.79	.65*	.53*	.81	.58*	.74*	.50	.55*	.55*
10	.70	.76*	.63	.57	.70*	.37	.74	.43	.50*	.47*
11	.78*	.64*	.65*	.53*	.51*	.84*	.57*	.66	.79*	.62*
12	.55*	.64*	.25	.79*	.74*	.54*	.68*	.48*	.72*	.41*
13	.91*	.74*	.75*	.44	.67	.69*	.78*	.64*	.71*	.85*
14	.92	.85*	.71*	.52*	.80*	.55*	.66*	.57	.48*	.54*
15	.68*	.35*	.56	.33*	.63	.67*	.58*	.57*	.47*	.62*
16	.75	.83	.39	.37	.53	.65*	.44*	.68*	.55*	.80*
17	.77*	.79*	.50*	.67*	.57	.86	.88*	.48*	.32*	.75*
18	.66*	.61*	.42*	.62	.71	.82*	.78*	.69	.60*	.59*
19	.73*	.54*	.61	.56	.82*	.61*	.59*	.60*	.73*	.86*
20	.89	.71*	.57*	.54*	.46	.84*	.38*	.45*	.65*	.81*
21	.77*	.50	.35	.62*	.24*	.61*	.58*	.52*	.75*	.43*
22	.87*	.45*	.56*	.49*	.71*	.44*	.75*	.59*	.86*	.77*
23	.78	.79*	.45*	.57*	.63*	.71*	.56*	.49*	.77*	.65*
24	.80*	.60*	.48	.33*	.70*	.61*	.59*	.56*	.47	.52*
25	.54*	.78*	.48*	.53	.77*	.58	.68	.70	.61	.59*
26	.75	.77*	.66*	.58·	.73*	.55*	.74*	.36*	.70	.48*
27	.85*	.60*	.52	.48	.75*	.60*	.30*	.75*	.61*	.60*
28	.71	.61	.47	.52*	.77	.59*	.62*	.39	.67*	.65*
29	.73*	.61*	.52*	.38*	.68	.75*	.45*	.61*	.82*	.81*
30	.68*	.79*	.47	.39	.48*	.83	.39	.55*	.61*	.85*
31		.52	.63*	.44*	.59*	.66	.85*	.52*	.76	.44*
32		.54	.44*	.44*	.70*	.62*	.48*	.75	.75	.57
33		.37	.62*		.68*	.32*	.59*	.50*	.77	.59
34		.72	.28		.55*	.61	.72	.69*	.47*	.41
35		.54	.45*		.76*	.77	.52	.39*	.75*	.72
36		.43	.58*		.54*	.61	.57	.54*	.52*	.33
37		.73	.41		.70	.72*	.35*	.61*	.80	.53
38		.60	.79*		.65	.47	.33*	.16	.32	.57
39		.75	.69*		.76	.68	.70	.44*	.67	.46
40		.55	.55*		.71	.49*	.54*	.69*	.81	.57
41		.80	.72		.59					.54
42		.75	.45		.84*					.68
43		.56	.51*		.83*					.56
44		.77	.36*		.61					.76
45		.54	.46*		.60					.73
46		.60	.50		.61*					.54
47		.50	.50		.57*					.51
48		.41	.58*		.64*					.55
49		.77*								
50		.60*								
51		.40								
52		.31*								
53		.50*								
54		.58*								

* Item is included in the Abbreviated Battery.

Appendix G

Table G–15. Item p–Values for Intermediate 3, Forms A/D, Spring, Grade 7

Item Number	Reading Vocabulary	Reading Comprehension	Mathematics Problem Solving	Mathematics Procedures	Language A	Spelling	Science	Social Science	Listening	Language D
1	.87*	.85*	.70*	.71	.69*	.86*	.89*	.85*	.84*	.72*
2	.78*	.76*	.57*	.71*	.87*	.78*	.75*	.79*	.88*	.73*
3	.65	.70*	.44	.63	.69*	.67*	.64	.64*	.46*	.66*
4	.79*	.70*	.67	.55*	.78*	.62*	.74*	.65*	.47*	.57*
5	.84	.68*	.55*	.77	.84*	.67*	.75*	.72*	.69* ·	.61*
6	.77*	.84	.58*	.77*	.71	.77*	.60	.84*	.64*	.60*
7	.70*	.90*	.41	.54*	.53	.80*	.88*	.31	.61*	.56*
8	.69	.80*	.69*	.72*	.82*	.80*	.70	.56*	.91*	.85
9	.59*	.77	.70*	.60*	.78	.64*	.73*	.51	.62*	.61*
10	.67	.76*	.67	.60	.67*	.42	.76	.45	.60*	.54*
11	.78*	.65*	.69*	.54*	.56*	.83*	.62*	.65	.77*	.63*
12	.58*	.66*	.29*	.75*	.69*	.57*	.73*	.46*	.75*	.47*
13	.90*	.75*	.77*	.58	.65	.64*	.79*	.55*	.75*	.85*
14	.89	.85*	.71*	.58*	.75*	.54*	.71*	.58	.55*	.43*
15	.67*	.37*	.57	.38*	.59	.65*	.61*	.61*	.57*	.68*
16	.79	.81	.46	.42	.59	.66*	.50*	.68*	.62*	.80*
17	.75*	.75*	.51*	.67*	.58	.84	.82*	.52*	.40*	.79*
18	.63	.61*	.45*	.63	.68	.81*	.79*	.68	.60*	.68*
19	.71*	.62*	.62	.59	.73*	.60*	.55*	.60*	.74*	.88*
20	.84	.73*	.59*	.57*	.45	.79*	.43*	.50*	.63*	.86*
21	.77*	.55	.40	.66*	.36*	.62*	.61*	.55*	.76*	.44*
22	.88*	.48*	.57*	.50*	.66*	.45*	.68*	.62*	.86*	.80*
23	.76	.79*	.50*	.57*	.60*	.67*	.61*	.49*	.78*	.67*
24	.76*	.62*	.54	.34*	.65*	.64*	.63*	.61*	.44	.59*
25	.65*	.76*	.55*	.51	.72*	.60	.69	.67	.63	.60*
26	.74	.72*	.73*	.60	.65*	.56*	.74*	.43*	.64	.44*
27	.81*	.61*	.66	.52	.67*	.62*	.35*	.77*	.67*	.61*
28	.72	.61	.47	.55*	.70	.63*	.57*	.39	.65*	.72*
29	.68*	.63*	.59*	.39*	.62	.73*	.51*	.58*	.81*	.85*
30	.68*	.72*	.53	.42	.47*	.82	.45	.59*	.62*	.84*
31		.53	.65*	.43*	.57*	.68	.83*	.49*	.76	.53*
32		.56	.51*	.46*	.67*	.68*	.56*	.72	.71	.64
33		.36	.59*		.63*	.33*	.61*	.50*	.78	.63
34		.70	.36		.56*	.63	.73	.67*	.51*	.36
35		.55	.47*		.73*	.77	.58	.39*	.75*	.72
36		.44	.62*		.60*	.60	.60	.60*	.55*	.42
37		.70	.46		.70	.74*	.43*	.62*	.80	.46
38		.60	.75*		.65	.52	.37*	.20	.32	.61
39		.72	.65*		.74	.65	.67	.41*	.73	.49
40		.57	.61*		.65	.53*	.58*	.68*	.83	.59
41		.77	.68		.60					.59
42		.71	.47		.75*					.69
43		.56	.54*		.74*					.51
44		.68	.42*		.59					.77
45		.57	.52*		.59					.68
46		.59	.57		.60*					.45
47		.49	.50		.48*					.48
48		.44	.58*		.61*					.54
49		.72*								
50		.63*								
51		.43								
52		.38*								
53		.49*								
54		.57*								

* Item is included in the Abbreviated Battery.

Table G–16. Item *p*–Values for Advanced 1, Forms A/D, Spring, Grade 7

Item Number	Reading Vocabulary	Reading Compre-hension	Mathematics Problem Solving	Mathematics Procedures	Language A	Spelling	Science	Social Science	Listening	Language D
1	.82*	.88*	.61	.52	.81*	.80*	.83*	.82*	.80*	.25
2	.74*	.80*	.55	.65*	.76	.86*	.78	.51	.74*	.76
3	.49	.49*	.43*	.55*	.67	.68*	.69	.58*	.46*	.82
4	.79*	.40*	.51*	.81	.67*	.60*	.52	.45*	.83*	.72
5	.72	.66*	.47*	.59*	.75*	.50*	.74*	.56*	.32*	.44
6	.57*	.84	.66	.43	.79*	.50	.74*	.63*	.48*	.33
7	.48*	.76*	.40*	.25*	.74*	.69*	.55*	.50*	.45*	.36
8	.57	.76	.44	.56*	.86*	.80*	.34*	.72*	.67*	.46
9	.48*	.46*	.46*	.66*	.69	.56*	.64	.60*	.43*	.65
10	.63	.59*	.40*	.50	.55*	.35*	.87*	.40*	.50*	.58
11	.77*	.65*	.60	.38*	.52	.72*	.64*	.69*	.47*	.56
12	.49*	.59*	.66*	.40	.70*	.78	.47*	.58*	.54*	.59
13	.77*	.51*	.50*	.37*	.82*	.44	.86	.41*	.73*	.23
14	.81	.75*	.26*	.45*	.63*	.37*	.68*	.49*	.55*	.50
15	.62*	.74	.55*	.45	.45*	.68*	.46*	.48*	.47*	.30
16	.80*	.56*	.44	.58*	.31*	.48*	.53*	.38	.70*	.65
17	.80	.71*	.28*	.63*	.70	.71*	.56*	.28*	.41*	.61*
18	.51*	.71*	.64	.45*	.78*	.74*	.56*	.65*	.82*	.48*
19	.71*	.84*	.42*	.49*	.72*	.71*	.55*	.51*	.49*	.40*
20	.80	.65*	.38	.69	.70*	.65*	.56*	.44*	.63*	.41*
21	.65*	.56	.37*	.65*	.69	.52*	.76*	.37*	.73*	.40*
22	.74*	.69*	.57*	.61*	.70	.79	.42*	.44*	.80*	.63*
23	.67	.65*	.42	.46*	.41	.24*	.50	.65*	.82*	.44
24	.58*	.78*	.72*	.30*	.54	.81*	.53*	.46*	.79*	.43*
25	.61*	.70	.43*	.46	.69	.58	.62*	.42	.35*	.48*
26	.61*	.66	.64	.37	.65*	.81	.46*	.59	.66*	.45*
27	.77*	.65	.35*	.50	.76*	.40	.68*	.51*	.60*	.41*
28	.66	.50	.18	.48*	.55*	.67*	.54*	.32	.67*	.47*
29	.50	.52	.34	.31*	.65	.76*	.49*	.59*	.75*	.48*
30	.70*	.76	.43*	.39	.66	.77*	.37*	.41	.73*	.57*
31		.67*	.23*	.51*	.63	.67	.47*	.52	.75	.51*
32		.61*	.58*	.46	.46*	.47*	.40*	.37*	.69	.57*
33		.71	.31*		.56	.57*	.39	.29*	.66	.41*
34		.69*	.46*		.54*	.78*	.70*	.61	.71	.59*
35		.59*	.39		.68*	.62	.60	.42*	.63	.48*
36		.41*	.27		.72*	.38*	.57	.44	.68	.42*
37		.72	.65		.69*	.49*	.60	.59	.43	.67*
38		.47	.47*		.67*	.56*	.31*	.34*	.61	.32*
39		.71	.66		.53*	.42*	.26*	.34*	.75	.52*
40		.39	.63*		.71	.59	.70*	.56*	.83	.45*
41		.47	.56*		.61					.58*
42		.47	.73*		.61					.33*
43		.44*	.42*		.55					.59*
44		.40*	.30*		.73*					.57*
45		.48	.51		.68*					.63
46		.45*	.49*		.56*					.34*
47		.51*	.49*		.56*					.55*
48		.21*	.35		.70*					.38*
49		.53								
50		.39								
51		.35								
52		.49								
53		.40								
54		.31								

* Item is included in the Abbreviated Battery.

Appendix G

Table G–17. Item *p*–Values for Advanced 1, Forms A/D, Spring, Grade 8

Item Number	Reading Vocabulary	Reading Compre-hension	Mathematics Problem Solving	Mathematics Procedures	Language A	Spelling	Science	Social Science	Listening	Language D
1	.88*	.89*	.66	.58	.81*	.83*	.83*	.85*	.81*	.47
2	.79*	.84*	.58	.65*	.78	.90*	.85	.58	.77*	.86
3	.57	.53*	.56*	.66*	.65	.69*	.75	.62*	.55*	.85
4	.82*	.52*	.60*	.82	.77*	.67*	.57	.56*	.83*	.81
5	.78	.72*	.51*	.64*	.79*	.56*	.75*	.61*	.37*	.56
6	.62*	.85	.73	.49	.80*	.60	.75*	.66*	.52*	.58
7	.56*	.71*	.54*	.38*	.75*	.77*	.53*	.49*	.58*	.44
8	.64	.77	.46	.59*	.86*	.84*	.38*	.77*	.70*	.53
9	.49*	.51*	.55*	.79*	.68	.61*	.67	.65*	.52*	.78
10	.72	.65*	.48*	.58	.55*	.35*	.89*	.42*	.59*	.69
11	.82*	.70*	.63	.43*	.50	.73*	.68*	.75*	.41*	.78
12	.55*	.60*	.73*	.50	.78*	.79	.60*	.62*	.58*	.75
13	.84*	.56*	.57*	.43*	.83*	.53	.86	.42*	.71*	.22
14	.86	.77*	.31*	.50*	.71*	.38*	.73*	.51*	.57*	.67
15	.68*	.78	.59*	.52	.45*	.74*	.49*	.53*	.54*	.35
16	.86*	.56*	.52	.65*	.31*	.55*	.59*	.42	.67*	.76
17	.88	.75*	.37*	.72*	.73	.80*	.55*	.43*	.50*	.76*
18	.56*	.79*	.70	.53*	.80*	.77*	.59*	.72*	.80*	.64*
19	.79*	.84*	.50*	.54*	.75*	.76*	.62*	.60*	.50*	.44*
20	.84	.70*	.39	.75	.74*	.70*	.61*	.52*	.63*	.63*
21	.75*	.62	.47*	.72*	.64	.56*	.76*	.39*	.71*	.53*
22	.81*	.74*	.75*	.69*	.70	.83	.45*	.45*	.79*	.78*
23	.76	.71*	.48	.52*	.48	.29*	.62	.72*	.84*	.60
24	.61*	.78*	.78*	.35*	.54	.84*	.55*	.53*	.82*	.59*
25	.67*	.72	.50*	.59	.75	.65	.69*	.41	.37*	.62*
26	.72*	.68	.79	.43	.65*	.86	.57*	.68	.60*	.56*
27	.86*	.67	.42*	.65	.74*	.46	.74*	.53*	.67*	.42*
28	.71	.56	.22	.55*	.59*	.72*	.58*	.42	.71*	.65*
29	.54	.55	.45	.37*	.65	.81*	.56*	.61*	.76*	.59*
30	.79*	.80	.55*	.44	.73	.81*	.38*	.49	.74*	.70*
31		.71*	.30*	.55*	.65	.72	.46*	.56	.75	.65*
32		.70*	.65*	.53	.45*	.53*	.43*	.44*	.70	.70*
33		.74	.28*		.60	.67*	.40	.31*	.76	.62*
34		.73*	.53*		.59*	.82*	.77*	.69	.73	.71*
35		.65*	.48		.68*	.63	.62	.56*	.61	.58*
36		.45*	.33		.73*	.41*	.61	.56	.69	.47*
37		.76	.72		.70*	.51*	.68	.71	.42	.80*
38		.54	.55*		.69*	.57*	.31*	.36*	.63	.48*
39		.77	.69		.57*	.49*	.31*	.39*	.76	.73*
40		.44	.73*		.67	.64	.76*	.66*	.83	.53*
41		.56	.61*		.59					.78*
42		.55	.75*		.66					.41*
43		.50*	.48*		.56					.73*
44		.43*	.35*		.71*					.68*
45		.60	.58		.64*					.74
46		.49*	.53*		.61*					.44*
47		.59*	.55*		.54*					.66*
48		.29*	.43		.71*					.43*
49		.57								
50		.47								
51		.39								
52		.61								
53		.45								
54		.28								

* Item is included in the Abbreviated Battery.

Table G–18. Item *p*–Values for Advanced 2, Forms A/D, Spring, Grade 8

Item Number	Reading Vocabulary	Reading Comprehension	Mathematics Problem Solving	Mathematics Procedures	Language A	Spelling	Science	Social Science	Listening	Language D
1	.81*	.93*	.63	.52	.83	.83*	.73*	.61	.61*	.89
2	.82	.93*	.51*	.54*	.72	.85*	.78*	.56*	.46*	.76
3	.68*	.54*	.54*	.36*	.78*	.55*	.65*	.64*	.87*	.84
4	.67*	.56	.47	.59	.80	.48*	.57*	.70*	.47*	.73
5	.79*	.51*	.37*	.52*	.86*	.79	.75*	.46	.73*	.74
6	.61*	.76*	.56	.35*	.78*	.55*	.44*	.66*	.60*	.84
7	.50*	.66	.42*	.39	.75	.65	.56*	.66*	.47*	.43
8	.54	.79	.54	.63*	.53*	.82*	.60*	.27*	.80*	.53
9	.69	.72	.38*	.61*	.81*	.67*	.54*	.38*	.43*	.55
10	.67*	.66	.40	.63*	.63	.39	.44*	.43*	.72*	.74
11	.77	.56	.40*	.37	.83	.68	.64*	.46*	.52	.89
12	.65*	.42	.29*	.52	.42	.65*	.68*	.52*	.85	.88
13	.89	.64	.66*	.43*	.53	.78*	.38*	.39*	.78	.71
14	.89*	.63	.45*	.55	.62*	.59	.54*	.47*	.67*	.55
15	.76*	.64	.59	.42*	.61*	.57	.69	.54*	.87*	.78
16	.73*	.50	.43*	.43*	.72*	.49*	.51*	.21*	.80*	.52
17	.82	.55	.29	.64	.56*	.75*	.43*	.54*	.13*	.67*
18	.66*	.63	.47*	.39*	.75	.85*	.70	.30*	.34*	.82*
19	.79*	.58*	.48*	.29	.66*	.90*	.66	.40*	.60*	.87*
20	.73	.39*	.58	.47*	.53*	.81*	.50	.45*	.35*	.90*
21	.59*	.88*	.58*	.45	.62	.54	.48*	.55	.55*	.81*
22	.81*	.73*	.48	.40	.32*	.36	.71*	.46*	.82*	.54*
23	.80*	.64	.39*	.39*	.52*	.66*	.35	.32	.61*	.81*
24	.71	.47*	.67*	.51	.67*	.66*	.19	.42*	.55	.74*
25	.59*	.70	.37*	.45	.70	.36*	.45	.50*	.37	.70*
26	.67	.61*	.53	.37*	.63*	.84*	.70*	.44*	.54	.63*
27	.83*	.48*	.59*	.43*	.62	.48*	.53*	.53	.33*	.50*
28	.55*	.58*	.40*	.52*	.60	.62*	.34*	.49*	.69*	.42*
29	.60	.73*	.43*	.33*	.60	.87	.43*	.59*	.66*	.79*
30	.47*	.69*	.58	.52*	.54*	.86*	.28	.08	.71*	.54*
31		.57*	.30	.32	.58*	.49*	.55*	.39*	.47*	.74
32		.68*	.63*	.48*	.56*	.82*	.33*	.42*	.74*	.79*
33		.68*	.33		.66	.52*	.46*	.60*	.84*	.84*
34		.42*	.25		.57*	.42*	.67*	.54*	.57*	.85*
35		.74	.66*		.65*	.41*	.40	.51	.62*	.68*
36		.64*	.41*		.55*	.72*	.26*	.43*	.44*	.84*
37		.59	.61*		.76*	.35*	.41*	.35	.62	.41*
38		.76	.54*		.64*	.69	.52*	.44	.52	.76*
39		.65	.59		.52*	.19*	.40*	.42*	.81	.85
40		.55	.48*		.78	.55*	.49	.44	.49	.69*
41		.54	.51*		.64					.47*
42		.60	.35		.56					.79*
43		.74*	.31*		.66					.77*
44		.59	.37*		.49*					.32*
45		.59*	.36*		.71*					.75*
46		.41*	.32		.63*					.67*
47		.64*	.25		.57*					.80*
48		.37*	.42*		.66*					.80*
49		.63*								
50		.60*								
51		.66								
52		.59*								
53		.57*								
54		.56*								

* Item is included in the Abbreviated Battery.

Appendix G

Table G–19. Item p–Values for Advanced 2, Forms A/D, Spring, Grade 9

Item Number	Reading Vocabulary	Reading Comprehension	Mathematics Problem Solving	Mathematics Procedures	Language A	Spelling	Science	Social Science	Listening	Language D
1	.86*	.98*	.68	.55	.83	.90*	.78*	.70	.73*	.84
2	.78	.95*	.57*	.61*	.76	.87*	.84*	.59*	.60*	.59
3	.69*	.62*	.58*	.46*	.83*	.59*	.72*	.74*	.92*	.74
4	.73*	.65	.57	.77	.84	.61*	.63*	.71*	.60*	.62
5	.79*	.56*	.52*	.65*	.86*	.83	.84*	.54	.82*	.75
6	.68*	.82*	.68	.43*	.85*	.53*	.43*	.69*	.69*	.80
7	.61*	.74	.48*	.45	.78	.76	.61*	.74*	.62*	.38
8	.59	.89	.67	.67*	.67*	.77*	.67*	.40*	.87*	.40
9	.67	.81	.43*	.73*	.85*	.73*	.59*	.36*	.51*	.48
10	.70*	.69	.52	.74*	.67	.41	.41*	.48*	.80*	.60
11	.85	.64	.55*	.47	.84	.75	.68*	.51*	.63	.81
12	.67*	.49	.34*	.58	.44	.70*	.71*	.53*	.88	.85
13	.90	.69	.74*	.46*	.59	.86*	.39*	.48*	.81	.68
14	.93*	.71	.54*	.70	.70*	.66	.57*	.52*	.69*	.53
15	.80*	.67	.67	.61*	.66*	.61	.72	.57*	.90*	.71
16	.77*	.53	.58*	.49*	.75*	.54*	.55*	.30*	.81*	.54
17	.82	.66	.31	.70	.58*	.76*	.44*	.59*	.13*	.64*
18	.69*	.75	.50*	.45*	.77*	.91*	.78	.30*	.41*	.78*
19	.76*	.63*	.49*	.31*	.74*	.93*	.73	.42*	.55*	.74*
20	.73	.39*	.71	.59*	.62*	.80*	.56	.54*	.37*	.79*
21	.60*	.94*	.63*	.44	.66	.64	.58*	.57	.56*	.67*
22	.79*	.81*	.57	.46	.33*	.42	.72*	.47*	.82*	.53*
23	.83*	.70	.62*	.38*	.53*	.77*	.41	.37	.60*	.73*
24	.71	.56*	.77*	.59	.71*	.76*	.16	.46*	.61	.61*
25	.63*	.79	.45*	.56	.74	.43*	.53	.46*	.42	.57*
26	.66	.70*	.69	.45*	.72*	.87*	.70*	.53*	.51	.65*
27	.85*	.51*	.73*	.48*	.71	.45*	.51*	.57	.42*	.46*
28	.67*	.69*	.47*	.54*	.67	.69*	.35*	.50*	.69*	.34*
29	.71	.79*	.48*	.44*	.66	.91	.49*	.61*	.66*	.73*
30	.51*	.77*	.65	.64*	.60*	.88*	.32	.28	.71*	.55*
31		.58*	.32	.42	.70*	.51*	.62*	.44*	.48*	.68
32		.77*	.76*	.58*	.58*	.90*	.42*	.45*	.75*	.62*
33		.78*	.37		.74	.49*	.46*	.64*	.81*	.78*
34		.49*	.33		.67*	.48*	.70*	.53*	.66*	.73*
35		.85	.72*		.73*	.51*	.44	.57	.70*	.59*
36		.72*	.44*		.62*	.75*	.30*	.48*	.54*	.73*
37		.62	.68*		.84*	.29*	.45*	.40	.68	.46*
38		.81	.66*		.75*	.72	.59*	.52	.58	.69*
39		.69	.72		.64*	.22*	.46*	.43*	.82	.77
40		.63	.54*		.85	.60*	.58	.47	.53	.63*
41		.59	.56*		.69					.49*
42		.64	.36		.62					.72*
43		.82*	.42*		.77					.65*
44		.66	.51*		.56*					.40*
45		.63*	.43*		.79*					.65*
46		.42*	.33		.74*					.55*
47		.72*	.30		.68*					.70*
48		.40*	.42*		.71*					.78*
49		.75*								
50		.69*								
51		.73								
52		.67*								
53		.63*								
54		.64*								

* Item is included in the Abbreviated Battery.

Table G–20. Item *p*–Values for TASK 1, Forms A/D, Spring, Grade 9

Item Number	Reading Vocabulary	Reading Comprehension	Mathematics	Language A	Spelling	Science	Social Science	Language D
1	.90	.69*	.48*	.90*	.79*	.92*	.55	.71
2	.73*	.81*	.52*	.79	.84*	.67*	.62	.63
3	.63	.47*	.36	.58	.77*	.54*	.49	.65
4	.69*	.85*	.14	.73*	.70	.53*	.67*	.80
5	.90*	.77*	.74	.70*	.74	.87	.63	.85
6	.59*	.49*	.61	.54	.84	.81*	.40*	.50
7	.59	.62*	.41*	.73*	.88	.75	.56*	.69
8	.49	.61*	.81*	.62*	.76*	.55	.60*	.87
9	.55*	.78*	.58*	.66	.72*	.76*	.47*	.74
10	.54*	.86*	.70	.73*	.65	.52	.27*	.72
11	.57*	.86*	.78*	.74*	.29*	.60*	.71	.91
12	.42*	.88*	.39*	.40*	.79*	.56*	.45*	.81
13	.89	.53*	.59	.43*	.58*	.82*	.30	.88
14	.87*	.69*	.56*	.68	.55	.64*	.19*	.75
15	.69*	.57*	.78*	.49*	.40*	.43*	.43*	.41
16	.75*	.53*	.53*	.58*	.73*	.59*	.40*	.62
17	.89*	.52	.36*	.86	.57*	.63*	.36	.83*
18	.85	.85*	.53	.34	.83*	.35*	.61*	.75*
19	.85	.55*	.53*	.70*	.75*	.59*	.56*	.66*
20	.82*	.48	.52	.80	.72*	.44*	.35*	.39*
21	.63	.40*	.67	.26*	.48*	.28*	.43*	.83*
22	.80*	.47	.26*	.74*	.43*	.67*	.52*	.48*
23	.90*	.60	.47	.53*	.53*	.39	.60*	.67*
24	.79*	.69*	.27*	.60	.71	.55	.37*	.82*
25	.72*	.71*	.50*	.72*	.51*	.37*	.35*	.74
26	.86*	.50*	.48	.72*	.65*	.55	.52	.75*
27	.83	.41	.39*	.68*	.27	.26*	.46*	.63*
28	.62*	.78	.53*	.72*	.71*	.55*	.53*	.56*
29	.53	.72	.38*	.49*	.70*	.47*	.45	.42*
30	.77*	.50	.34	.73	.71*	.33	.34*	.80*
31		.65	.47	.57*	.56	.57*	.57*	.50*
32		.63	.55*	.47*	.65*	.33*	.51*	.82*
33		.71	.26	.62*	.33*	.30*	.45*	.62*
34		.32	.40*	.57	.74	.53	.41	.67*
35		.62	.16*	.66*	.41*	.41*	.39*	.64*
36		.33	.29	.63*	.62*	.26*	.40*	.61*
37		.76	.45*	.75	.51*	.47*	.43*	.76*
38		.80	.31*	.77	.42*	.25*	.33*	.53*
39		.31	.19	.65	.33*	.25	.36*	.60*
40		.72	.51*	.68*	.46*	.50*	.37*	.62*
41		.66	.24	.74*				.59*
42		.59	.30*	.51				.65*
43		.64	.35	.48				.68*
44		.65	.34*	.48				.44*
45		.71	.35*	.45				.54*
46		.50*	.16	.67*				.61*
47		.53	.26*	.69*				.61*
48		.62*	.30*	.58*				.37
49		.48*	.27					
50		.63*	.41*					
51		.26*						
52		.41*						
53		.40*						
54		.51*						

* Item is included in the Abbreviated Battery.

Appendix G

Table G–21. Item p–Values for TASK 2, Forms A/D, Spring, Grade 10

Item Number	Reading Vocabulary	Reading Comprehension	Mathematics	Language A	Spelling	Science	Social Science	Language D
1	.89	.65*	.58*	.89*	.82*	.70*	.62*	.72
2	.87*	.63*	.51	.74	.83*	.73	.57*	.70
3	.80*	.87*	.47*	.70	.75*	.74*	.65*	.78
4	.70	.77*	.62*	.54*	.75*	.65	.49*	.73
5	.75*	.67*	.53	.43*	.59*	.47*	.54	.67
6	.76	.83*	.56*	.65	.68*	.55*	.29*	.90
7	.50	.89*	.62*	.53*	.78	.55*	.33	.56
8	.57*	.77*	.15	.67*	.82*	.54*	.60*	.75
9	.42*	.51*	.51	.54*	.54	.85*	.42*	.88
10	.58*	.95	.74*	.63*	.63*	.73*	.36*	.88
11	.82*	.85	.65*	.50	.66*	.58	.61*	.86
12	.51*	.70	.64	.81	.43	.42	.57*	.89
13	.93	.80	.51	.55*	.78*	.78*	.32	.74
14	.94*	.69	.77*	.31*	.43*	.78*	.30*	.81
15	.87*	.74	.47*	.57*	.68	.59*	.24*	.71
16	.75	.88	.51	.67*	.66*	.44	.58	.61
17	.92*	.41	.23*	.69*	.65	.74	.31*	.83*
18	.56*	.70	.60*	.64*	.71	.63*	.38*	.80*
19	.92	.65*	.66*	.78	.84*	.47	.46*	.81*
20	.91*	.70*	.51	.76*	.80	.49*	.42*	.93*
21	.82	.60*	.42*	.43*	.54*	.70	.29*	.70*
22	.83	.68*	.36	.36	.50*	.26*	.40	.40*
23	.93*	.78*	.46*	.63	.17*	.40*	.57*	.43*
24	.77*	.42*	.36	.28	.68*	.42*	.40	.67*
25	.58*	.72*	.51	.69	.56*	.55*	.35*	.63*
26	.94*	.68*	.32*	.67*	.67	.49*	.41*	.56*
27	.87*	.65	.29*	.57	.50*	.60*	.50*	.75
28	.45	.81	.45*	.71*	.52*	.46*	.36	.43*
29	.71*	.52	.63*	.69*	.84*	.49*	.67	.58*
30	.62*	.47	.34	.60	.80*	.45*	.27*	.75*
31		.56	.44	.79	.35*	.68	.39*	.56*
32		.79	.58*	.63*	.58	.40*	.51*	.53*
33		.73	.61	.55*	.53*	.64*	.36*	.43*
34		.70	.32*	.78	.69*	.21*	.28*	.75*
35		.83	.42*	.78*	.70	.49*	.41*	.49*
36		.72	.24	.75*	.80*	.35	.29	.64*
37		.66*	.57	.78*	.58*	.71*	.38*	.51*
38		.61*	.44*	.69*	.83*	.33*	.25*	.46*
39		.83*	.16*	.71*	.41*	.37*	.39*	.56*
40		.67*	.56	.75	.62*	.72*	.49	.40
41		.39	.45*	.69				.69*
42		.76*	.46*	.70				.68*
43		.80*	.26	.66				.64*
44		.65*	.29	.60*				.41*
45		.44*	.37*	.54*				.49*
46		.34*	.17*	.59*				.73*
47		.49*	.34*	.61*				.43*
48		.55*	.27*	.57*				.61*
49		.69	.35*					
50		.46	.29					
51		.55						
52		.60*						
53		.35*						
54		.34						

* Item is included in the Abbreviated Battery.

204

Table G–22. Item *p*–Values for TASK 3, Forms A/D, Spring, Grade 11

Item Number	Reading Vocabulary	Reading Comprehension	Mathematics	Language A	Spelling	Science	Social Science	Language D
1	.84*	.62*	.55*	.76	.83*	.75	.69*	.70
2	.75*	.66*	.54	.75*	.65*	.69*	.48*	.62
3	.54*	.61*	.34*	.76*	.84*	.47	.42*	.76
4	.83	.75*	.74	.62*	.51*	.62*	.28	.35
5	.71*	.65*	.46	.46*	.78*	.63*	.45	.55
6	.77*	.66*	.44*	.39	.48*	.58	.45*	.47
7	.45*	.41*	.40*	.52*	.69*	.55*	.29*	.60
8	.62*	.73*	.50*	.43*	.79*	.63*	.32	.72
9	.43	.55*	.49	.67	.64*	.46*	.39	.80
10	.65*	.74*	.40	.68*	.64	.43*	.58	.63
11	.73	.58*	.44*	.34*	.14*	.42*	.23*	.75
12	.37	.44*	.62*	.71*	.80	.20*	.45*	.70
13	.81	.63	.38*	.59	.68*	.65*	.28*	.72
14	.85*	.65	.42	.65*	.48*	.60*	.17*	.78
15	.78*	.73	.35*	.65*	.71	.32*	.41	.65
16	.73	.63*	.24	.68	.70*	.40*	.47*	.45
17	.81*	.73*	.49*	.30*	.63*	.31*	.29*	.70*
18	.67*	.48	.48*	.39*	.60*	.31	.41*	.79*
19	.81*	.66	.42*	.62	.74*	.28*	.27*	.73*
20	.85*	.86	.38	.36	.74	.54*	.38*	.62*
21	.70	.67	.60*	.67*	.56*	.70*	.35	.57*
22	.74*	.79	.24	.50	.63	.42*	.37*	.61*
23	.86	.50	.48	.30	.23	.27*	.40*	.45*
24	.67*	.64	.29*	.59*	.62	.47	.35	.63*
25	.53	.69	.50*	.70*	.44*	.27*	.28*	.74*
26	.80*	.79	.37*	.62*	.55	.51*	.38*	.45*
27	.88	.66	.32*	.57*	.39*	.32*	.39*	.48*
28	.62*	.47	.41*	.54	.45	.45*	.40*	.65*
29	.61*	.52	.46	.50	.82*	.31*	.40*	.73*
30	.84*	.35	.36	.58	.71*	.58*	.27*	.65*
31		.65	.43*	.63	.64*	.37	.48	.42*
32		.29	.45	.51*	.27*	.49	.38*	.38
33		.64	.23*	.65*	.67*	.30	.46*	.70*
34		.68	.41	.44	.43*	.29*	.48*	.42*
35		.67	.32	.60*	.55*	.35*	.45	.55*
36		.71	.52*	.64*	.65*	.52	.35*	.71*
37		.66*	.42*	.56*	.56*	.34	.30*	.57*
38		.56*	.53*	.42*	.68	.32*	.31*	.39*
39		.46*	.49	.67*	.51*	.52*	.27*	.35
40		.58*	.45*	.65	.46*	.25*	.45*	.61*
41		.53*	.48	.64				.44*
42		.38	.41*	.48*				.58*
43		.47*	.24*	.67*				.40*
44		.59*	.35*	.59				.58*
45		.53*	.34	.64				.55*
46		.50*	.26*	.45*				.30*
47		.59*	.23	.60*				.39*
48		.44*	.30*	.55*				.51*
49		.56*	.26*					
50		.44*	.28					
51		.47*						
52		.64*						
53		.67*						
54		.36						

* Item is included in the Abbreviated Battery.

Appendix G

Table G–23. Item p–Values for TASK 3, Forms A/D, Spring, Grade 12

Item Number	Reading Vocabulary	Reading Comprehension	Mathematics	Language A	Spelling	Science	Social Science	Language D
1	.84*	.58*	.53*	.74	.88*	.74	.67*	.70
2	.75*	.70*	.53	.73*	.67*	.65*	.52*	.60
3	.59*	.59*	.35*	.71*	.86*	.49	.42*	.87
4	.80	.72*	.75	.60*	.52*	.56*	.26	.40
5	.75*	.63*	.48	.44*	.77*	.54*	.42	.75
6	.81*	.64*	.43*	.44	.55*	.58	.38*	.54
7	.48*	.45*	.44*	.47*	.73*	.50*	.29*	.65
8	.68*	.71*	.52*	.43*	.81*	.62*	.30	.77
9	.46	.53*	.51	.64	.65*	.51*	.37	.85
10	.70*	.75*	.41	.65*	.65	.43*	.57	.64
11	.77	.60*	.44*	.33*	.13*	.43*	.27*	.81
12	.38	.37*	.58*	.67*	.78	.22*	.40*	.79
13	.85	.59	.37*	.59	.74*	.63*	.27*	.72
14	.89*	.65	.44	.62*	.51*	.58*	.19*	.80
15	.82*	.67	.33*	.61*	.66	.36*	.39	.65
16	.76	.61*	.26	.65	.75*	.39*	.46*	.50
17	.78*	.67*	.45*	.38*	.63*	.32*	.30*	.74*
18	.65*	.47	.48*	.38*	.60*	.35	.40*	.83*
19	.81*	.65	.45*	.63	.74*	.33*	.29*	.72*
20	.84*	.80	.35	.42	.75	.46*	.35*	.73*
21	.74	.62	.51*	.62*	.61*	.66*	.30	.62*
22	.75*	.77	.21	.50	.64	.44*	.31*	.54*
23	.85	.46	.46	.33	.23	.24*	.35*	.64*
24	.67*	.57	.25*	.58*	.64	.49	.30	.67*
25	.53	.69	.47*	.65*	.49*	.29*	.28*	.74*
26	.81*	.74	.34*	.63*	.61	.48*	.38*	.50*
27	.85	.61	.33*	.55*	.40*	.32*	.41*	.44*
28	.65*	.41	.41*	.49	.51	.44*	.46*	.69*
29	.65*	.53	.40	.48	.81*	.30*	.37*	.77*
30	.84*	.38	.35	.58	.70*	.57*	.23*	.69*
31		.61	.39*	.57	.65*	.36	.41	.45*
32		.27	.39	.54*	.34*	.48	.34*	.36
33		.62	.19*	.60*	.67*	.28	.46*	.75*
34		.64	.34	.50	.44*	.30*	.44*	.50*
35		.63	.30	.58*	.55*	.32*	.47	.57*
36		.69	.44*	.59*	.72*	.47	.39*	.77*
37		.61*	.35*	.51*	.57	.33	.32*	.57*
38		.50*	.45*	.45*	.73	.31*	.35*	.34*
39		.49*	.40	.62*	.53*	.50*	.23*	.37
40		.57*	.45*	.59	.45*	.22*	.42*	.66*
41		.48*	.41	.56				.44*
42		.32	.36*	.48*				.64*
43		.47*	.28*	.63*				.46*
44		.56*	.30*	.54				.61*
45		.47*	.31	.58				.63*
46		.47*	.28*	.41*				.29*
47		.54*	.23	.58*				.41*
48		.46*	.29*	.50*				.52*
49		.53*	.25*					
50		.41*	.28					
51		.42*						
52		.61*						
53		.63*						
54		.34						

* Item is included in the Abbreviated Battery.

Table G–24. Item *p*–Values for Primary 1, Forms B/E, Spring, Grade 1

Item Number	Word Study Skills	Word Reading	Sentence Reading	Reading Comprehension	Mathematics Problem Solving	Mathematics Procedures	Language B	Spelling	Environment	Listening	Language E
1	.82*	.86*	.96*	.92*	.94*	.75*	.71*	.90	.96*	.75*	.80*
2	.69*	.74*	.88*	.88*	.67*	.87*	.83*	.83*	.88*	.66*	.91*
3	.67*	.77*	.91	.89*	.90*	.74*	.65	.85	.61*	.64*	.66
4	.82	.84*	.93	.88	.51	.59	.71*	.55	.87	.65*	.79*
5	.94*	.83*	.85*	.69*	.50*	.67	.64	.70*	.88	.44*	.70*
6	.93	.68*	.81	.86*	.43	.59*	.51*	.87*	.85*	.60*	.76*
7	.90	.81*	.87*	.78	.72	.41*	.67*	.91*	.51	.75*	.90*
8	.84*	.74*	.92*	.83*	.84*	.42	.54	.81*	.96*	.58*	.85
9	.94*	.87*	.84*	.73*	.83*	.76	.70*	.74*	.94*	.46*	.72
10	.96*	.76*	.77	.88	.82*	.71*	.32*	.58*	.64*	.57*	.75
11	.90*	.80*	.81*	.90*	.61*	.47	.60*	.81*	.76*	.97*	.73*
12	.95	.70*	.88	.86*	.71	.72*	.61*	.62*	.57*	.84*	.58*
13	.88*	.81*	.91*	.87*	.36*	.73*	.71*	.55*	.80*	.83	.52*
14	.71*	.66*	.70*	.92*	.86	.31*	.59*	.64*	.45	.53	.75*
15	.82*	.85*	.80*	.81*	.61*	.56	.71*	.74*	.76*	.78	.61*
16	.87	.88	.85*	.85*	.47*	.69*	.64	.92*	.78*	.89*	.38*
17	.49*	.70	.73*	.89*	.88*	.53*	.68*	.92*	.88	.47*	.53*
18	.57*	.85	.70*	.86*	.70	.60*	.43*	.73*	.41*	.80*	.75*
19	.58	.82*	.83*	.60*	.70*	.44*	.73	.53*	.92*	.49*	.64*
20	.62*	.73*	.85	.94	.86*	.51	.50*	.79*	.97*	.65*	.50*
21	.84	.68*	.88*	.82	.45*	.64*	.73*	.75	.94	.63*	.41*
22	.58*	.85	.79*	.75	.64*	.77*	.64	.45	.79*	.58*	.62*
23	.64*	.86	.76*	.71*	.76*	.48*	.55*	.58*	.87*	.71	.53*
24	.15*	.79	.73	.75*	.57	.43	.61*	.55*	.46*	.67	.72*
25	.63	.87	.88	.74*	.80	.68*	.55*	.78*	.39*	.67	.18*
26	.66*	.72	.81*	.52*	.70*	.46*	.73	.73*	.88*	.75*	.46*
27	.36	.75	.87*	.76*	.65*	.50	.68*	.90*	.74*	.60*	.34*
28	.61*	.69*	.88*	.55*	.60*	.29*	.39*	.65*	.86*	.42*	.52*
29	.70*	.78*	.85	.43*	.55*	.39*	.66	.53*	.86*	.58*	.66*
30	.19	.72	.89	.57*	.85*	.32	.54	.66*	.67	.51*	.57*
31				.46*	.73		.48*	.70*	.95*	.63*	.72*
32				.50*	.69*		.81*	.40*	.85*	.79*	.55*
33				.66*	.79*		.77*	.67*	.90	.67*	.65*
34				.53*	.88*		.67	.51*	.66*	.41*	.52*
35				.61*	.86*		.41*	.53	.93*	.71*	.49
36				.53*	.82		.73*	.57*	.59	.62*	.60
37				.56	.59*		.60*		.46	.48	.73
38				.51	.59		.79*		.30*	.71	.38
39				.72	.68		.53*		.34*	.63	.66
40				.62	.66*		.61*		.84*	.55	.41
41					.80*						
42					.45*						

* Item is included in the Abbreviated Battery.

Appendix G

Table G–25. Item _p_–Values for Primary 1, Forms B/E, Spring, Grade 2

Item Number	Word Study Skills	Word Reading	Sentence Reading	Reading Compre-hension	Mathematics Problem Solving	Mathematics Procedures	Language B	Spelling	Environment	Listening	Language E
1	.95*	.98*	.98*	.98*	.98*	.92*	.92*	.99	.98*	.92*	.90*
2	.93*	.89*	.96*	.96*	.89*	.96*	.94*	.96*	.92*	.81*	.95*
3	.88*	.97*	.98	.97*	.97*	.91*	.88	.97	.72*	.75*	.84
4	.96	.96*	.98	.96	.75	.90	.85*	.88	.94	.81*	.93*
5	.98*	.96*	.94*	.85*	.81*	.81	.92	.93*	.95	.64*	.85*
6	.98	.86*	.92	.97*	.70	.74*	.70*	.98*	.89*	.78*	.86*
7	.95	.96*	.96*	.95	.90	.79*	.89*	.97*	.64	.91*	.98*
8	.91*	.95*	.98*	.95*	.94*	.70	.84	.95*	.98*	.82*	.85
9	.98*	.98*	.96*	.90*	.93*	.93	.90*	.90*	.98*	.75*	.89
10	.99*	.97*	.94	.96	.93*	.93*	.64*	.92*	.85*	.69*	.86
11	.98*	.94*	.92*	.97*	.77*	.79	.81*	.95*	.83*	.97*	.88*
12	.98	.92*	.97	.97*	.87	.94*	.88*	.89*	.76*	.90*	.77*
13	.94*	.96*	.96*	.97*	.59*	.88*	.92*	.86*	.89*	.89	.76*
14	.87*	.88*	.91*	.99*	.94	.75*	.82*	.94*	.64	.76	.85*
15	.89*	.97*	.96*	.92*	.85*	.89	.88*	.95*	.88*	.84	.89*
16	.90	.97	.97*	.98*	.77*	.94*	.83	.96*	.86*	.94*	.77*
17	.83*	.88	.89*	.98*	.93*	.90*	.85*	.97*	.96	.62*	.77*
18	.85*	.95	.92*	.97*	.86	.81*	.76*	.93*	.39*	.87*	.94*
19	.78	.97*	.96*	.83*	.87*	.87*	.89	.88*	.94*	.61*	.89*
20	.78*	.93*	.98	.99	.94*	.88	.70*	.89*	.98*	.82*	.72*
21	.91	.91*	.98*	.97	.76*	.75*	.84*	.90	.97	.83*	.68*
22	.83*	.98	.95*	.95	.82*	.94*	.87	.70	.88*	.73*	.82*
23	.75*	.97	.88*	.93*	.86*	.80*	.81*	.88*	.93*	.84	.66*
24	.26*	.90	.95	.93*	.78	.77	.85*	.86*	.60*	.75	.84*
25	.78	.97	.96	.90*	.91	.86*	.82*	.89*	.60*	.89	.20*
26	.79*	.93	.96*	.84*	.84*	.82*	.86	.94*	.87*	.80*	.56*
27	.43	.93	.96*	.95*	.81*	.77	.79*	.97*	.84*	.77*	.57*
28	.76*	.90*	.96*	.71*	.81*	.75*	.60*	.90*	.90*	.60*	.71*
29	.78*	.94*	.97	.70*	.66*	.79*	.80	.80*	.95*	.70*	.82*
30	.45	.95	.96	.82*	.97*	.77	.73	.88*	.83	.73*	.71*
31				.77*	.89		.66*	.88*	.97*	.83*	.89*
32				.79*	.86*		.93*	.72*	.83*	.88*	.74*
33				.84*	.87*		.88*	.92*	.94	.79*	.89*
34				.77*	.94*		.78	.91*	.76*	.56*	.81*
35				.84*	.94*		.53*	.70	.97*	.89*	.69
36				.82*	.92		.85*	.85*	.73	.72*	.82
37				.84	.80*		.73*		.66	.63	.92
38				.82	.84		.91*		.36*	.88	.51
39				.92	.73		.69*		.50*	.75	.87
40				.87	.88*		.79*		.93*	.73	.76
41				.95*							
42				.80*							

* Item is included in the Abbreviated Battery.

Table G–26. Item *p*–Values for Primary 2, Forms B/E, Spring, Grade 2

Item Number	Word Study Skills	Reading Vocabulary	Reading Compre-hension	Mathematics Problem Solving	Mathematics Procedures	Language B	Spelling	Environment	Listening	Language E
1	.97*	.93	.88*	.91*	.95	.84*	.86	.86*	.84*	.80
2	.95*	.95*	.84*	.89	.93*	.92*	.89*	.88*	.58*	.96*
3	.76*	.86*	.86*	.83	.91*	.83*	.84*	.69*	.79*	.94*
4	.95	.87*	.55*	.91*	.85*	.84	.78*	.76*	.64*	.68*
5	.97*	.89	.59*	.90*	.61*	.78*	.77	.62*	.58*	.92*
6	.98	.90*	.80*	.87*	.63	.80*	.87*	.73	.58*	.82*
7	.95	.78*	.78*	.82*	.74	.63	.92*	.72	.58*	.90
8	.97*	.60	.70*	.68*	.68*	.72*	.86*	.54*	.71*	.62
9	.98*	.74*	.38*	.82	.85*	.75*	.78*	.86*	.51*	.49*
10	.91*	.59	.80*	.67*	.84	.81	.86*	.55*	.68*	.53
11	.95	.89*	.81	.64*	.75*	.84*	.71*	.75	.80*	.81*
12	.96*	.81*	.67	.77	.67*	.80*	.81*	.91*	.90*	.60*
13	.97*	.67*	.74	.63*	.75	.72	.70	.48	.93*	.67
14	.86*	.63*	.59	.46*	.82*	.77*	.79*	.65*	.97*	.70*
15	.55*	.54*	.83	.70	.75*	.58*	.71*	.97*	.65*	.85
16	.75*	.73*	.72	.61*	.57	.75*	.82*	.68*	.97*	.65*
17	.78	.85	.38	.68*	.83*	.84	.90*	.76*	.85*	.75
18	.58	.87	.70*	.84	.88	.68	.82*	.51	.78*	.82*
19	.50*	.70*	.50*	.85*	.82*	.76	.88	.86*	.70*	.68*
20	.64*	.73*	.83*	.53*	.76*	.77*	.78*	.69*	.46*	.65*
21	.45	.66*	.63*	.89*	.84*	.78	.82*	.94	.83	.75*
22	.49	.64*	.69*	.55	.48*	.76*	.83*	.75*	.68	.66*
23	.49*	.55	.73*	.57	.80	.83	.87	.42*	.61	.65*
24	.74	.63	.52*	.37*	.77*	.66*	.83*	.72*	.68	.64*
25	.46*	.90*	.61*	.68*	.35*	.72*	.82*	.60*	.74	.74*
26	.35*	.85*	.73*	.78*	.69	.76*	.91	.51*	.55	.78*
27	.59*	.74	.47*	.73	.67*	.68*	.87*	.38*	.85	.47*
28	.49*	.75*	.84*	.70*	.65*	.71*	.86*	.91*	.56*	.77*
29	.24	.60	.70*	.81*	.54*	.78	.79*	.78*	.56*	.54*
30	.23*	.59*	.76*	.87*	.59	.76*	.81*	.23*	.53*	.77*
31			.40*	.60*		.76*	.61*	.93*	.67*	.86*
32			.61*	.51		.75	.92*	.83*	.63*	.63*
33			.57*	.74		.70	.63*	.41	.83*	.70
34			.56*	.60*		.71*	.83*	.93	.59*	.66*
35			.48*	.90*		.61	.61*	.84*	.40*	.74*
36			.69*	.60*		.61*	.77*	.41*	.80*	.85
37			.69*	.85		.90*		.42	.71*	.61*
38			.64	.63*		.65*		.97	.74	.57*
39			.61	.69*		.73*		.57*	.39	.63
40			.65	.76*		.60		.92*	.62	.82*
41				.69		.69				
42				.37		.70				
43				.28*		.91				
44				.86*		.54*				
45						.53*				
46						.75*				
47						.80				
48						.57*				

* Item is included in the Abbreviated Battery.

Appendix G

Table G–27. Item *p*–Values for Primary 2, Forms B/E, Spring, Grade 3

Item Number	Word Study Skills	Reading Vocabulary	Reading Comprehension	Mathematics Problem Solving	Mathematics Procedures	Language B	Spelling	Environment	Listening	Language E
1	.98*	.98	.94*	.93*	.96	.93*	.94	.90*	.86*	.89
2	.96*	.98*	.93*	.95	.94*	.97*	.96*	.90*	.66*	.95*
3	.80*	.93*	.93*	.89	.95*	.93*	.90*	.79*	.84*	.94*
4	.98	.93*	.74*	.97*	.91*	.95	.88*	.78*	.79*	.66*
5	.97*	.96	.72*	.94*	.66*	.89*	.87	.66*	.68*	.92*
6	.97	.95*	.89*	.92*	.73	.89*	.93*	.79	.67*	.83*
7	.98	.91*	.90*	.86*	.85	.84	.97*	.83	.71*	.92
8	.98*	.81	.82*	.80*	.73*	.90*	.95*	.57*	.76*	.61
9	.98*	.81*	.56*	.88	.91*	.84*	.85*	.92*	.64*	.58*
10	.95*	.73	.86*	.74*	.93	.92	.93*	.52*	.73*	.55
11	.98	.95*	.89	.67*	.85*	.96*	.83*	.83	.84*	.80*
12	.97*	.90*	.75	.89	.85*	.88*	.91*	.96*	.95*	.57*
13	.98*	.84*	.82	.80*	.86	.92	.83	.67	.96*	.75
14	.93*	.79*	.73	.57*	.89*	.90*	.90*	.72*	.95*	.72*
15	.70*	.70*	.89	.82	.86*	.60*	.83*	.99*	.71*	.87
16	.88*	.85*	.74	.77*	.77	.85*	.90*	.80*	.97*	.73*
17	.86	.92	.49	.80*	.92*	.92	.96*	.74*	.88*	.80
18	.72	.95	.82*	.92	.95	.88	.93*	.68	.88*	.91*
19	.53*	.80*	.63*	.95*	.91*	.78	.96	.93*	.76*	.72*
20	.65*	.82*	.92*	.67*	.85*	.89*	.87*	.79*	.53*	.70*
21	.58	.80*	.75*	.92*	.90*	.88	.89*	.97	.90	.74*
22	.62	.71*	.76*	.76	.66*	.91*	.92*	.84*	.70	.65*
23	.57*	.68	.83*	.76	.90	.95	.94	.50*	.62	.72*
24	.80	.77	.65*	.61*	.87*	.76*	.88*	.74*	.73	.75*
25	.58*	.97*	.74*	.87*	.51*	.89*	.91*	.73*	.79	.78*
26	.50*	.93*	.80*	.93*	.83	.85*	.96	.53*	.65	.83*
27	.74*	.89	.61*	.88	.85*	.81*	.95*	.44*	.91	.48*
28	.54*	.86*	.92*	.85*	.92*	.89*	.96*	.94*	.70*	.82*
29	.34	.70	.77*	.87*	.86*	.88	.89*	.82*	.64*	.58*
30	.27*	.77*	.88*	.93*	.90	.90*	.92*	.31*	.68*	.82*
31			.52*	.77*		.92*	.82*	.94*	.75*	.89*
32			.75*	.63		.87	.97*	.88*	.67*	.71*
33			.66*	.82		.86	.81*	.50	.85*	.74
34			.68*	.80*		.80*	.90*	.97	.72*	.66*
35			.53*	.97*		.78	.75*	.91*	.50*	.76*
36			.77*	.67*		.77*	.90*	.48*	.85*	.89
37			.78*	.94		.96*		.46	.75*	.62*
38			.75	.74*		.73*		.97	.81	.65*
39			.74	.71*		.84*		.64*	.55	.73
40			.79	.86*		.70		.93*	.76	.82*
41				.78		.83				
42				.52		.83				
43				.32*		.95				
44				.91*		.72*				
45						.69*				
46						.85*				
47						.90				
48						.67*				

* Item is included in the Abbreviated Battery.

Table G–28. Item *p*–Values for Primary 3, Forms B/E, Spring, Grade 3

Item Number	Word Study Skills	Reading Vocabulary	Reading Compre-hension	Mathematics Problem Solving	Mathematics Procedures	Language B	Spelling	Science	Social Science	Listening	Language E
1	.96	.94	.76*	.88	.84*	.90*	.84	.96*	.89*	.96*	.48*
2	.92*	.86*	.82*	.89	.78*	.90	.65*	.89*	.84*	.64*	.69*
3	.88*	.88	.51*	.60*	.78	.77	.56*	.53*	.84*	.37*	.73*
4	.94*	.54	.87*	.59*	.65	.78*	.68*	.73*	.78*	.92*	.53*
5	.80*	.93*	.67*	.85*	.73*	.66	.50	.88*	.68	.35*	.52*
6	.65	.73*	.64	.81	.64*	.45	.69*	.80*	.58*	.81*	.66*
7	.70*	.91*	.77	.89	.62*	.48*	.82*	.70*	.80*	.64*	.75*
8	.59*	.63	.81	.89*	.81*	.72*	.47*	.36	.82*	.50*	.54*
9	.63	.57*	.64	.83*	.49	.66*	.65*	.43*	.82*	.42*	.62*
10	.62*	.79*	.80	.55	.47*	.33*	.66*	.63*	.88*	.54*	.30*
11	.61	.58*	.64	.74*	.47	.72*	.36*	.35	.68*	.73	.35*
12	.41*	.64*	.81	.60*	.67	.65*	.55*	.48*	.62	.81	.72*
13	.93*	.75	.89	.67*	.56*	.81	.53*	.87	.87	.67	.66*
14	.54*	.49*	.74	.48*	.63*	.59	.70	.71*	.50*	.74	.29*
15	.61*	.60*	.89	.80	.66*	.64*	.42*	.58	.64	.62	.62*
16	.79	.81	.69	.80*	.60	.77*	.70*	.82	.76*	.77	.45*
17	.67	.76*	.35	.72*	.77*	.68*	.73*	.65*	.58	.79	.76*
18	.59*	.69	.83	.74	.44	.62*	.47*	.53*	.46*	.90*	.23*
19	.76*	.64*	.69	.56	.48*	.81	.33	.54*	.63*	.48*	.68*
20	.79	.72*	.48	.79*	.50	.71	.77*	.58*	.75	.51*	.42*
21	.75*	.67*	.80	.64	.33*	.76*	.55*	.65*	.56*	.65*	.45*
22	.77*	.80*	.59	.51*	.58*	.69*	.45*	.44	.80*	.66*	.48*
23	.53*	.60*	.67	.84*	.42	.27	.73*	.62	.40*	.57*	.51*
24	.48*	.77*	.57	.73*	.50*	.28*	.69	.74*	.52*	.81*	.38*
25	.62*	.78	.85*	.67*	.51*	.65	.53*	.57	.74*	.58*	.69*
26	.51	.57	.51*	.76*	.45*	.80*	.57*	.54*	.44*	.80*	.73*
27	.53	.71*	.64	.59	.55*	.67*	.79*	.76*	.62*	.51*	.43*
28	.76	.70	.62*	.89*	.59	.70	.74*	.51*	.77*	.86*	.57*
29	.52*	.48*	.73*	.85	.49*	.62*	.49*	.52*	.76*	.75*	.51*
30	.32*	.75*	.49*	.62*	.55*	.75	.63	.69	.60*	.30*	.47*
31			.65*	.72		.59*	.44*	.42*	.72	.37*	.63
32			.56	.75*		.79*	.43*	.48*	.25*	.48*	.47
33			.44*	.67*		.67*	.61*	.76*	.78*	.64*	.37
34			.56*	.65*		.42*	.40	.42*	.71*	.77*	.52
35			.68*	.42*		.59*	.81	.78*	.78*	.58*	.73
36			.61*	.70		.73	.37*	.80	.58*	.75*	.43
37			.72*	.81*		.79*	.53*	.48*	.57*	.78*	.37
38			.42*	.14*		.57	.50*	.43*	.56	.54	.45
39			.57*	.74		.70		.79*	.70	.73	.38
40			.33*	.79*		.35*		.70*	.72*	.75	.47
41			.58	.70		.66					.39
42			.67*	.55*		.65*					.50
43			.62	.69*		.76*					.38
44			.58*	.75*		.65*					.72
45			.34*	.71		.58					.46
46			.70*	.68*		.54*					
47			.51*			.42					
48			.55*			.47*					
49			.56*								
50			39*								
51			.49*								
52			.34*								
53			.46*								
54			.73								

* Item is included in the Abbreviated Battery.

Table G–29. Item *p*–Values for Primary 3, Forms B/E, Spring, Grade 4

Item Number	Word Study Skills	Reading Vocab-ulary	Reading Compre-hension	Mathematics Problem Solving	Mathematics Procedures	Language B	Spelling	Science	Social Science	Listening	Language E
1	.94	.96	.78*	.88	.90*	.93*	.88	.94*	.91*	.97*	.53*
2	.93*	.89*	.88*	.88	.83*	.92	.80*	.92*	.92*	.72*	.67*
3	.90*	.89	.58*	.67*	.83	.82	.65*	.43*	.86*	.46*	.78*
4	.94*	.65	.89*	.70*	.76	.87*	.77*	.76*	.77*	.96*	.64*
5	.81*	.93*	.66*	.89*	.84*	.74	.67	.85*	.75	.47*	.48*
6	.73	.84*	.69	.85	.77*	.60	.80*	.79*	.68*	.90*	.65*
7	.68*	.93*	.80	.90	.65*	.64*	.81*	.74*	.84*	.68*	.71*
8	.62*	.74	.86	.89*	.86*	.76*	.58*	.48	.74	.65*	.51*
9	.72	.70*	.75	.86*	.62	.77*	.77*	.44*	.82*	.57*	.60*
10	.69*	.84*	.83	.60	.63*	.57*	.70*	.60*	.89*	.56*	.26*
11	.69	.67*	.69	.79*	.61	.79*	.47*	.42	.72*	.78	.48*
12	.57*	.74*	.86	.78*	.81	.76*	.61*	.54*	.68	.87	.73*
13	.95*	.83	.89	.72*	.76*	.89	.65*	.84	.86	.71	.65*
14	.58*	.62*	.84	.56*	.81*	.72	.75	.73*	.59*	.74	.37*
15	.64*	.67*	.90	.86	.83*	.81*	.57*	.69	.75	.71	.56*
16	.77	.84	.76	.86*	.79	.81*	.83*	.82	.82*	.76	.57*
17	.69	.85*	.45	.79*	.82*	.83*	.82*	.74*	.64	.82	.78*
18	.70*	.72	.85	.80	.57	.61*	.61*	.68*	.55*	.89*	.25*
19	.76*	.72*	.70	.66	.62*	.88	.42	.66*	.61*	.58*	.69*
20	.83	.81*	.63	.82*	.57	.82	.82*	.57*	.78	.59*	.44*
21	.77*	.73*	.84	.73	.52*	.82*	.69*	.67*	.64*	.72*	.40*
22	.78*	.86*	.63	.58*	.74*	.76*	.53*	.46	.82*	.65*	.48*
23	.51*	.69*	.74	.88*	.60	.50	.76*	.61	.51*	.63*	.57*
24	.56*	.79*	.62	.75*	.63*	.41*	.77	.73*	.62*	.83*	.44*
25	.60*	.84	.89*	.71*	.61*	.76	.60*	.62	.79*	.62*	.57*
26	.58	.65	.58*	.82*	.55*	.83*	.73*	.61*	.39*	.82*	.67*
27	.56	.76*	.67	.65	.70*	.80*	.82*	.73*	.70*	.59*	.46*
28	.80	.75	.70*	.85*	.72	.73	.81*	.64*	.72*	.87*	.65*
29	.56*	.63*	.78*	.86	.66*	.68*	.56*	.61*	.81*	.76*	.63*
30	.38*	.80*	.47*	.69*	.70*	.78	.74	.76	.69*	.33*	.45*
31			.69*	.79		.69*	.56*	.49*	.76	.38*	.69
32			.66	.76*		.85*	.51*	.51*	.28*	.56*	.39
33			.53*	.74*		.75*	.67*	.74*	.82*	.71*	.52
34			.62*	.75*		.46*	.57	.49*	.78*	.82*	.58
35			.75*	.46*		.69*	.76	.76*	.82*	.56*	.73
36			.65*	.75		.79	.50*	.78	.67*	.81*	.48
37			.79*	.83*		.83*	.54*	.57*	.68*	.84*	.30
38			.48*	.21*		.61	.65*	.50*	.61	.60	.44
39			.69*	.79		.73		.81*	.75	.77	.25
40			.40*	.67*		.42*		.73*	.79*	.76	.42
41			.72	.72		.79					.34
42			.73*	.59*		.75*					.62
43			.71	.77*		.82*					.33
44			.61*	.79*		.73*					.65
45			.40*	.74		.70					.47
46			.76*	.71*		.69*					
47			.59*			.52					
48			.63*			.54*					
49			.60*								
50			.52*								
51			.52*								
52			.47*								
53			.55*								
54			.76								

* Item is included in the Abbreviated Battery.

Table G–30. Item *p*–Values for Intermediate 1, Forms B/E, Spring, Grade 4

Item Number	Word Study Skills	Reading Vocabulary	Reading Comprehension	Mathematics Problem Solving	Mathematics Procedures	Language B	Spelling	Science	Social Science	Listening	Language E
1	.92	.83	.82*	.81	.77*	.91*	.74*	.90*	.84*	.75*	.84*
2	.83*	.73*	.82*	.72*	.80	.52*	.69*	.76*	.84*	.66*	.79*
3	.87*	.57	.67	.74	.85*	.80*	.68*	.47	.63	.84*	.81*
4	.88*	.85*	.63*	.77	.52*	.74*	.61	.38	.70*	.36*	.56*
5	.75*	.66*	.71*	.65*	.58	.77	.56*	.80	.69*	.23*	.43*
6	.74	.59*	.48*	.72*	.70*	.42*	.52*	.45*	.65*	.70*	.55*
7	.63	.59*	.80	.65	.65*	.81	.69*	.74*	.78*	.57*	.47
8	.64*	.67*	.61	.76*	.59	.73	.80*	.75*	.57	.44*	.72*
9	.70*	.59*	.45	.86*	.69	.51*	.73*	.55	.66*	.48*	.48*
10	.65	.62	.44	.24*	.74*	.61*	.68*	.30	.70*	.81*	.55*
11	.70*	.66*	.75	.76	.81*	.47*	.37*	.80*	.66*	.62	.69*
12	.38*	.59	.78	.62*	.47*	.51*	.78*	.63*	.47*	.52	.59*
13	.62*	.87*	.84	.74	.65	.73*	.43*	.75*	.83	.70	.70*
14	.64*	.84	.67	.83	.72*	.72*	.72*	.35*	.55	.38*	.50*
15	.64*	.67	.84	.64	.58*	.39	.68*	.65*	.69*	.61*	.81*
16	.51*	.80*	.45	.64*	.61	.75	.47	.66*	.58	.91*	.57*
17	.47	.78*	.53	.71*	.65*	.55	.68*	.74*	.47*	.67*	.74*
18	.54	.75	.72	.61	.52	.69*	.66*	.53*	.49*	.41*	.65*
19	.73*	.76*	.61*	.54*	.49*	.58*	.72*	.63*	.52*	.65*	.53*
20	.74	.73*	.65*	.79*	.68	.52*	.42	.57	.51	.89*	.56*
21	.74*	.60*	.53*	.45	.58*	.59	.36	.77*	.66*	.81*	.72*
22	.64	.78	.83*	.65*	.60*	.67*	.75*	.74*	.69*	.79*	.56*
23	.57	.79	.82	.38*	.59*	.63	.55	.58*	.52*	.57*	.52*
24	.56	.72*	.70*	.84*	.69*	.45	.82*	.62	.44*	.47*	.74*
25	.51*	.62*	.58*	.43	.61	.65*	.47*	.40*	.58*	.55	.52*
26	.68*	.74*	.32*	.75*	.64*	.46*	.52	.54*	.54*	.45	.66*
27	.49*	.73*	.48*	.48*	.39*	.60	.45*	.54*	.55*	.48	.59
28	.45*	.76*	.61*	.71	.45*	.49*	.49*	.73*	.61*	.61	.62*
29	.60*	.66	.55*	.64*	.54*	.63	.79	.72*	.76	.32*	.33*
30	.39*	.66*	.65	.66*	.43	.65*	.70	.48*	.37*	.75*	.69*
31			.71*	.55	.59	.55	.87*	.50*	.51	.60*	.83*
32			.57	.58*	.46	.52*	.33*	.47*	.64*	.62*	.72*
33			.36*	.75*		.63	.62	.25	.69	.70*	.51
34			.68*	.61*		.70	.31*	.60*	.68*	.63*	.67
35			.54*	.58		.56*	.62	.81*	.36*	.75*	.74
36			.60*	.69		.56	.42*	.68	.66	.72*	.59
37			.60	.57*		.56*	.65*	.42*	.59*	.67*	.56
38			.54	.64*		.71*	.39*	.64*	.62*	.79	.69
39			.40	.42*		.55*	.68*	.37*	.69*	.33	.41
40			.41	.59		.79*	.46*	.63	.70*	.33	.41
41			.74	.40*		.64*					.46
42			.67	.23*		.58*					.53
43			.39*	.80*		.64*					.51
44			.60*	.60*		.63*					.49
45			.61*	.38		.47*					.47
46			.67*	.60*		.70					.71
47			.38	.60		.61					.58
48			.60*	.59*		.42					.37
49			.48								
50			.47*								
51			.52*								
52			.61*								
53			.50*								
54			.71*								

* Item is included in the Abbreviated Battery.

213

Appendix G

Table G–31. Item p–Values for Intermediate 1, Forms B/E, Spring, Grade 5

Item Number	Word Study Skills	Reading Vocabulary	Reading Comprehension	Mathematics Problem Solving	Mathematics Procedures	Language B	Spelling	Science	Social Science	Listening	Language E
1	.93	.89	.87*	.87	.80*	.96*	.82*	.94*	.89*	.82*	.78*
2	.87*	.84*	.87*	.77*	.80	.53*	.77*	.84*	.88*	.77*	.78*
3	.90*	.72	.74	.81	.85*	.91*	.74*	.62	.68	.92*	.79*
4	.91*	.91*	.75*	.85	.68*	.85*	.67	.46	.83*	.53*	.58*
5	.76*	.81*	.79*	.73*	.67	.90	.66*	.86	.76*	.30*	.38*
6	.79	.71*	.61*	.81*	.74*	.59*	.65*	.39*	.72*	.80*	.62*
7	.68	.72*	.88	.71	.72*	.93	.73*	.84*	.85*	.67*	.56
8	.74*	.73*	.66	.83*	.71	.81	.87*	.86*	.64	.53*	.76*
9	.78*	.71*	.51	.89*	.79	.69*	.79*	.72	.75*	.53*	.44*
10	.71	.75	.59	.42*	.84*	.69*	.75*	.33	.78*	.90*	.62*
11	.80*	.80*	.82	.83	.86*	.59*	.46*	.87*	.74*	.67	.67*
12	.45*	.66	.80	.70*	.61*	.73*	.85*	.74*	.52*	.56	.61*
13	.67*	.91*	.89	.83	.73	.88*	.43*	.80*	.87	.76	.70*
14	.68*	.89	.75	.88	.78*	.84*	.77*	.43*	.60	.49*	.52*
15	.66*	.79	.90	.77	.70*	.58	.74*	.74*	.78*	.76*	.77*
16	.55*	.84*	.53	.75*	.72	.84	.53	.77*	.73	.95*	.56*
17	.51	.86*	.63	.78*	.74*	.71	.72*	.80*	.58*	.76*	.69*
18	.61	.83	.80	.72	.69	.80*	.77*	.63*	.61*	.48*	.64*
19	.79*	.82*	.68*	.63*	.63*	.70*	.85*	.75*	.62*	.76*	.51*
20	.78	.81*	.73*	.84*	.77	.59*	.56	.61	.60	.90*	.58*
21	.82*	.71*	.61*	.58	.71*	.80	.39	.84*	.76*	.89*	.72*
22	.70	.87	.89*	.74*	.67*	.77*	.78*	.82*	.77*	.89*	.57*
23	.63	.85	.90	.53*	.71*	.75	.67	.66*	.67*	.67*	.50*
24	.59	.84*	.80*	.88*	.76*	.62	.88*	.70	.59*	.60*	.73*
25	.54*	.72*	.69*	.48	.71	.82*	.60*	.49*	.75*	.69	.50*
26	.72*	.85*	.41*	.84*	.74*	.63*	.59	.66*	.71*	.58	.69*
27	.55*	.79*	.59*	.52*	.48*	.77	.55*	.71*	.61*	.57	.60
28	.52*	.84*	.71*	.82	.60*	.66*	.56*	.80*	.70*	.73	.58*
29	.65*	.79	.66*	.74*	.69*	.73	.84	.81*	.84	.41*	.36*
30	.45*	.79*	.76	.78*	.57	.79*	.78	.56*	.46*	.83*	.74*
31			.80*	.63	.70	.71	.90*	.53*	.63	.68*	.79*
32			.69	.69*	.58	.70*	.44*	.48*	.74*	.72*	.70*
33			.45*	.83*		.77	.72	.35	.76	.75*	.55
34			.75*	.70*		.83	.38*	.73*	.76*	.71*	.69
35			.66*	.69		.70*	.75	.87*	.47*	.79*	.76
36			.69*	.76		.73	.55*	.77	.78	.81*	.62
37			.73	.68*		.73*	.75*	.46*	.75*	.74*	.60
38			.68	.72*		.84*	.51*	.74*	.74*	.82	.71
39			.51	.58*		.70*	.76*	.50*	.79*	.38	.37
40			.50	.67		.89*	.55*	.70	.77*	.40	.45
41			.81	.43*		.80*					.52
42			.80	.24*		.71*					.53
43			.47*	.84*		.78*					.46
44			.72*	.69*		.79*					.53
45			.69*	.46		.62*					.42
46			.77*	.65*		.85					.71
47			.45	.65		.77					.54
48			.73*	.65*		.53					.31
49			.56								
50			.57*								
51			.61*								
52			.70*								
53			.61*								
54			.80*								

* Item is included in the Abbreviated Battery.

214

Table G–32. Item _p_–Values for Intermediate 2, Forms B/E, Spring, Grade 5

Item Number	Reading Vocabulary	Reading Comprehension	Mathematics Problem Solving	Mathematics Procedures	Language B	Spelling	Science	Social Science	Listening	Language E
1	.85	.75	.58*	.84	.89*	.83*	.86	.92*	.89*	.70*
2	.65*	.75*	.73	.81*	.94*	.84*	.90	.48*	.64*	.89*
3	.52*	.80*	.16	.86	.65*	.79*	.70*	.80*	.65*	.94*
4	.70*	.60*	.66*	.53*	.56*	.70	.59*	.64	.46*	.72*
5	.78*	.47*	.48	.71*	.56*	.71*	.64	.58	.57*	.75*
6	.56	.72*	.60	.74*	.76*	.56	.44	.51*	.56*	.56*
7	.51*	.78*	.32*	.86	.70	.79*	.49*	.73*	.49*	.72*
8	.52*	.50*	.79*	.28*	.58*	.66*	.65*	.61	.66*	.65*
9	.63*	.71*	.49*	.68*	.52	.65*	.57*	.56	.69*	.68
10	.67	.56	.47*	.51	.61	.41	.52*	.41*	.71*	.51*
11	.68	.66*	.41	.33*	.18*	.79*	.55	.51	.82	.68*
12	.67*	.55*	.60*	.33	.60	.66*	.58*	.59*	.37	.76*
13	.84*	.41	.78*	.40*	.76*	.55*	.54*	.42	.80	.56*
14	.85	.74	.71*	.66*	.63*	.63	.44*	.75	.61*	.41*
15	.65*	.87	.27	.61	.58*	.59	.81*	.45*	.61*	.62*
16	.78*	.54	.61	.68*	.81*	.73*	.47*	.34*	.63*	.69*
17	.83	.50	.60*	.78*	.52	.61*	.49*	.62*	.83*	.87*
18	.55*	.76	.22*	.66*	.74	.79*	.44	.29*	.83*	.71*
19	.78*	.66	.40*	.59	.76	.70*	.58*	.72*	.40*	.72*
20	.85*	.88*	.53	.58	.48	.59*	.59*	.43	.77*	.75*
21	.78	.81*	.51	.45	.51*	.52*	.78*	.56*	.82*	.32*
22	.82	.33*	.47*	.57*	.66*	.82	.60	.40*	.74*	.35*
23	.79*	.62*	.50	.65*	.68	.47*	.62*	.77*	.41*	.73*
24	.80*	.48*	.53*	.65*	.53*	.70*	.59*	.40*	.30*	.28*
25	.65*	.65	.63*	.32*	.70	.59*	.29*	.41*	.44*	.68*
26	.70	.68	.48*	.43*	.57	.44*	.82*	.67*	.55*	.77*
27	.74*	.55	.51	.45	.66	.58*	.50*	.35*	.28*	.51*
28	.59	.55	.53*	.41	.63	.62*	.38	.61	.70*	.48*
29	.66*	.62	.70*	.52	.49*	.84*	.36*	.71	.47*	.50*
30	.70*	.72	.48	.52*	.53*	.74	.49*	.66*	.82*	.70
31		.60*	.73*	.39*	.54	.69	.66*	.48*	.68*	.57*
32		.80*	.71	.42*	.55*	.82*	.33*	.38*	.70*	.53*
33		.60	.61		.65*	.65*	.67	.61*	.75*	.54
34		.66*	.80*		.68*	.45	.38	.45*	.56	.55
35		.74*	.36*		.63*	.72*	.35*	.64*	.76	.68
36		.50*	.58*		.56*	.28*	.65*	.33*	.53	.82
37		.77	.38*		.64*	.52*	.59*	.66*	.53	.56
38		.74	.41		.56	.76	.47*	.35*	.54	.59
39		.44	.46		.63*	.57*	.47*	.50*	.53	.54
40		.57	.55*		.67*	.79*	.72*	.52*	.33	.61
41		.75	.43*		.54					.47
42		.63	.63*		.67					.69
43		.45*	.62*		.61					.54
44		.58*	.22*		.44*					.70
45		.37*	.45*		.58*					.74
46		.55*	.32		.60*					.51
47		.47*	.31*		.43*					.61
48		.59	.63		.42*					.81
49		.63*								
50		.47*								
51		.59*								
52		.48								
53		.37*								
54		.61*								

* Item is included in the Abbreviated Battery.

Table G–33. Item *p*–Values for Intermediate 2, Forms B/E, Spring, Grade 6

Item Number	Reading Vocabulary	Reading Comprehension	Mathematics Problem Solving	Mathematics Procedures	Language B	Spelling	Science	Social Science	Listening	Language E
1	.94	.83	.69*	.82	.88*	.84*	.90	.94*	.94*	.68*
2	.75*	.78*	.80	.79*	.93*	.87*	.94	.65*	.72*	.88*
3	.59*	.86*	.38	.83	.77*	.80*	.81*	.82*	.70*	.92*
4	.78*	.67*	.71*	.63*	.59*	.72	.61*	.70	.48*	.70*
5	.89*	.59*	.58	.73*	.58*	.71*	.77	.65	.73*	.74*
6	.64	.76*	.65	.81*	.83*	.59	.58	.67*	.66*	.60*
7	.55*	.84*	.36*	.85	.76	.80*	.57*	.75*	.62*	.74*
8	.60*	.49*	.89*	.37*	.66*	.69*	.77*	.72	.67*	.65*
9	.77*	.80*	.58*	.70*	.60	.71*	.66*	.65	.71*	.77
10	.77	.59	.55*	.48	.64	.49	.60*	.49*	.89*	.45*
11	.75	.71*	.55	.31*	.26*	.84*	.64	.60	.90	.64*
12	.76*	.58*	.72*	.35	.70	.68*	.74*	.72*	.34	.74*
13	.88*	.47	.85*	.48*	.85*	.64*	.66*	.58	.86	.64*
14	.88	.79	.69*	.68*	.69*	.66	.57*	.76	.62*	.35*
15	.72*	.88	.34	.65	.61*	.60	.85*	.52*	.71*	.55*
16	.88*	.64	.69	.69*	.85*	.79*	.57*	.42*	.72*	.68*
17	.87	.56	.61*	.78*	.52	.67*	.49*	.72*	.83*	.83*
18	.61*	.78	.25*	.73*	.74	.83*	.48	.44*	.88*	.64*
19	.80*	.76	.60*	.67	.84	.70*	.70*	.81*	.47*	.71*
20	.91*	.92*	.66	.66	.61	.63*	.65*	.60	.81*	.71*
21	.83	.87*	.61	.53	.57*	.62*	.81*	.68*	.88*	.38*
22	.86	.44*	.60*	.64*	.67*	.84	.67	.42*	.80*	.44*
23	.85*	.65*	.63	.69*	.70	.52*	.71*	.84*	.50*	.67*
24	.86*	.62*	.59*	.72*	.56*	.73*	.65*	.53*	.40*	.37*
25	.75*	.73	.71*	.40*	.72	.61*	.39*	.50*	.54*	.67*
26	.78	.71	.56*	.56*	.64	.54*	.86*	.73*	.62*	.80*
27	.86*	.62	.64	.53	.68	.61*	.59*	.39*	.33*	.54*
28	.71	.63	.65*	.49	.68	.66*	.39	.74	.80*	.50*
29	.72*	.66	.70*	.58	.48*	.89*	.51*	.78	.57*	.49*
30	.79*	.78	.63	.57*	.61*	.74	.46*	.76*	.89*	.71
31		.74*	.79*	.47*	.53	.77	.75*	.57*	.79*	.59*
32		.84*	.78	.47*	.63*	.88*	.41*	.42*	.76*	.55*
33		.70	.69		.53*	.73*	.72	.71*	.82*	.60
34		.77*	.86*		.75*	.46	.40	.47*	.58	.64
35		.78*	.45*		.69*	.80*	.38*	.76*	.81	.62
36		.64*	.74*		.65*	.34*	.72*	.54*	.61	.77
37		.78	.51*		.68*	.58*	.69*	.73*	.52	.54
38		.78	.46		.60	.80	.58*	.38*	.60	.66
39		.48	.56		.63*	.63*	.57*	.70*	.59	.63
40		.66	.48*		.68*	.81*	.77*	.65*	.35	.66
41		.78	.57*		.65					.49
42		.68	.70*		.75					.72
43		.52*	.69*		.67					.52
44		.66*	.33*		.48*					.69
45		.47*	.54*		.64*					.77
46		.61*	.43		.63*					.53
47		.56*	.37*		.50*					.70
48		.65	.68		.48*					.83
49		.70*								
50		.52*								
51		.67*								
52		.59								
53		.45*								
54		.67*								

* Item is included in the Abbreviated Battery.

Table G–34. Item *p*–Values for Intermediate 3, Forms B/E, Spring, Grade 6

Item Number	Reading Vocabulary	Reading Comprehension	Mathematics Problem Solving	Mathematics Procedures	Language B	Spelling	Science	Social Science	Listening	Language E
1	.86*	.81*	.61	.83	.81	.90*	.91*	.50*	.78*	.80*
2	.73*	.66*	.62*	.64*	.85*	.64*	.68*	.79	.84*	.51*
3	.57	.52*	.63*	.61*	.89	.71*	.61	.44*	.60*	.56*
4	.67*	.55*	.79	.74	.70*	.70*	.55*	.58*	.30*	.39*
5	.83*	.90*	.55*	.73	.60	.61	.84	.67	.61*	.40*
6	.46*	.81	.57	.72*	.63	.70*	.64*	.80*	.69*	.30*
7	.61*	.51*	.65*	.59*	.76*	.74*	.58*	.49	.39*	.56*
8	.70	.91*	.42	.73*	.74*	.78*	.58	.54	.91*	.55
9	.48*	.56*	.57*	.53	.65	.58	.57	.68*	.44*	.56*
10	.84	.86*	.62*	.46*	.71*	.66	.69*	.84*	.69*	.55*
11	.71	.35*	.68*	.49*	.62*	.59*	.59*	.57	.71*	.46*
12	.48*	.76	.82*	.82*	.80	.65*	.60*	.42*	.71*	.59*
13	.91	.62	.48	.44	.52	.49*	.74*	.66*	.83*	.57*
14	.90*	.58	.51	.55*	.69*	.80*	.68*	.51*	.81*	.58*
15	.77	.47	.50*	.35*	.84*	.40	.69*	.41*	.48*	.14*
16	.81*	.74	.35*	.34	.81*	.83*	.44*	.59*	.78*	.71*
17	.87	.48	.49*	.73*	.69*	.63*	.53*	.49*	.63*	.62*
18	.65*	.75	.40	.68	.49*	.63*	.76	.32*	.63*	.75*
19	.68*	.87	.58	.71*	.61*	.83*	.45*	.44*	.56*	.47*
20	.81*	.87	.48*	.54	.33*	.80*	.41*	.63*	.59*	.65*
21	.74*	.78	.51*	.49	.83	.63	.68	.54*	.83*	.43*
22	.87	.64	.36*	.48*	.71*	.46*	.84*	.51*	.35*	.52*
23	.78*	.66	.46*	.57*	.56*	.67	.46*	.58*	.46*	.48*
24	.71*	.66	.42*	.63*	.68	.78*	.67*	.66*	.34	.37*
25	.70	.47*	.48	.50	.79	.63*	.34*	.66*	.76	.57*
26	.82	.73*	.54*	.38	.58*	.46	.56	.28*	.83	.40*
27	.81*	.72	.50	.30*	.65	.47*	.40	.48*	.92*	.57*
28	.58*	.30*	.37	.42*	.68*	.59*	.34*	.77*	.70*	.68
29	.74*	.65*	.63*	.25*	.60*	.66	.63	.40	.69*	.61*
30	.59*	.56*	.46*	.37*	.70*	.79*	.43*	.40*	.41*	.61*
31		.48*	.33	.36	.63*	.56*	.72*	.60	.60	.40*
32		.50*	.65*	.44*	.62*	.59	.53	.58*	.78	.43*
33		.58*	.55		.66*	.29*	.35*	.32	.53	.47
34		.56*	.41*		.66	.88	.63*	.45	.56*	.69
35		.54	.49*		.76	.42*	.47*	.69*	.54*	.38
36		.50*	.37		.64	.49*	.51*	.34*	.34*	.66
37		.49*	.40*		.74*	.62*	.62*	.46*	.59	.30
38		.49*	.55*		.75*	.38*	.38*	.63	.63	.52
39		.50*	.69		.54*	.50*	.46*	.44*	.82	.63
40		.63*	.35*		.81*	.53*	.70*	.75*	.54	.38
41		.55*	.57		.79*					.55
42		.73	.78*		.69					.52
43		.56	.38		.76					.28
44		.69	.55*		.58					.54
45		.48	.68*		.51					.69
46		.75	.52		.56*					.51
47		.69	.56*		.57*					.50
48		.52	.54*		.69*					.70
49		.69*								
50		.36								
51		.28*								
52		.53*								
53		.50*								
54		.44*								

* Item is included in the Abbreviated Battery.

Appendix G

Table G–35. Item *p*–Values for Intermediate 3, Forms B/E, Spring, Grade 7

Item Number	Reading Vocabulary	Reading Comprehension	Mathematics Problem Solving	Mathematics Procedures	Language B	Spelling	Science	Social Science	Listening	Language E
1	.87*	.79*	.69	.83	.83	.89*	.89*	.57*	.79*	.83*
2	.76*	.70*	.72*	.68*	.85*	.67*	.72*	.81	.85*	.63*
3	.55	.49*	.64*	.66*	.90	.69*	.67	.51*	.61*	.68*
4	.72*	.52*	.78	.78	.72*	.70*	.62*	.64*	.31*	.37*
5	.83*	.86*	.58*	.72	.70	.63	.85	.67	.64*	.48*
6	.49*	.81	.56	.74*	.72	.69*	.68*	.81*	.65*	.31*
7	.64*	.55*	.67*	.60*	.78*	.69*	.57*	.52	.46*	.70*
8	.74	.89*	.43	.78*	.76*	.80*	.57	.62	.89*	.64
9	.50*	.60*	.67*	.60	.63	.60	.57	.68*	.49*	.60*
10	.84	.82*	.64*	.57*	.71*	.65	.67*	.83*	.69*	.65*
11	.70	.36*	.75*	.51*	.74*	.64*	.63*	.62	.70*	.40*
12	.51*	.73	.83*	.84*	.80	.69*	.66*	.44*	.71*	.67*
13	.89	.64	.70	.53	.55	.50*	.76*	.70*	.82*	.64*
14	.87*	.61	.67	.56*	.73*	.84*	.69*	.52*	.77*	.65*
15	.75	.49	.57*	.35*	.86*	.50	.75*	.45*	.50*	.23*
16	.78*	.72	.45*	.38	.79*	.79*	.50*	.65*	.71*	.81*
17	.84	.48	.63*	.77*	.67*	.64*	.57*	.59*	.59*	.69*
18	.66*	.79	.47	.74	.62*	.66*	.77	.42*	.62*	.76*
19	.74*	.83	.65	.71*	.69*	.81*	.51*	.47*	.60*	.51*
20	.78*	.86	.51*	.63	.46*	.80*	.41*	.66*	.62*	.78*
21	.75*	.79	.56*	.54	.83	.65	.70	.57*	.81*	.41*
22	.85	.68	.45*	.54*	.73*	.49*	.82*	.49*	.37*	.57*
23	.82*	.64	.50*	.60*	.61*	.73	.51*	.58*	.51*	.57*
24	.72*	.69	.49*	.68*	.74	.81*	.67*	.65*	.36	.34*
25	.69	.52*	.52	.52	.72	.63*	.34*	.67*	.70	.61*
26	.80	.73*	.69*	.41	.63*	.49	.58	.36*	.81	.45*
27	.82*	.67	.62	.39*	.76	.45*	.40	.55*	.89*	.59*
28	.63*	.31*	.49	.43*	.72*	.62*	.39*	.78*	.69*	.77
29	.75*	.65*	.72*	.27*	.67*	.72	.58	.45	.70*	.67*
30	.65*	.56*	.52*	.37*	.71*	.78*	.47*	.48*	.42*	.65*
31		.53*	.47	.39	.61*	.58*	.70*	.65	.62	.46*
32		.50*	.65*	.46*	.63*	.63	.56	.61*	.78	.51*
33		.61*	.63		.73*	.36*	.38*	.36	.51	.56
34		.57*	.49*		.73	.88	.64*	.55	.64*	.76
35		.56	.54*		.77	.46*	.50*	.75*	.52*	.50
36		.58*	.45		.64	.50*	.50*	.33*	.38*	.74
37		.49*	.46*		.77*	.60*	.66*	.55*	.58	.36
38		.52*	.57*		.73*	.44*	.42*	.65	.59	.55
39		.48*	.70		.55*	.50*	.55*	.48*	.75	.66
40		.63*	.40*		.77*	.54*	.66*	.79*	.55	.42
41		.58*	.64		.77*					.67
42		.71	.79*		.70					.59
43		.53	.42		.71					.36
44		.67	.64*		.58					.62
45		.53	.68*		.52					.76
46		.71	.58		.55*					.53
47		.66	.61*		.65*					.47
48		.52	.57*		.65*					.73
49		.66*								
50		.38								
51		.27*								
52		.51*								
53		.46*								
54		.44*								

* Item is included in the Abbreviated Battery.

Table G–36. Item *p*–Values for Advanced 1, Forms B/E, Spring, Grade 7

Item Number	Reading Vocabulary	Reading Comprehension	Mathematics Problem Solving	Mathematics Procedures	Language B	Spelling	Science	Social Science	Listening	Language E
1	.86*	.87	.67	.55	.79	.84*	.84*	.75*	.81*	.54
2	.57	.88	.46*	.60*	.61*	.68*	.74	.48*	.63*	.81
3	.48*	.86	.65*	.62*	.72	.49	.73*	.45*	.51*	.38
4	.69*	.82	.60*	.66	.71*	.56	.80*	.57*	.94*	.57
5	.74*	.64	.57	.64*	.72	.45*	.80*	.24	.45*	.59
6	.54*	.79	.41*	.54	.64*	.72	.52*	.48*	.68*	.61
7	.46*	.87	.65	.31*	.53*	.77*	.62*	.52*	.38*	.60
8	.70	.68	.51	.55*	.80	.64*	.38*	.57	.50*	.64
9	.48	.86	.56*	.76*	.70*	.64	.68*	.71*	.39*	.50
10	.66*	.75	.56*	.58	.63*	.42*	.84*	.48*	.64*	.75
11	.77	.72	.47	.38*	.73*	.30*	.45	.48	.69*	.67
12	.53*	.65	.72	.48	.71	.56*	.55*	.70*	.74*	.64
13	.88	.61*	.71*	.37*	.64	.54*	.73	.51*	.40*	.47
14	.80*	.64	.33*	.42*	.71	.57*	.78*	.44	.41*	.47
15	.77*	.60*	.56	.51	.54	.32	.56*	.32*	.73*	.64
16	.78*	.71*	.64*	.43*	.73*	.48*	.69*	.40*	.67*	.75
17	.91	.70*	.32*	.61	.67*	.49*	.53	.44*	.76*	.79*
18	.51*	.42*	.69*	.65*	.55*	.71*	.58*	.51*	.69*	.36*
19	.75*	.54	.62*	.50*	.49*	.75*	.43*	.49	.72*	.23*
20	.91	.59*	.47*	.60	.73*	.84*	.79*	.49*	.28*	.39*
21	.53*	.46*	.45	.56	.67*	.27*	.80*	.42	.82*	.40
22	.81*	.34*	.30*	.28	.77*	.68	.62	.25	.73*	.63*
23	.83*	.56*	.71	.56*	.25*	.49	.37	.59*	.70*	.81*
24	.76	.66*	.49*	.63	.59	.65*	.42*	.22*	.78	.62*
25	.50*	.84*	.73*	.39*	.70*	.44*	.59*	.48*	.74	.59*
26	.71	.66*	.35	.44*	.49*	.80	.71	.77*	.66	.63*
27	.74*	.37*	.65*	.51*	.62*	.24*	.53*	.47*	.57	.64*
28	.50*	.35*	.45	.30*	.53	.62*	.52	.59	.67	.80*
29	.49	.75*	.65*	.51*	.57*	.87	.30*	.49*	.48	.34*
30	.62*	.50	.53*	.24*	.55*	.87*	.37*	.69*	.67	.54*
31		.73*	.29*	.57	.58*	.69*	.69	.38	.62	.43*
32		.49*	.53	.41*	.62	.61*	.54*	.42*	.64	.41*
33		.78*	.41		.56	.73*	.53	.28*	.65	.70*
34		.74*	.38		.57	.67*	.40*	.58*	.67*	.66*
35		.65*	.46*		.61*	.39*	.41*	.45*	.63*	.61*
36		.42	.31*		.56*	.74*	.44*	.53	.40*	.50*
37		.57*	.70*		.62*	.32	.41*	.61*	.50*	.55*
38		.64*	.59*		.76	.80*	.29*	.34*	.64*	.35*
39		.43*	.52*		.67*	.51*	.35*	.45*	.79*	.56*
40		.69*	.60		.70	.59*	.54*	.60*	.63*	.53*
41		.70*	.57*		.55					.54*
42		.66	.64		.61*					.52*
43		.41*	.26*		.52*					.63*
44		.66*	.42*		.57					.45*
45		.59*	.41*		.53					.52
46		.53*	.36*		.66*					.65*
47		.60	.52		.59*					.63*
48		.69*	.40		.62*					.43*
49		.29								
50		.56								
51		.47								
52		.55								
53		.52								
54		.57								

* Item is included in the Abbreviated Battery.

Appendix G

Table G–37. Item *p*–Values for Advanced 1, Forms B/E, Spring, Grade 8

Item Number	Reading Vocabulary	Reading Comprehension	Mathematics Problem Solving	Mathematics Procedures	Language B	Spelling	Science	Social Science	Listening	Language E
1	.87*	.85	.71	.57	.78	.89*	.85*	.76*	.83*	.54
2	.69	.86	.57*	.63*	.65*	.70*	.71	.52*	.67*	.81
3	.55*	.84	.64*	.63*	.75	.58	.74*	.46*	.61*	.39
4	.69*	.82	.64*	.63	.70*	.69	.80*	.62*	.93*	.64
5	.81*	.65	.61	.64*	.70	.54*	.79*	.26	.58*	.61
6	.69*	.80	.50*	.52	.67*	.76	.61*	.57*	.76*	.70
7	.53*	.85	.67	.45*	.54*	.79*	.66*	.53*	.48*	.68
8	.70	.72	.63	.60*	.79	.63*	.43*	.59	.58*	.57
9	.53	.85	.62*	.77*	.72*	.71	.71*	.75*	.51*	.57
10	.73*	.77	.65*	.58	.66*	.43*	.85*	.57*	.74*	.75
11	.81	.76	.54	.41*	.75*	.43*	.55	.53	.67*	.70
12	.63*	.66	.77	.52	.70	.56*	.61*	.68*	.75*	.71
13	.90	.58*	.73*	.38*	.61	.58*	.71	.56*	.40*	.44
14	.78*	.66	.40*	.41*	.66	.59*	.77*	.45	.41*	.50
15	.76*	.68*	.61	.51	.61	.43	.58*	.34*	.74*	.71
16	.81*	.68*	.70*	.45*	.70*	.50*	.70*	.46*	.70*	.76
17	.91	.71*	.42*	.63	.68*	.56*	.61	.49*	.75*	.80*
18	.59*	.45*	.76*	.67*	.53*	.74*	.61*	.56*	.73*	.39*
19	.65*	.60	.66*	.56*	.46*	.80*	.46*	.51	.75*	.26*
20	.90	.61*	.52*	.64	.74*	.85*	.75*	.50*	.31*	.41*
21	.53*	.46*	.55	.58	.69*	.38*	.78*	.49	.82*	.40
22	.81*	.40*	.47*	.31	.79*	.69	.58	.30	.75*	.67*
23	.83*	.61*	.75	.61*	.29*	.51	.49	.64*	.72*	.79*
24	.80	.68*	.56*	.62	.62	.66*	.43*	.21*	.78	.67*
25	.56*	.83*	.75*	.45*	.68*	.48*	.67*	.54*	.75	.69*
26	.73	.70*	.46	.49*	.47*	.79	.68	.75*	.62	.64*
27	.78*	.45*	.74*	.54*	.61*	.31*	.55*	.46*	.56	.65*
28	.55*	.42*	.55	.38*	.53	.61*	.53	.61	.68	.83*
29	.50	.75*	.70*	.56*	.58*	.86	.35*	.51*	.50	.48*
30	.65*	.59	.57*	.30*	.52*	.90*	.37*	.68*	.70	.64*
31		.75*	.39*	.59	.60*	.73*	.70	.42	.65	.43*
32		.54*	.59	.42*	.57	.64*	.56*	.46*	.64	.41*
33		.78*	.49		.53	.73*	.58	.37*	.68	.76*
34		.78*	.46		.56	.70*	.43*	.57*	.67*	.66*
35		.67*	.52*		.64*	.43*	.45*	.46*	.65*	.65*
36		.43	.33*		.57*	.71*	.50*	.55	.43*	.57*
37		.64*	.71*		.64*	.45	.45*	.67*	.52*	.59*
38		.67*	.66*		.70	.82*	.32*	.43*	.66*	.42*
39		.48*	.53*		.63*	.55*	.39*	.50*	.79*	.59*
40		.69*	.67		.62	.63*	.57*	.64*	.68*	.64*
41		.73*	.56*		.55					.54*
42		.68	.68		.58*					.67*
43		.47*	.33*		.49*					.68*
44		.74*	.46*		.58					.49*
45		.63*	.44*		.50					.62
46		.55*	.46*		.58*					.76*
47		.64	.60		.53*					.63*
48		.76*	.43		.59*					.53*
49		.33								
50		.61								
51		.54								
52		.57								
53		.61								
54		.63								

* Item is included in the Abbreviated Battery.

Table G–38. Item *p*–Values for Advanced 2, Forms B/E, Spring, Grade 8

Item Number	Reading Vocabulary	Reading Comprehension	Mathematics Problem Solving	Mathematics Procedures	Language B	Spelling	Science	Social Science	Listening	Language E
1	.83	.52*	.53	.54	.84*	.89*	.80*	.73*	.56*	.49
2	.86*	.47*	.57	.55*	.87	.71*	.65*	.50*	.44*	.80
3	.72	.34*	.29*	.43*	.81	.76*	.73	.49*	.78*	.78
4	.74*	.81*	.64	.61	.61*	.29	.63*	.57*	.37*	.78
5	.67*	.65	.44*	.47*	.70*	.54*	.59*	.58*	.68*	.75
6	.66*	.88*	.35*	.47	.72	.53*	.54*	.45*	.56*	.38
7	.63*	.58*	.64	.39*	.75*	.85*	.74	.54*	.41*	.44
8	.55	.62	.37*	.63*	.73*	.76*	.54*	.77	.70*	.71
9	.52*	.72*	.64*	.66*	.71*	.76*	.47	.39*	.30*	.68
10	.74	.79*	.55*	.61*	.77*	.47	.58*	.26*	.74*	.76
11	.56*	.82*	.40*	.42	.79*	.83*	.49	.27*	.66*	.58
12	.41*	.37*	.31	.58*	.63*	.63	.42*	.45*	.68*	.51
13	.91	.66*	.64*	.41	.64*	.55*	.34*	.58	.46*	.63
14	.92	.76*	.56	.48*	.79*	.74*	.49	.42*	.53	.67
15	.63*	.48*	.48*	.53	.40*	.68*	.64*	.40*	.74	.59
16	.83	.48	.49	.63*	.34	.63*	.57	.37*	.65	.74
17	.84*	.55*	.36*	.55	.77	.71	.42*	.60*	.23	.71*
18	.68*	.61*	.48*	.46*	.76*	.91	.80*	.50	.45	.77*
19	.65*	.70	.69*	.63	.65*	.55*	.31*	.46*	.62	.62*
20	.80*	.52*	.53*	.28*	.73	.52*	.58*	.55*	.54	.54*
21	.62*	.75*	.51	.43*	.66	.89*	.36*	.33*	.35*	.55*
22	.81*	.69*	.65*	.67*	.49	.24*	.61*	.50*	.45*	.32*
23	.72*	.84*	.46*	.39*	.49	.52*	.71*	.47	.72*	.62*
24	.70	.50*	.36*	.50	.74*	.43*	.38*	.43*	.52*	.46*
25	.45*	.46*	.42	.38*	.79	.64	.30	.39*	.66*	.42*
26	.72	.69*	.46*	.42*	.63*	.55*	.31*	.48*	.39*	.57*
27	.84*	.67	.43	.38*	.57*	.62*	.71*	.49*	.53*	.49*
28	.63	.67*	.63	.40	.74	.78*	.58*	.36*	.70*	.48*
29	.58*	.67*	.38*	.37	.57*	.79*	.35*	.36*	.53*	.65*
30	.49*	.82*	.34	.30	.59	.87*	.39*	.42	.78*	.52*
31		.46	.62*	.30*	.67*	.82	.43*	.42*	.70*	.44*
32		.79	.61*	.46*	.65	.30*	.55*	.43	.57*	.66
33		.58	.48*		.74	.60*	.31*	.27*	.71	.64*
34		.66	.39		.30*	.70	.58	.35	.58	.42*
35		.36	.39		.71*	.34*	.36*	.50	.64	.68*
36		.47	.35*		.80*	.70*	.28	.39	.62*	.74*
37		.70	.56		.65*	.64	.33*	.43*	.48*	.51*
38		.73	.34*		.65*	.55*	.46*	.39	.65*	.49
39		.45	.45*		.70*	.70*	.46*	.30*	.46*	.36*
40		.73	.34		.63*	.37	.41	.59*	.57*	.41*
41		.55	.50*		.76*					.36*
42		.38	.38*		.71					.44*
43		.58	.38		.43					.59*
44		.62	.47*		.72					.66*
45		.48	.25*		.47					.43*
46		.61	.26		.63*					.61*
47		.60	.41*		.39*					.57*
48		.68	.41*		.67*					.67*
49		.60*								
50		.47*								
51		.44*								
52		.61								
53		.39*								
54		.43*								

* Item is included in the Abbreviated Battery.

Appendix G

Table G–39. Item *p*–Values for Advanced 2, Forms B/E, Spring, Grade 9

Item Number	Reading Vocabulary	Reading Comprehension	Mathematics Problem Solving	Mathematics Procedures	Language B	Spelling	Science	Social Science	Listening	Language E
1	.83	.58*	.57	.54	.87*	.89*	.82*	.74*	.64*	.58
2	.89*	.54*	.60	.53*	.87	.65*	.65*	.60*	.47*	.82
3	.78	.33*	.36*	.45*	.76	.79*	.78	.44*	.81*	.87
4	.80*	.85*	.67	.64	.62*	.31	.66*	.59*	.44*	.80
5	.78*	.68	.50*	.56*	.74*	.54*	.64*	.63*	.71*	.83
6	.68*	.91*	.42*	.46	.75	.51*	.58*	.51*	.59*	.40
7	.65*	.67*	.64	.41*	.80*	.82*	.80	.56*	.52*	.48
8	.62	.70	.43*	.65*	.62*	.77*	.61*	.80	.83*	.73
9	.55*	.73*	.66*	.66*	.71*	.70*	.47	.45*	.30*	.67
10	.80	.82*	.61*	.62*	.75*	.53	.58*	.31*	.81*	.79
11	.65*	.87*	.39*	.41	.75*	.84*	.50	.30*	.71*	.62
12	.49*	.39*	.31	.55*	.55*	.67	.51*	.51*	.68*	.62
13	.94	.76*	.70*	.42	.60*	.60*	.40*	.56	.52*	.72
14	.95	.80*	.59	.46*	.72*	.73*	.49	.45*	.55	.75
15	.64*	.51*	.49*	.57	.41*	.69*	.65*	.43*	.68	.60
16	.84	.54	.50	.62*	.43	.59*	.59	.37*	.72	.86
17	.88*	.61*	.39*	.55	.74	.62	.43*	.61*	.28	.76*
18	.67*	.62*	.49*	.45*	.70*	.84	.80*	.47	.51	.86*
19	.72*	.75	.70*	.70	.63*	.51*	.33*	.46*	.59	.76*
20	.85*	.54*	.56*	.28*	.73	.55*	.67*	.54*	.58	.63*
21	.61*	.79*	.57	.47*	.69	.87*	.41*	.36*	.35*	.44*
22	.81*	.75*	.65*	.68*	.55	.29*	.67*	.54*	.47*	.32*
23	.79*	.83*	.51*	.40*	.43	.53*	.74*	.55	.75*	.60*
24	.75	.45*	.43*	.56	.68*	.47*	.39*	.41*	.56*	.49*
25	.53*	.46*	.42	.39*	.79	.59	.38	.45*	.70*	.47*
26	.75	.79*	.53*	.41*	.67*	.63*	.40*	.50*	.47*	.58*
27	.87*	.64	.46	.44*	.66*	.61*	.75*	.53*	.60*	.39*
28	.68	.73*	.63	.42	.77	.77*	.63*	.40*	.73*	.51*
29	.61*	.71*	.43*	.49	.55*	.79*	.42*	.34*	.51*	.65*
30	.55*	.81*	.37	.30	.62	.87*	.36*	.46	.80*	.59*
31		.53	.65*	.39*	.74*	.83	.44*	.42*	.76*	.45*
32		.83	.62*	.48*	.65	.28*	.55*	.46	.61*	.67
33		.60	.49*		.77	.59*	.40*	.29*	.77	.64*
34		.70	.40		.30*	.76	.54	.39	.60	.44*
35		.37	.38		.76*	.34*	.37*	.47	.72	.71*
36		.57	.32*		.83*	.73*	.34	.45	.72*	.82*
37		.72	.61		.66*	.61	.37*	.48*	.56*	.56*
38		.79	.43*		.65*	.55*	.52*	.43	.67*	.58
39		.52	.46*		.71*	.74*	.54*	.29*	.47*	.38*
40		.79	.39		.60*	.42	.43	.64*	.57*	.47*
41		.62	.52*		.72*					.48*
42		.39	.43*		.70					.59*
43		.61	.40		.45					.62*
44		.69	.51*		.70					.65*
45		.56	.29*		.57					.44*
46		.65	.26		.62*					.63*
47		.59	.38*		.52*					.59*
48		.75	.42*		.70*					.67*
49		.60*								
50		.49*								
51		.50*								
52		.60								
53		.41*								
54		.49*								

* Item is included in the Abbreviated Battery.

Table G–40. Item *p*–Values for TASK 1, Forms B/E, Spring, Grade 9

Item Number	Reading Vocabulary	Reading Comprehension	Mathematics	Language B	Spelling	Science	Social Science	Language E
1	.83	.88*	.67*	.73*	.85	.73*	.60*	.81
2	.65*	.51*	.61	.86*	.71	.81*	.65	.86
3	.54	.82*	.36*	.72	.81*	.56*	.53*	.28
4	.78*	.82*	.79*	.77*	.56*	.64	.37	.45
5	.86*	.62*	.51*	.71*	.70	.49*	.70*	.54
6	.65*	.65*	.66*	.77*	.58*	.27*	.43*	.66
7	.47	.49*	.69	.88*	.63*	.62	.23*	.59
8	.56	.76*	.44	.66*	.76*	.59*	.49*	.78
9	.38*	.86*	.18	.85	.82	.64*	.43*	.38
10	.58*	.92*	.51*	.38*	.53*	.49*	.61*	.53
11	.49*	.56*	.48*	.69*	.44*	.75*	.68	.43
12	.41*	.80*	.71*	.63*	.73*	.47*	.41	.58
13	.92	.79*	.76	.69	.48*	.54*	.32*	.44
14	.87*	.90*	.68	.69*	.62*	.72*	.51	.68
15	.67	.77*	.52	.77*	.67*	.53*	.38	.82
16	.73*	.68*	.41	.58	.67*	.67*	.53*	.56
17	.75*	.72*	.59	.50	.33*	.55*	.50*	.66*
18	.68*	.83	.71*	.27*	.75*	.37	.32*	.54*
19	.87	.65*	.53*	.69*	.82*	.44*	.29	.46*
20	.82*	.67*	.73*	.83	.63*	.41	.64*	.59*
21	.84*	.73*	.36*	.68*	.50	.65	.53	.55*
22	.83*	.73*	.51*	.69	.66*	.67	.24*	.62*
23	.91	.64	.38	.37	.54	.78	.54*	.63*
24	.73	.38*	.39*	.74	.47*	.40*	.37*	.54*
25	.55*	.73*	.28*	.63*	.53*	.45*	.70*	.48*
26	.79*	.71*	.41*	.54	.53*	.41*	.41*	.66*
27	.81*	.67*	.39	.60*	.83*	.59*	.44	.45
28	.70*	.88	.42	.76*	.50	.39*	.39*	.68*
29	.35*	.79	.58*	.55*	.74*	.33*	.36*	.66*
30	.53	.74	.50	.70*	.73*	.54	.61*	.36*
31		.71	.38*	.64*	.58	.31	.28*	.67*
32		.44	.29*	.60*	.45*	.57*	.36*	.45*
33		.89	.52*	.64	.66	.54	.38*	.74*
34		.65	.13*	.59*	.38*	.39*	.52*	.67*
35		.59	.28	.73*	.64	.20*	.33*	.72*
36		.58	.37	.72*	.66*	.53*	.36	.48*
37		.76	.46*	.74	.37*	.47*	.55	.53*
38		.87	.23	.67	.28*	.22*	.38*	.66*
39		.38	.58*	.72	.55*	.27*	.45*	.50*
40		.74	.44*	.66*	.23*	.43*	.36*	.41
41		.82	.37*	.69*				.48*
42		.62	.33	.65				.48*
43		.30	.32*	.57				.55*
44		.48	.33*	.58				.33*
45		.65	.42*	.58				.54*
46		.54*	.24	.55*				.47*
47		.77	.28	.65*				.63*
48		.50	.30	.49*				.50*
49		.40*	.42*					
50		.66	.31*					
51		.68*						
52		.52*						
53		.50						
54		.42*						

* Item is included in the Abbreviated Battery.

Appendix G

Table G–41. Item p–Values for TASK 2, Forms B/E, Spring, Grade 10

Item Number	Reading Vocabulary	Reading Comprehension	Mathematics	Language B	Spelling	Science	Social Science	Language E
1	.82	.83*	.41*	.90*	.82*	.81*	.56*	.43
2	.77*	.77*	.52*	.82*	.91*	.60*	.41*	.75
3	.68*	.40*	.35	.54*	.65*	.44*	.55*	.75
4	.66*	.43*	.51*	.49*	.70	.64	.29*	.42
5	.83*	.59*	.24*	.71	.62*	.57*	.54*	.63
6	.66*	.30*	.48*	.72	.65	.60*	.44*	.57
7	.40	.63*	.65	.77*	.71*	.56*	.26*	.41
8	.62	.48*	.41	.60	.84*	.67*	.36*	.29
9	.50*	.79*	.43	.66*	.80*	.41*	.53	.48
10	.62*	.68	.62	.57*	.50	.69*	.24	.80
11	.58	.76	.62	.56	.45*	.41*	.38*	.45
12	.46*	.77	.31*	.39	.85*	.66	.49*	.31
13	.96	.87	.62*	.75*	.60*	.56	.42*	.38
14	.86*	.55	.27*	.83	.65*	.67*	.38*	.77
15	.84*	.54	.66	.57*	.69*	.32*	.41*	.65
16	.84*	.69	.19*	.79*	.48*	.63*	.48	.59
17	.84	.46	.53*	.37*	.74	.43*	.24*	.87*
18	.56*	.66	.14*	.62	.62*	.60*	.44	.62*
19	.94	.53*	.23	.80*	.73	.44*	.49*	.69*
20	.90*	.72*	.31*	.87	.70	.35	.64*	.66*
21	.79*	.72*	.43*	.73*	.61*	.76*	.53*	.37*
22	.85*	.58*	.27	.38	.57	.40	.23*	.48*
23	.91*	.82	.50	.47*	.82*	.69*	.30*	.58*
24	.65*	.60*	.57*	.45*	.55*	.43*	.26*	.85*
25	.60*	.67*	.62*	.77*	.68*	.34*	.40*	.44*
26	.84	.56*	.51*	.71*	.50*	.65*	.56	.49
27	.87*	.20*	.37	.72*	.58	.32*	.36	.63*
28	.73*	.65	.39	.59*	.77*	.65*	.42*	.63*
29	.76	.40	.43*	.73	.85*	.63	.47*	.48*
30	.41	.74	.28	.73	.66*	.22*	.59	.78*
31		.46	.68*	.61*	.72*	.39	.30	.56*
32		.40	.49*	.67	.34*	.62*	.25*	.33*
33		.55	.22*	.67	.79*	.24*	.27*	.85*
34		.68	.39	.48*	.49*	.39	.40	.78*
35		.47	.43*	.74	.70	.27*	.42*	.33*
36		.75	.49	.74*	.24*	.42	.36	.80*
37		.82*	.46	.62*	.72*	.32*	.39*	.53*
38		.61*	.38*	.75*	.11*	.19*	.31*	.69*
39		.74*	.57	.64*	.43*	.42	.31*	.33*
40		.81*	.57*	.69	.60	.53*	.32*	.68*
41		.83	.28*	.77				.41*
42		.60*	.52*	.60				.56*
43		.78*	.44*	.54				.66
44		.38*	.27*	.62*				.56*
45		.64*	.35	.67*				.67*
46		.45*	.17*	.36*				.63*
47		.70	.24	.70*				.40*
48		.51*	.25	.68*				.61*
49		.63*	.29*					
50		.64*	.29*					
51		.32*						
52		.57						
53		.53						
54		.49						

* Item is included in the Abbreviated Battery.

Table G–42. Item *p*–Values for TASK 3, Forms B/E, Spring, Grade 11

Item Number	Reading Vocabulary	Reading Comprehension	Mathematics	Language B	Spelling	Science	Social Science	Language E
1	.80	.84*	.32	.54	.87*	.43*	.54*	.60
2	.61*	.78*	.39*	.74*	.67*	.74*	.29*	.83
3	.69*	.84*	.50*	.67	.74	.18*	.49*	.76
4	.71*	.57*	.46*	.60*	.33*	.59*	.35*	.43
5	.72*	.80*	.30*	.53	.62	.53*	.33*	.80
6	.48*	.52*	.67	.47	.55	.58*	.34*	.74
7	.37*	.82*	.42*	.66*	.69*	.53*	.35*	.46
8	.58*	.85*	.29	.36*	.71*	.49*	.31*	.68
9	.51*	.43*	.40	.72*	.69*	.50	.54*	.58
10	.72	.68*	.72	.55*	.50*	.55	.53	.49
11	.70	.69	.49	.76	.58*	.38	.37*	.67
12	.55	.75*	.27*	.50	.70*	.44	.22*	.83
13	.89	.53	.14*	.49	.52*	.40*	.45	.77
14	.85*	.70	.52*	.64*	.61*	.67*	.24	.72
15	.77*	.67*	.28	.52*	.81*	.55*	.33	.47
16	.77	.64*	.33	.45*	.72	.70	.35*	.59
17	.89*	.37	.23	.22*	.60*	.53*	.28	.36*
18	.61*	.71*	.46*	.43*	.74*	.26*	.37*	.61*
19	.83*	.76*	.24*	.56*	.72*	.36*	.41*	.55*
20	.83*	.43*	.32*	.66*	.74	.63	.44*	.71*
21	.79*	.73*	.35	.42*	.70	.55*	.51	.63*
22	.77*	.66*	.30*	.35	.53*	.65*	.53	.66
23	.88	.63*	.43*	.19*	.72*	.55	.43*	.63*
24	.59*	.61	.45*	.65	.42*	.36*	.24*	.66*
25	.66*	.43*	.36	.52*	.54*	.27	.30*	.52*
26	.85*	.65*	.58*	.53*	.68*	.60*	.33	.82*
27	.82	.49*	.28	.63	.31*	.32*	.29*	.68*
28	.64	.65	.28	.55	.72*	.46*	.49*	.61*
29	.83	.56	.28*	.43	.72	.24*	.30*	.34*
30	.39*	.52	.45*	.59*	.70*	.52	.29*	.40*
31		.72	.41*	.39*	.73*	.27*	.23	.60*
32		.57	.34*	.67	.27*	.53*	.29	.77*
33		.59	.51	.46*	.70*	.31*	.26*	.64*
34		.72	.18*	.46	.63*	.43*	.23*	.66*
35		.54	.42*	.62*	.57*	.17*	.43*	.79*
36		.42	.36	.55*	.30	.20*	.31*	.78*
37		.68*	.26*	.52*	.73	.49*	.37*	.68*
38		.58	.40	.50*	.49	.46*	.30*	.52*
39		.50*	.60*	.64*	.53*	.19*	.33*	.58*
40		.59*	.28*	.69	.49*	.32	.18*	.62*
41		.52*	.36	.51				.54*
42		.55*	.27*	.48*				.75*
43		.41*	.19*	.59*				.72*
44		.59*	.33	.62				.77*
45		.58*	.26	.54				.58*
46		.67	.33*	.61*				.57
47		.52	.25*	.56*				.57*
48		.49	.32*	.64*				.39*
49		.59	.25					
50		.45	.25*					
51		.38						
52		.63						
53		.44						
54		.45						

* Item is included in the Abbreviated Battery.

Appendix G

Table G–43. Item *p*–Values for TASK 3, Forms B/E, Spring, Grade 12

Item Number	Reading Vocabulary	Reading Comprehension	Mathematics	Language B	Spelling	Science	Social Science	Language E
1	.71	.72*	.30	.43	.78*	.38*	.53*	.60
2	.63*	.71*	.30*	.71*	.62*	.67*	.38*	.79
3	.65*	.73*	.44*	.68	.67	.20*	.42*	.78
4	.74*	.56*	.41*	.49*	.37*	.58*	.37*	.37
5	.68*	.71*	.20*	.47	.60	.44*	.36*	.64
6	.47*	.45*	.57	.50	.54	.59*	.31*	.63
7	.36*	.75*	.42*	.63*	.64*	.43*	.33*	.40
8	.48*	.81*	.34	.34*	.66*	.44*	.26*	.50
9	.46*	.42*	.38	.64*	.65*	.50	.50*	.46
10	.69	.56*	.64	.54*	.48*	.51	.46	.50
11	.63	.66	.45	.66	.51*	.43	.36*	.57
12	.52	.66*	.31*	.47	.74*	.41	.22*	.76
13	.78	.49	.18*	.54	.46*	.34*	.41	.57
14	.79*	.60	.41*	.55*	.59*	.57*	.29	.57
15	.76*	.62*	.29	.44*	.78*	.46*	.27	.37
16	.67	.69*	.39	.49*	.72	.61	.33*	.55
17	.78*	.34	.28	.34*	.53*	.50*	.31	.26*
18	.58*	.63*	.45*	.43*	.72*	.26*	.34*	.39*
19	.71*	.67*	.23*	.55*	.67*	.29*	.37*	.45*
20	.74*	.51*	.28*	.62*	.64	.57	.40*	.61*
21	.69*	.64*	.42	.39*	.69	.45*	.45	.53*
22	.72*	.63*	.29*	.36	.50*	.53*	.44	.59
23	.85	.55*	.44*	.17*	.64*	.46	.43*	.49*
24	.56*	.61	.35*	.57	.39*	.37*	.30*	.51*
25	.58*	.43*	.33	.53*	.46*	.26	.30*	.43*
26	.73*	.55*	.48*	.55*	.62*	.54*	.30	.59*
27	.81	.54*	.24	.64	.29*	.34*	.24*	.56*
28	.54	.58	.28	.48	.67*	.45*	.46*	.53*
29	.68	.51	.25*	.45	.73	.22*	.35*	.23*
30	.40*	.44	.41*	.57*	.63*	.44	.25*	.37*
31		.63	.38*	.37*	.63*	.30*	.31	.49*
32		.59	.24*	.68	.23*	.47*	.31	.57*
33		.56	.38	.37*	.66*	.29*	.24*	.53*
34		.61	.20*	.49	.56*	.37*	.20*	.53*
35		.53	.37*	.57*	.53*	.14*	.40*	.65*
36		.37	.31	.48*	.38	.22*	.29*	.60*
37		.60*	.25*	.50*	.69	.42*	.35*	.55*
38		.53	.33	.38*	.48	.42*	.33*	.41*
39		.45*	.55*	.64*	.45*	.21*	.30*	.44*
40		.51*	.28*	.64	.44*	.31	.20*	.47*
41		.47*	.35	.52				.46*
42		.50*	.24*	.49*				.65*
43		.39*	.23*	.58*				.61*
44		.49*	.33	.64				.61*
45		.52*	.27	.57				.46*
46		.58	.28*	.63*				.52
47		.47	.29*	.63*				.46*
48		.45	.36*	.63*				.39*
49		.52	.24					
50		.43	.28*					
51		.38						
52		.56						
53		.44						
54		.42						

* Item is included in the Abbreviated Battery.

226

Table G-44. Item *p*-Values for SESAT 1, Forms A/D, Fall, Grade K

Item Number	Sounds and Letters	Word Reading	Mathematics	Environment	Listening to Words and Stories
1	.86	.55	.92	.92	.85
2	.80	.56	.84	.75	.95
3	.80	.39	.89	.77	.80
4	.77	.62	.90	.48	.94
5	.86	.41	.58	.77	.93
6	.84	.57	.66	.30	.64
7	.81	.51	.58	.59	.94
8	.70	.30	.63	.88	.76
9	.68	.60	.85	.63	.55
10	.60	.48	.51	.64	.94
11	.65	.43	.64	.67	.84
12	.55	.31	.47	.70	.28
13	.78	.64	.71	.79	.65
14	.68	.36	.83	.67	.61
15	.80	.58	.17	.88	.74
16	.78	.52	.91	.71	.49
17	.68	.60	.55	.90	.85
18	.63	.51	.68	.84	.79
19	.89	.52	.31	.48	.78
20	.83	.72	.67	.78	.75
21	.59	.47	.44	.93	.87
22	.94	.22	.39	.64	.82
23	.89	.46	.62	.92	.55
24	.88	.57	.74	.88	.63
25	.95	.28	.65	.75	.51
26	.83	.49	.63	.66	.58
27	.89	.50	.36	.72	.54
28	.93	.65	.70	.71	.73
29	.84	.28	.38	.52	.87
30	.90	.43	.58	.84	.56
31	.76		.93	.63	.85
32	.83		.69	.78	.82
33	.72		.84	.58	.30
34	.60		.74	.42	.44
35	.63		.70	.79	.60
36	.69		.94	.58	.64
37	.74		.94	.59	.62
38	.74		.70	.71	.63
39	.61		.78	.34	.72
40	.73		.65	.65	.80

Appendix G

Table G-45. Item *p*-Values for SESAT 2, Forms A/D, Fall, Grade 1

Item Number	Sounds and Letters	Word Reading	Sentence Reading	Mathematics	Environment	Listening to Words and Stories
1	.93	.97	.82	.94	.95	.95
2	.89	.87	.84	.95	.84	.95
3	.91	.68	.90	.86	.89	.68
4	.95	.77	.87	.82	.80	.98
5	.90	.76	.83	.83	.93	.89
6	.91	.82	.79	.84	.92	.88
7	.91	.77	.82	.88	.75	.83
8	.94	.77	.84	.97	.82	.81
9	.97	.75	.83	.87	.87	.93
10	.92	.70	.46	.88	.83	.91
11	.92	.49	.61	.96	.36	.95
12	.91	.81	.82	.98	.80	.57
13	.94	.74	.70	.85	.81	.67
14	.91	.69	.80	.60	.96	.55
15	.90	.69	.80	.81	.79	.91
16	.80	.68	.56	.57	.40	.64
17	.83	.79	.87	.86	.89	.66
18	.80	.54	.77	.78	.91	.72
19	.96	.56	.79	.85	.74	.43
20	.99	.54	.77	.91	.89	.70
21	.94	.62	.70	.89	.92	.66
22	.95	.67	.61	.75	.90	.45
23	.98	.88	.73	.91	.86	.69
24	.96	.81	.66	.54	.83	.42
25	.95	.80	.76	.86	.85	.70
26	.95	.71	.65	.67	.15	.53
27	.89	.66	.45	.43	.75	.34
28	.92	.74	.60	.94	.49	.79
29	.91	.64	.54	.95	.86	.33
30	.91	.80	.76	.81	.57	.59
31	.87			.96	.55	.73
32	.70			.91	.95	.78
33	.98			.94	.86	.62
34	.98			.94	.38	.85
35	.80			.50	.58	.45
36	.97			.88	.70	.56
37	.97			.49	.64	.39
38	.80			.96	.41	.51
39	.79			.89	.93	.88
40	.67			.55	.90	.87

228

Table G-46. Item *p*-Values for Primary 1, Forms A/D, Fall, Grade 2

Item Number	Word Study Skills	Word Reading	Sentence Reading	Reading Compre-hension	Mathematics Problem Solving	Mathematics Procedures	Language A	Spelling	Environ-ment	Listening	Language D
1	.89*	.98*	.97*	.95	.77	.84*	.70*	.97*	.93*	.79*	.76*
2	.87*	.96*	.95*	.95*	.57	.94*	.94*	.93*	.82	.73*	.85*
3	.83*	.84*	.93	.79*	.91*	.60*	.79	.76*	.68*	.79*	.87
4	.85	.93	.96	.88	.83	.77	.63*	.75*	.89*	.50*	.97*
5	.92	.89	.84*	.93*	.72*	.81*	.77*	.83*	.46*	.53*	.92*
6	.96*	.89	.95*	.86*	.85*	.62	.59*	.79*	.81*	.86*	.93
7	.91*	.94	.84*	.95*	.87*	.64*	.85*	.96*	.68*	.61*	.79
8	.95	.88	.95	.84*	.69	.71	.58	.84*	.94*	.73*	.76
9	.97*	.87	.95*	.93*	.90	.91	.80*	.83*	.86*	.82*	.88*
10	.97	.81	.95*	.91	.56*	.81*	.44*	.88*	.64*	.61*	.86*
11	.95*	.79	.89	.97*	.90*	.83	.71*	.62	.80*	.86*	.63*
12	.97*	.75	.93*	.84*	.52*	.66*	.69	.71	.75	.64*	.76*
13	.95	.94*	.92*	.69*	.80*	.63	.76*	.90*	.91	.94*	.73*
14	.81*	.86*	.96*	.88	.90*	.68*	.69*	.65*	.91*	.89*	.54*
15	.82*	.95*	.89	.88	.72*	.85*	.80*	.78*	.49	.85*	.65*
16	.80*	.75*	.92*	.94	.76*	.59*	.64*	.89*	.77*	.76*	.85*
17	.61*	.90*	.66	.92*	.84*	.77*	.84	.94*	.95	.85	.62*
18	.77*	.89*	.68*	.85*	.80*	.70*	.61	.89	.38*	.73	.60*
19	.57	.90*	.93*	.88*	.91*	.30*	.73*	.74*	.88*	.76	.87*
20	.79	.83*	.80*	.90*	.55	.53	.45*	.52*	.94*	.49*	.44*
21	.84*	.79*	.86*	.93*	.79*	.84*	.83	.80*	.97	.67*	.78*
22	.39	.93*	.71*	.90*	.88	.86*	.73*	.49*	.93*	.58*	.45*
23	.65	.91*	.86	.93*	.46*	.68*	.73*	.77*	.81*	.63	.86*
24	.37*	.70*	.81	.89*	.80*	.71	.75*	.67*	.81*	.70	.29*
25	.75*	.84*	.89*	.69*	.85*	.67	.82*	.87*	.77*	.76	.66*
26	.72*	.93*	.91	.84*	.70*	.54*	.76	.83*	.59	.81*	.72*
27	.53*	.92*	.92*	.77*	.69	.59	.71*	.83	.97*	.76*	.86*
28	.58	.85*	.86*	.74*	.75*	.39*	.66	.74*	.89*	.60*	.81*
29	.59*	.79*	.86*	.54*	.81*	.44*	.70	.69*	.51	.83*	.46*
30	.23*	.85	.87	.61*	.91*	.50*	.64*	.66*	.61*	.85*	.82*
31				.60*	.77		.65	.76*	.77*	.65*	.79*
32				.64	.68*		.87*	.58	.61*	.65*	.75*
33				.59	.96		.69*	.80*	.96	.73*	.56*
34				.76	.94*		.76*	.80*	.83*	.56*	.55*
35				.81*	.61*		.69*	.77*	.72*	.80*	.79
36				.67*	.90*		.71*	.69	.85*	.77*	.65
37				.62*	.57*		.57*		.73*	.75	.80
38				.74*	.56*		.70*		.95	.74	.68
39				.53*	.77*		.64*		.62*	.70	.70
40				.51	.87*		.47*		.97*	.54	.62
41					.78						
42					.67						

* Item is included in the Abbreviated Battery.

Appendix G

Table G-47. Item *p*-Values for Primary 2, Forms A/D, Fall, Grade 3

Item Number	Word Study Skills	Reading Vocabulary	Reading Compre- hension	Mathematics Problem Solving	Mathematics Procedures	Language A	Spelling	Environment	Listening	Language D
1	.96*	.95	.91	.88	.95*	.88*	.90	.94*	.95*	.92*
2	.94*	.94*	.92	.86*	.95	.99*	.86*	.92*	.68*	.89*
3	.92*	.90*	.75	.85*	.92*	.85	.82*	.79	.33*	.91
4	.95	.86*	.66*	.90*	.90*	.92	.72*	.80*	.53*	.89
5	.99*	.94*	.77*	.74	.71*	.85*	.85*	.95	.69*	.69
6	.95	.95*	.86*	.85*	.80	.80*	.94*	.84	.85*	.96*
7	.95*	.82	.75*	.89*	.71*	.78*	.90*	.67*	.56*	.93
8	.99	.77*	.73*	.79*	.77	.88	.86*	.88	.74*	.82*
9	.99*	.66	.86*	.85	.88	.85*	.94*	.87*	.57*	.91*
10	.98	.73*	.84*	.71*	.89*	.83*	.82*	.89*	.67*	.71*
11	.98*	.91*	.79*	.70*	.79*	.80*	.88*	.50*	.98*	.85*
12	.91*	.93	.65*	.59*	.75*	.93*	.79*	.81	.87*	.64*
13	.90*	.85*	.47*	.80*	.69	.88	.86*	.94*	.96*	.65*
14	.74*	.79*	.85*	.57	.82*	.82*	.86*	.96*	.73*	.87*
15	.71*	.71*	.85*	.69	.84*	.56*	.95*	.85*	.85*	.86*
16	.75*	.76	.90*	.78	.61	.86	.91*	.64*	.76*	.83*
17	.89*	.76*	.48*	.91	.84*	.88	.89*	.93*	.68*	.62*
18	.68	.94	.92*	.91*	.81*	.74	.83*	.43*	.66*	.48
19	.45*	.85*	.88	.42*	.85	.87*	.77*	.62*	.84*	.65*
20	.62*	.77*	.90	.49	.75*	.61*	.87*	.83*	.88	.56*
21	.41	.81*	.84	.93*	.76*	.90	.68*	.99	.91	.88*
22	.66*	.84*	.44*	.67*	.46*	.86*	.60*	.82*	.90	.86
23	.87	.73	.73*	.51*	.78*	.91*	.94	.70*	.69*	.73*
24	.87	.79	.76*	.56*	.83	.45	.89	.74*	.55*	.84
25	.65	.93*	.53*	.78*	.41	.91	.65*	.79*	.57*	.87*
26	.50	.85*	.83*	.89*	.66*	.87*	.88	.88*	.66*	.66*
27	.33*	.86*	.68*	.59*	.77*	.86*	.92*	.84*	.74*	.44*
28	.88*	.70*	.80*	.79	.70*	.88	.87*	.58	.65*	.86*
29	.67*	.81	.73*	.81	.77	.89*	.85*	.80*	.69*	.74*
30	.36*	.75	.49*	.65*	.57*	.88	.77*	.91*	.46*	.83*
31			.69*	.68*		.81*	.89	.78*	.38*	.85*
32			.58*	.84*		.92*	.83*	.49*	.93*	.44*
33			.41*	.83		.74	.86*	.83*	.65*	.91
34			.85*	.78*		.68*	.84*	.93	.64	.94*
35			.75*	.81*		.72	.85*	.86	.78	.82*
36			.76*	.95		.87*	.77	.97*	.63	.60
37			.68	.55*		.87		.53*	.81	.41*
38			.70	.68*		.66		.90*	.68	.62*
39			.70	.74		.61*		.35	.76	.62
40			.54	.49		.89*		.94*	.88	.86*
41				.67*		.72*				
42				.86*		.75				
43				.37*		.77*				
44				.76*		.86*				
45						.84				
46						.91*				
47						.82*				
48						.60*				

* Item is included in the Abbreviated Battery.

Table G-48. Item *p*-Values for Primary 3, Forms A/D, Fall, Grade 4

Item Number	Word Study Skills	Reading Vocab- ulary	Reading Compre- hension	Mathematics Problem Solving	Mathematics Procedures	Language A	Spelling	Science	Social Science	Listening	Language D
1	.93*	.87	.70*	.66	.79*	.91*	.82*	.92*	.90	.88*	.45*
2	.93	.81*	.76	.55*	.84*	.84*	.71	.91*	.89*	.77*	.87*
3	.92*	.83*	.86*	.84	.65	.41*	.55*	.73	.86*	.45*	.62*
4	.91*	.70	.73*	.51*	.73*	.67*	.59*	.66	.62*	.91*	.62*
5	.83*	.70*	.87*	.72*	.83	.67*	.47*	.72	.75	.37*	.46*
6	.84*	.67	.83*	.56	.64*	.73	.65*	.68*	.67	.76*	.37*
7	.76	.88*	.73*	.81*	.60*	.51	.85*	.70*	.86*	.52*	.42*
8	.74	.84*	.78*	.55	.77	.71*	.74*	.55*	.55*	.56*	.50*
9	.65*	.66*	.52	.86	.49*	.68	.48*	.79*	.71*	.77*	.44*
10	.42*	.49*	.86*	.41*	.56*	.76	.46	.72*	.69*	.71*	.61*
11	.59*	.89*	.66*	.82*	.57	.60	.61*	.90*	.63	.84*	.49*
12	.61	.47	.58*	.62*	.83*	.56*	.35	.50*	.45*	.67*	.44*
13	.81*	.65*	.81	.76*	.73*	.77	.70	.52*	.57	.82*	.43*
14	.78*	.74	.75	.62*	.36	.82*	.39*	.77*	.38*	.61*	.41*
15	.57*	.58*	.65	.49*	.68*	.64	.64*	.62*	.48	.85*	.69*
16	.54*	.67	.84	.72*	.60	.75*	.82*	.70*	.61*	.58*	.71*
17	.77	.71*	.70	.68	.66	.76*	.77*	.66	.69	.68*	.64*
18	.39	.63*	.59	.73*	.57	.80*	.44*	.46	.61*	.77*	.60*
19	.77*	.79*	.46*	.80*	.56	.63*	.59*	.84	.78*	.63	.40*
20	.70	.72*	.76*	.60	.51*	.59*	.44*	.52*	.64*	.68	.56*
21	.92*	.61	.88*	.67*	.50*	.72	.73*	.69*	.71*	.71	.38*
22	.77*	.81*	.52	.75*	.56*	.74	.43*	.58	.42*	.80	.55*
23	.54	.70	.71*	.66	.54*	.70*	.46*	.78*	.80*	.66	.42*
24	.36*	.75*	.69*	.45*	.48*	.45*	.65*	.78*	.70*	.72	.36*
25	.71*	.77	.72	.68	.56*	.75	.52*	.59*	.65*	.74*	.51*
26	.61*	.58*	.49	.94*	.45	.80*	.66*	.71*	.75*	.67*	.56*
27	.61	.73*	.61	.81*	.61*	.60*	.68*	.59*	.52*	.54*	.56*
28	.38*	.67*	.72	.61	.50*	.69	.57*	.50*	.64*	.75*	.49*
29	.59	.57*	.69	.64*	.58*	.73	.69	.48	.80*	.55*	.53*
30	.31*	.63	.74	.75	.32*	.58*	.59*	.59*	.49*	.68*	.66*
31			.52*	.68*		.69*	.48*	.62*	.55	.57*	.66
32			.74*	.74*		.80*	.44*	.44*	.61*	.73*	.27
33			.81*	.69*		.67*	.71*	.75*	.69	.45*	.29
34			.79*	.73		.60*	.51	.57*	.35*	.74*	.34
35			.81*	.60*		.44*	.54	.81*	.68*	.80*	.59
36			.78	.55		.71*	.63	.36*	.62*	.67*	.37
37			.58	.76		.75	.51*	.70	.40*	.60	.51
38			.57	.60*		.70	.51*	.43*	.72	.63	.60
39			.72	.56*		.77		.72	.47*	.80	.46
40			.63	.66*		.73*		.78*	.66*	.61	.40
41			.64	.62*		.75*					.42
42			.52*	.58		.51*					.70
43			.62*	.54*		.75					.48
44			.54*	.65		.53					.28
45			.65*	.81*		.62*					.35
46			.45	.69*		.73*					
47			.32*			.73					
48			.59*			.59*					
49			.63*								
50			.53								
51			.64*								
52			.66*								
53			.54*								
54			.50								

* Item is included in the Abbreviated Battery.

Appendix G

Table G-49. Item *p*-Values for Intermediate 1, Forms A/D, Fall, Grade 5

Item Number	Word Study Skills	Reading Vocabulary	Reading Comprehension	Mathematics Problem Solving	Mathematics Procedures	Language A	Spelling	Science	Social Science	Listening	Langua D
1	.98*	.88	.95	.79	.74*	.81*	.81*	.82*	.84*	.62*	.64*
2	.89*	.69*	.61	.80*	.78*	.64	.66	.77*	.82*	.92*	.75*
3	.95*	.52*	.72	.62	.79	.86*	.69*	.64	.78*	.53*	.81*
4	.93	.86*	.88	.64	.57*	.67*	.39*	.71	.52*	.75*	.67*
5	.83	.73*	.57	.65*	.74*	.48*	.79*	.80*	.63*	.31*	.55*
6	.73	.63	.63	.28*	.64	.57*	.53*	.58*	.66	.68*	.60
7	.88*	.69*	.80*	.71*	.75	.83	.77*	.75*	.55*	.40*	.66*
8	.73*	.71	.68*	.75	.72*	.53*	.83*	.87*	.57*	.65*	.82*
9	.73*	.59*	.80*	.68	.73*	.63	.70*	.48	.73*	.56*	.76*
10	.58	.70*	.64	.83*	.75	.79*	.62*	.59	.61*	.72*	.44*
11	.64*	.74*	.73*	.44*	.80*	.42	.49*	.48*	.68*	.89*	.45*
12	.66*	.59	.41*	.78*	.55*	.63*	.84	.79*	.57	.54*	.55*
13	.78*	.90*	.88*	.73*	.50	.67	.60	.69	.46*	.73*	.61*
14	.84*	.81*	.80	.61	.74*	.73*	.54*	.76*	.55*	.48	.68*
15	.58	.76*	.79*	.92*	.46*	.72	.63	.55*	.73	.65	.66*
16	.52*	.84	.67*	.61	.54	.75*	.81*	.67*	.47*	.57	.49*
17	.39*	.76*	.52*	.68*	.78*	.74	.60	.78*	.69*	.68	.67*
18	.56*	.70	.78*	.73	.71*	.77*	.73*	.85*	.72*	.44*	.54*
19	.51	.65*	.90*	.81*	.58*	.53*	.87*	.57*	.54	.41*	.50*
20	.64*	.81	.63*	.11*	.64	.46*	.59*	.74*	.65*	.65*	.75*
21	.72*	.73*	.47*	.51*	.61*	.75*	.42*	.74	.40*	.79	.43*
22	.58	.84	.42*	.51*	.57*	.39	.47*	.71*	.63	.81	.63*
23	.45*	.77*	.90*	.49	.70*	.53*	.71*	.50	.76*	.90	.65*
24	.55	.79*	.56	.43	.52*	.48	.48*	.55*	.72*	.67*	.55*
25	.57*	.59*	.67*	.84	.62	.60*	.49*	.39	.51	.85*	.55*
26	.66*	.72*	.56*	.80*	.65*	.69*	.39*	.78*	.45*	.25*	.62*
27	.48*	.79	.46*	.75*	.50	.80*	.62*	.79	.58*	.66*	.36*
28	.70	.70*	.64*	.61*	.69	.55	.40	.52*	.69*	.63*	.36*
29	.58	.67	.64	.79*	.61	.59	.66*	.52*	.69	.73*	.64*
30	.37*	.75*	.48*	.81	.61*	.54*	.77*	.55*	.54	.31*	.69
31			.76	.65	.53	.69	.83*	.80*	.72*	.69*	.62*
32			.69	.52*	.59*	.55*	.35	.73*	.74	.43*	.64*
33			.57	.42*		.59*	.57	.73	.74*	.84*	.68
34			.81	.64*		.67	.42*	.51*	.44*	.66*	.63
35			.69	.54		.68	.85	.55*	.56*	.66*	.29
36			.63	.66*		.77*	.47*	.72*	.72*	.84*	.52
37			.58*	.76		.72	.57*	.60*	.61*	.79*	.41
38			.71*	.61*		.84*	.50*	.45*	.70	.63	.44
39			.59*	.62		.59*	.63*	.79*	.22*	.67	.53
40			.58*	.33*		.70	.60	.64*	.74*	.54	.57
41			.61	.41*		.76					.44
42			.50*	.40*		.69					.57
43			.62	.44		.82*					.32
44			.73	.76*		.76*					.69
45			.53	.57		.70*					.30
46			.33	.26*		.76*					.65
47			.40	.61*		.61*					.52
48			.46	.76*		.59*					.45
49			.46*								
50			.65*								
51			.64*								
52			.33*								
53			.54*								
54			.49								

* Item is included in the Abbreviated Battery.

Table G-50. Item _p_-Values for Intermediate 2, Forms A/D, Fall, Grade 6

Item Number	Reading Vocabulary	Reading Comprehension	Mathematics Problem Solving	Mathematics Procedures	Language A	Spelling	Science	Social Science	Listening	Language D
1	.92*	.71*	.77	.85	.86*	.88	.91	.86*	.91*	.77*
2	.81*	.89*	.74*	.83*	.75*	.73	.70*	.67	.62*	.60*
3	.73	.60*	.34	.71	.85	.75*	.57*	.72*	.91*	.48*
4	.72*	.77	.76*	.56*	.73*	.73*	.64*	.86*	.23*	.83*
5	.87	.84*	.62*	.78*	.71*	.52*	.61*	.84*	.64*	.53*
6	.58*	.74*	.23*	.66*	.68*	.77*	.58*	.78*	.31*	.44*
7	.63*	.70*	.76*	.83	.66	.79	.87*	.72*	.60*	.65*
8	.58*	.77	.70*	.28*	.83*	.86*	.58	.65*	.84*	.80
9	.65*	.71*	.74	.65*	.66*	.69*	.45*	.67*	.47*	.66*
10	.75	.56*	.45*	.33*	.68	.52*	.42*	.83*	.74*	.67*
11	.74*	.66*	.44	.50	.72	.58*	.82	.49*	.85	.50*
12	.58	.85*	.49	.71	.56*	.61*	.63*	.67	.64	.72*
13	.87*	.66	.68*	.21*	.85*	.73*	.50*	.48	.44	.53*
14	.85	.68	.63	.64	.60*	.54*	.69*	.63	.75*	.80*
15	.79*	.84	.85*	.60*	.64	.66	.56	.77*	.83*	.50*
16	.83	.75	.61*	.56*	.18*	.55*	.40*	.71	.30*	.57*
17	.80*	.68	.51*	.77*	.82	.54*	.80*	.66*	.66*	.64*
18	.69*	.68	.56	.73*	.64	.85	.56*	.48*	.74*	.60*
19	.76*	.62*	.56	.65	.80	.77*	.52*	.59*	.57*	.64*
20	.77*	.42*	.50*	.48	.66*	.56*	.59*	.78*	.49*	.65*
21	.62	.63	.58*	.53	.73*	.78*	.55	.48*	.48*	.60*
22	.80	.36*	.55*	.67*	.73	.59*	.45	.57	.59*	.43*
23	.81*	.66*	.68	.54	.66*	.37	.66*	.80*	.80*	.53*
24	.77*	.65*	.71	.36	.71*	.75*	.52*	.49*	.49*	.42*
25	.77	.89	.79*	.73*	.78*	.36*	.54*	.52	.48*	.30*
26	.80*	.54	.49*	.29*	.70	.75*	.70*	.80*	.53*	.73*
27	.85*	.47	.70	.55*	.78*	.65*	.45*	.64*	.68*	.73*
28	.71*	.68	.72	.43*	.72*	.78*	.62*	.45	.54*	.75*
29	.68*	.38	.67*	.46*	.64*	.70*	.67*	.62	.73*	.54*
30	.76	.49	.71*	.53*	.67	.77*	.38	.77*	.80*	.60*
31		.85*	.89*	.45	.43*	.64*	.41	.57	.88*	.74*
32		.62*	.64*	.55*	.63	.56	.72*	.54*	.78*	.29
33		.46*	.75*		.55	.47*	.50	.82*	.83*	.85
34		.64*	.60		.78*	.75*	.73*	.45*	.42	.45
35		.54*	.65		.71*	.58	.55*	.58*	.63	.66
36		.66	.54*		.66	.79	.72	.59*	.78	.70
37		.51	.70*		.59	.51	.40*	.53*	.51	.52
38		.61	.50		.77	.82*	.37*	.70*	.69	.48
39		.72	.58*		.85*	.56*	.83*	.33*	.55	.63
40		.71	.66*		.70*	.53*	.79*	.42*	.71	.46
41		.37	.51		.71*					.47
42		.59	.51*		.60*					.60
43		.74	.56*		.61*					.48
44		.68*	.34*		.72					.53
45		.43*	.63*		.56					.57
46		.68*	.69*		.61*					.41
47		.62*	.80		.55*					.69
48		.46*	.84		.48*					.14
49		.36*								
50		.67								
51		.67*								
52		.48*								
53		.74*								
54		.63*								

* Item is included in the Abbreviated Battery.

Appendix G

Table G-51. Item *p*-Values for Intermediate 3, Forms A/D, Fall, Grade 7

Item Number	Reading Vocabulary	Reading Compre-hension	Mathematics Problem Solving	Mathematics Procedures	Language A	Spelling	Science	Social Science	Listening	Language D
1	.80*	.84*	.54*	.75	.62*	.82*	.88*	.87*	.76*	.73*
2	.77*	.72*	.48*	.69*	.86*	.75*	.75*	.78*	.84*	.69*
3	.63	.68*	.38	.57	.73*	.65*	.61	.60*	.46*	.69*
4	.77*	.69*	.60	.46*	.77*	.56*	.73*	.59*	.37*	.60*
5	.78	.62*	.61*	.77	.84*	.66*	.75*	.73*	.66*	.57*
6	.77*	.85	.57*	.75*	.72	.77*	.58	.82*	.57*	.59*
7	.64*	.88*	.40	.42*	.48	.78*	.86*	.32	.53*	.55*
8	.63	.78*	.54*	.67*	.79*	.76*	.70	.49*	.90*	.83
9	.52*	.76	.64*	.54*	.79	.58*	.73*	.49	.56*	.58*
10	.71	.76*	.63	.57	.64*	.39	.80	.43	.54*	.53*
11	.76*	.63*	.62*	.41*	.46*	.81*	.57*	.61	.76*	.64*
12	.50*	.63*	.24*	.76*	.64*	.44*	.74*	.45*	.71*	.44*
13	.91*	.73*	.76*	.45	.58	.63*	.77*	.57*	.71*	.85*
14	.86	.82*	.68*	.51*	.74*	.53*	.67*	.59	.49*	.48*
15	.65*	.36*	.53	.25*	.54	.64*	.58*	.56*	.46*	.64*
16	.71	.78	.42	.30	.49	.64*	.45*	.70*	.54*	.77*
17	.72*	.78*	.45*	.68*	.53	.82	.82*	.42*	.30*	.75*
18	.63*	.61*	.41*	.61	.68	.81*	.80*	.65	.55*	.66*
19	.68*	.56*	.60	.51	.74*	.60*	.55*	.58*	.66*	.86*
20	.86	.72*	.57*	.57*	.45	.79*	.39*	.46*	.61*	.83*
21	.76*	.49	.34	.63*	.24*	.56*	.62*	.54*	.70*	.41*
22	.84*	.50*	.50*	.48*	.63*	.44*	.70*	.55*	.85*	.81*
23	.74	.78*	.47*	.56*	.59*	.69*	.56*	.44*	.73*	.63*
24	.73*	.55*	.48	.28*	.64*	.59*	.57*	.51*	.46	.58*
25	.58*	.73*	.43*	.50	.77*	.56	.68	.62	.58	.57*
26	.71	.71*	.67*	.59	.71*	.55*	.71*	.37*	.66	.39*
27	.79*	.60*	.60	.46	.70*	.63*	.34*	.69*	.60*	.61*
28	.67	.59	.43	.51*	.72	.52*	.60*	.35	.62*	.73*
29	.68*	.64*	.54*	.33*	.60	.72*	.49*	.59*	.74*	.82*
30	.64*	.71*	.50	.34	.48*	.78	.43	.50*	.58*	.85*
31		.49	.64*	.34*	.54*	.61	.84*	.47*	.71	.47*
32		.52	.46*	.40*	.64*	.65*	.51*	.72	.68	.59
33		.39	.62*		.61*	.30*	.59*	.47*	.73	.59
34		.67	.31		.49*	.57	.70	.67*	.48*	.37
35		.50	.45*		.71*	.74	.46	.36*	.71*	.74
36		.41	.58*		.53*	.58	.55	.57*	.48*	.42
37		.70	.46		.66	.69*	.41*	.61*	.76	.47
38		.56	.75*		.63	.49	.32*	.17	.32	.57
39		.71	.65*		.69	.60	.68	.38*	.62	.53
40		.53	.55*		.68	.48*	.53*	.65*	.75	.52
41		.77	.69		.60					.53
42		.69	.48		.75*					.68
43		.55	.52*		.77*					.54
44		.71	.41*		.59					.77
45		.51	.45*		.53					.67
46		.58	.51		.55*					.51
47		.51	.45		.50*					.46
48		.40	.57*		.62*					.54
49		.72*								
50		.59*								
51		.40								
52		.32*								
53		.45*								
54		.58*								

* Item is included in the Abbreviated Battery.

Table G-52. Item *p*-Values for Advanced 1, Forms A/D, Fall, Grade 8

Item Number	Reading Vocabulary	Reading Comprehension	Mathematics Problem Solving	Mathematics Procedures	Language A	Spelling	Science	Social Science	Listening	Language D
1	.87*	.88*	.63	.48	.78*	.82*	.83*	.84*	.82*	.51
2	.77*	.82*	.56	.57*	.74	.88*	.82	.52	.75*	.85
3	.52	.53*	.51*	.58*	.63	.65*	.74	.62*	.56*	.84
4	.80*	.48*	.54*	.81	.69*	.65*	.55	.52*	.87*	.85
5	.76	.70*	.46*	.55*	.76*	.55*	.76*	.58*	.35*	.58
6	.62*	.84	.74	.41	.77*	.54	.75*	.65*	.51*	.61
7	.54*	.76*	.50*	.33*	.71*	.71*	.54*	.51*	.56*	.45
8	.61	.78	.48	.56*	.82*	.79*	.40*	.74*	.69*	.53
9	.48*	.56*	.44*	.72*	.67	.56*	.65	.64*	.53*	.80
10	.67	.66*	.43*	.51	.53*	.33*	.87*	.42*	.56*	.67
11	.78*	.70*	.61	.36*	.49	.71*	.66*	.71*	.47*	.78
12	.51*	.59*	.74*	.42	.73*	.76	.53*	.59*	.59*	.77
13	.81*	.57*	.52*	.41*	.83*	.50	.86	.43*	.70*	.20
14	.85	.79*	.27*	.41*	.62*	.38*	.71*	.50*	.60*	.66
15	.65*	.79	.55*	.45	.40*	.69*	.46*	.51*	.51*	.37
16	.83*	.58*	.47	.63*	.34*	.53*	.58*	.40	.71*	.81
17	.84	.75*	.33*	.65*	.66	.74*	.60*	.35*	.46*	.81*
18	.55*	.76*	.67	.46*	.75*	.77*	.58*	.69*	.83*	.66*
19	.77*	.85*	.47*	.47*	.70*	.72*	.59*	.57*	.50*	.47*
20	.82	.68*	.40	.73	.68*	.66*	.59*	.50*	.65*	.64*
21	.69*	.62	.44*	.68*	.64	.54*	.71*	.39*	.75*	.47*
22	.77*	.71*	.66*	.65*	.67	.81	.40*	.43*	.79*	.80*
23	.71	.68*	.48	.49*	.43	.30*	.56	.70*	.84*	.63
24	.61*	.81*	.75*	.29*	.52	.81*	.54*	.52*	.81*	.59*
25	.65*	.73	.46*	.52	.72	.62	.65*	.39	.39*	.63*
26	.68*	.69	.74	.39	.65*	.80	.53*	.68	.64*	.63*
27	.80*	.68	.39*	.56	.74*	.43	.70*	.51*	.62*	.45*
28	.68	.56	.20	.50*	.56*	.68*	.59*	.39	.70*	.68*
29	.54	.55	.41	.31*	.62	.78*	.53*	.61*	.76*	.61*
30	.76*	.76	.47*	.42	.68	.78*	.39*	.44	.74*	.66*
31		.72*	.27*	.53*	.62	.68	.46*	.55	.77	.69*
32		.68*	.64*	.48	.49*	.47*	.41*	.40*	.67	.72*
33		.73	.32*		.59	.61*	.42	.31*	.68	.63*
34		.72*	.53*		.53*	.81*	.74*	.65	.72	.75*
35		.63*	.42		.69*	.58	.63	.49*	.65	.61*
36		.43*	.29		.72*	.40*	.59	.47	.70	.51*
37		.74	.69		.68*	.52*	.65	.65	.45	.79*
38		.50	.49*		.66*	.56*	.32*	.36*	.63	.49*
39		.75	.67		.51*	.47*	.28*	.39*	.76	.73*
40		.43	.69*		.72	.61	.72*	.63*	.82	.54*
41		.53	.56*		.58					.77*
42		.54	.75*		.62					.43*
43		.51*	.44*		.57					.77*
44		.41*	.34*		.67*					.72*
45		.54	.54		.65*					.76
46		.47*	.49*		.57*					.46*
47		.52*	.52*		.55*					.67*
48		.27*	.40		.70*					.43*
49		.57								
50		.45								
51		.39								
52		.56								
53		.45								
54		.31								

* Item is included in the Abbreviated Battery.

Table G-53. Item *p*-Values for Advanced 2, Forms A/D, Fall, Grade 9

Item Number	Reading Vocabulary	Reading Comprehension	Mathematics Problem Solving	Mathematics Procedures	Language A	Spelling	Science	Social Science	Listening	Language D
1	.80*	.94*	.62	.50	.79	.82*	.75*	.60	.65*	.82
2	.79	.96*	.53*	.56*	.68	.83*	.77*	.56*	.54*	.65
3	.65*	.57*	.55*	.38*	.81*	.58*	.57*	.65*	.91*	.71
4	.70*	.61	.43	.64	.82	.51*	.59*	.70*	.56*	.62
5	.77*	.52*	.40*	.55*	.80*	.74	.73*	.48	.77*	.77
6	.62*	.77*	.57	.33*	.76*	.57*	.42*	.64*	.64*	.80
7	.56*	.61	.45*	.35	.65	.72	.60*	.67*	.52*	.39
8	.52	.84	.58	.65*	.58*	.80*	.57*	.28*	.87*	.45
9	.65	.75	.35*	.67*	.81*	.70*	.57*	.37*	.47*	.47
10	.66*	.67	.41	.63*	.63	.38	.35*	.42*	.73*	.63
11	.75	.59	.43*	.33	.85	.71	.66*	.47*	.60	.84
12	.61*	.44	.25*	.49	.37	.69*	.71*	.47*	.85	.87
13	.87	.67	.67*	.40*	.47	.85*	.33*	.43*	.85	.71
14	.91*	.65	.53*	.59	.70*	.61	.57*	.47*	.66*	.52
15	.80*	.63	.65	.48*	.61*	.55	.70	.50*	.87*	.76
16	.72*	.52	.49*	.36*	.70*	.48*	.52*	.30*	.76*	.61
17	.83	.61	.31	.64	.62*	.73*	.42*	.53*	.14*	.65*
18	.68*	.61	.43*	.36*	.77*	.88*	.70	.31*	.41*	.83*
19	.75*	.55*	.44*	.26*	.68*	.92*	.67	.38*	.57*	.79*
20	.70	.44*	.62	.47*	.47*	.78*	.51	.42*	.41*	.81*
21	.59*	.90*	.58*	.42	.64	.66	.50*	.51	.58*	.77*
22	.81*	.74*	.52	.39	.24*	.36	.71*	.46*	.82*	.59*
23	.78*	.68	.51*	.37*	.50*	.73*	.40	.29	.64*	.71*
24	.65	.52*	.72*	.50	.71*	.67*	.17	.44*	.59	.68*
25	.63*	.72	.40*	.45	.63	.38*	.48	.46*	.38	.64*
26	.66	.61*	.61	.34*	.62*	.88*	.66*	.49*	.50	.64*
27	.80*	.45*	.65*	.40*	.63	.42*	.46*	.53	.36*	.50*
28	.59*	.61*	.40*	.52*	.59	.65*	.29*	.45*	.70*	.38*
29	.59	.72*	.45*	.37*	.57	.89	.34*	.54*	.66*	.78*
30	.45*	.69*	.64	.53*	.56*	.86*	.30	.56	.71*	.53*
31		.54*	.34	.28	.61*	.53*	.56*	.46*	.40*	.76
32		.71*	.66*	.45*	.51*	.85*	.37*	.44*	.73*	.68*
33		.68*	.35		.69	.50*	.45*	.59*	.80*	.81*
34		.39*	.31		.58*	.44*	.67*	.52*	.65*	.78*
35		.75	.67*		.64*	.43*	.44	.51	.63*	.65*
36		.66*	.40*		.51*	.77*	.25*	.39*	.48*	.78*
37		.57	.58*		.74*	.35*	.36*	.34	.63	.54*
38		.75	.59*		.63*	.71	.54*	.43	.55	.80*
39		.64	.59		.48*	.17*	.37*	.43*	.82	.83
40		.55	.47*		.73	.58*	.49	.39	.47	.67*
41		.50	.53*		.61					.49*
42		.56	.35		.55					.73*
43		.70*	.33*		.66					.65*
44		.61	.31*		.51*					.40*
45		.57*	.39*		.69*					.71*
46		.38*	.29		.63*					.59*
47		.61*	.26		.55*					.77*
48		.38*	.34*		.64*					.81*
49		.61*								
50		.62*								
51		.62								
52		.57*								
53		.57*								
54		.59*								

* Item is included in the Abbreviated Battery.

Table G-54. Item *p*-Values for TASK 1, Forms A/D, Fall, Grade 9

Item Number	Reading Vocabulary	Reading Comprehension	Mathematics	Language A	Spelling	Science	Social Science	Language D
1	.86	.68*	.46*	.89*	.78*	.93*	.54	.70
2	.71*	.82*	.48*	.82	.88*	.67*	.60	.60
3	.60	.48*	.34	.60	.79*	.54*	.49	.66
4	.68*	.82*	.17	.75*	.67	.55*	.65*	.81
5	.84*	.74*	.73	.67*	.76	.86	.61	.88
6	.58*	.53*	.50	.50	.84	.80*	.37*	.58
7	.54	.64*	.38*	.72*	.88	.78	.52*	.66
8	.49	.62*	.76*	.57*	.80*	.51	.58*	.87
9	.52*	.77*	.60*	.66	.69*	.75*	.46*	.74
10	.50*	.86*	.70	.70*	.68	.49	.24*	.72
11	.58*	.80*	.75*	.70*	.27*	.60*	.73	.92
12	.34*	.88*	.37*	.37*	.78*	.57*	.41*	.82
13	.89	.56*	.56	.48*	.52*	.85*	.24	.86
14	.87*	.67*	.51*	.50	.53	.60*	.18*	.74
15	.70*	.56*	.76*	.46*	.33*	.39*	.44*	.44
16	.66*	.51*	.49*	.55*	.73*	.60*	.33*	.66
17	.88*	.57	.29*	.85	.54*	.60*	.36	.81*
18	.81	.83*	.46	.35	.85*	.34*	.58*	.74*
19	.80	.52*	.50*	.69*	.70*	.56*	.59*	.70*
20	.79*	.46	.49	.76	.73*	.43*	.36*	.42*
21	.59	.41*	.60	.17*	.52*	.26*	.41*	.81*
22	.82*	.45	.20*	.69*	.40*	.65*	.45*	.50*
23	.91*	.65	.46	.52*	.52*	.40	.56*	.68*
24	.76*	.69*	.26*	.54	.75	.52	.36*	.80*
25	.64*	.68*	.43*	.66*	.49*	.36*	.32*	.73
26	.82*	.50*	.42	.70*	.64*	.53	.54	.76*
27	.83	.39	.38*	.67*	.28	.25*	.41*	.63*
28	.59*	.74	.51*	.72*	.73*	.55*	.48*	.60*
29	.51	.70	.37*	.42*	.72*	.45*	.45	.36*
30	.76*	.49	.36	.67	.71*	.28	.36*	.83*
31		.61	.34	.57*	.58	.52*	.54*	.56*
32		.64	.51*	.49*	.62*	.31*	.46*	.82*
33		.63	.23	.58*	.33*	.32*	.40*	.68*
34		.32	.41*	.59	.73	.50	.38	.62*
35		.60	.16*	.65*	.40*	.40*	.39*	.67*
36		.31	.28	.65*	.56*	.21*	.48*	.64*
37		.76	.50*	.76	.50*	.47*	.43*	.77*
38		.77	.28*	.75	.46*	.23*	.37*	.54*
39		.31	.18	.61	.32*	.26	.41*	.60*
40		.67	.42*	.58*	.42*	.51*	.41*	.66*
41		.60	.24	.68*				.55*
42		.56	.29*	.51				.63*
43		.59	.35	.50				.67*
44		.62	.31*	.47				.42*
45		.68	.32*	.42				.55*
46		.49*	.16	.64*				.65*
47		.48	.24*	.62*				.62*
48		.61*	.30*	.53*				.42
49		.49*	.28					
50		.64*	.40*					
51		.26*						
52		.42*						
53		.44*						
54		.48*						

* Item is included in the Abbreviated Battery.

Appendix G

Table G-55. Item *p*-Values for TASK 2, Forms A/D, Fall, Grade 10

Item Number	Reading Vocabulary	Reading Comprehension	Mathematics	Language A	Spelling	Science	Social Science	Language D
1	.68	.68*	.50*	.87*	.69*	.65*	.40*	.52
2	.68*	.69*	.39	.64	.66*	.62	.44*	.50
3	.57*	.75*	.45*	.64	.58*	.67*	.47*	.63
4	.59	.67*	.50*	.47*	.59*	.59	.40*	.69
5	.60*	.56*	.42	.40*	.47*	.46*	.43	.63
6	.54	.73*	.49*	.51	.56*	.56*	.26*	.91
7	.37	.76*	.58*	.36*	.65	.47*	.29	.57
8	.38*	.68*	.21	.61*	.69*	.40*	.54*	.72
9	.38*	.39*	.39	.51*	.45	.79*	.41*	.80
10	.46*	.78	.62*	.50*	.57*	.57*	.33*	.80
11	.69*	.72	.68*	.44	.51*	.48	.54*	.78
12	.32*	.49	.54	.65	.23	.29	.44*	.87
13	.95	.80	.45	.50*	.70*	.73*	.29	.66
14	.79*	.71	.62*	.27*	.34*	.71*	.32*	.68
15	.71*	.61	.36*	.54*	.55	.54*	.29*	.67
16	.63	.72	.42	.52*	.48*	.47	.57	.62
17	.90*	.29	.17*	.60*	.68	.64	.29*	.86*
18	.54*	.70	.41*	.58*	.60	.56*	.32*	.78*
19	.78	.57*	.55*	.70	.71*	.34	.38*	.53*
20	.76*	.56*	.35	.63*	.71	.42*	.32*	.91*
21	.65	.49*	.37*	.36*	.42*	.57	.26*	.49*
22	.72	.49*	.31	.29	.38*	.35*	.37	.41*
23	.80*	.61*	.29*	.58	.12*	.33*	.46*	.43*
24	.68*	.31*	.38	.25	.65*	.37*	.35	.59*
25	.45*	.72*	.59	.54	.36*	.46*	.31*	.58*
26	.90*	.56*	.30*	.57*	.57	.42*	.25*	.44*
27	.70*	.49	.23*	.53	.35*	.51*	.43*	.68
28	.35	.74	.41*	.55*	.34*	.38*	.31	.38*
29	.59*	.49	.53*	.56*	.82*	.47*	.56	.49*
30	.52*	.48	.25	.60	.69*	.42*	.21*	.76*
31		.55	.40	.76	.23*	.59	.37*	.56*
32		.69	.41*	.56*	.43	.30*	.47*	.59*
33		.60	.48	.42*	.37*	.59*	.31*	.38*
34		.55	.23*	.67	.52*	.23*	.25*	.59*
35		.76	.32*	.58*	.71	.47*	.32*	.55*
36		.60	.22	.62*	.69*	.30	.29	.53*
37		.53*	.47	.62*	.40*	.66*	.32*	.39*
38		.49*	.31*	.66*	.72*	.20*	.23*	.40*
39		.68*	.13*	.57*	.40*	.36*	.31*	.65*
40		.65*	.45	.62	.54*	.74*	.39	.28
41		.32	.29*	.58				.58*
42		.66*	.37*	.57				.52*
43		.80*	.22	.53				.56*
44		.51*	.24	.55*				.45*
45		.28*	.34*	.45*				.27*
46		.27*	.13*	.48*				.66*
47		.40*	.35*	.59*				.41*
48		.49*	.26*	.48*				.52*
49		.67	.30*					
50		.41	.21					
51		.40						
52		.54*						
53		.30*						
54		.27						

* Item is included in the Abbreviated Battery.

Table G-56. Item *p*-Values for TASK 3, Forms A/D, Fall, Grade 11

Item Number	Reading Vocabulary	Reading Comprehension	Mathematics	Language A	Spelling	Science	Social Science	Language D
1	.72*	.55*	.45*	.62	.77*	.79	.54*	.66
2	.63*	.61*	.22	.60*	.57*	.61*	.44*	.52
3	.42*	.53*	.36*	.71*	.79*	.35	.27*	.74
4	.78	.79*	.63	.71*	.37*	.49*	.21	.38
5	.64*	.63*	.38	.49*	.71*	.41*	.36	.64
6	.65*	.53*	.36*	.29	.44*	.65	.34*	.46
7	.41*	.38*	.35*	.27*	.61*	.44*	.23*	.51
8	.52*	.78*	.46*	.29*	.78*	.56*	.29	.77
9	.33	.50*	.49	.53	.61*	.31*	.28	.86
10	.56*	.73*	.42	.55*	.58	.42*	.43	.56
11	.63	.63*	.46*	.46*	.11*	.32*	.21*	.80
12	.34	.35*	.52*	.65*	.72	.22*	.32*	.72
13	.71	.62	.38*	.57	.71*	.54*	.16*	.72
14	.79*	.60	.34	.58*	.41*	.49*	.09*	.69
15	.75*	.61	.20*	.53*	.71	.29*	.44	.51
16	.58	.64*	.26	.58	.71*	.30*	.42*	.47
17	.73*	.63*	.37*	.28*	.49*	.23*	.29*	.72*
18	.65*	.43	.39*	.35*	.52*	.33	.28*	.71*
19	.63*	.65	.35*	.54	.63*	.25*	.24*	.72*
20	.76*	.81	.37	.25	.65	.48*	.30*	.59*
21	.62	.55	.48*	.58*	.49*	.64*	.40	.51*
22	.59*	.73	.22	.28	.63	.30*	.33*	.52*
23	.80	.44	.34	.29	.16	.21*	.38*	.56*
24	.60*	.64	.22*	.40*	.62	.46	.26	.61*
25	.37	.62	.52*	.57*	.30*	.29*	.28*	.66*
26	.71*	.74	.38*	.50*	.45	.47*	.41*	.44*
27	.80	.60	.31*	.43*	.27*	.24*	.30*	.42*
28	.57*	.36	.40*	.40	.43	.41*	.34*	.63*
29	.35*	.46	.40	.38	.79*	.23*	.22*	.71*
30	.74*	.38	.33	.48	.66*	.55*	.22*	.60*
31		.66	.43*	.52	.58*	.38	.38	.46*
32		.22	.38	.61*	.21*	.48	.33*	.27
33		.62	.24*	.63*	.62*	.35	.39*	.62*
34		.64	.39	.30	.43*	.25*	.37*	.35*
35		.56	.29	.43*	.43*	.36*	.43	.48*
36		.66	.44*	.59*	.70*	.51	.26*	.56*
37		.59*	.33*	.44*	.50*	.41	.26*	.32*
38		.45*	.39*	.45*	.66	.29*	.32*	.28*
39		.36*	.45	.65*	.41*	.43*	.27*	.31
40		.49*	.38*	.55	.40*	.24*	.34*	.58*
41		.45*	.37	.54				.27*
42		.36	.34*	.38*				.50*
43		.31*	.24*	.62*				.41*
44		.46*	.23*	.53				.44*
45		.47*	.28	.60				.48*
46		.47*	.22*	.42*				.16*
47		.53*	.19	.64*				.34*
48		.33*	.25*	.43*				.47*
49		.46*	.20*					
50		.35*	.28					
51		.41*						
52		.56*						
53		.58*						
54		.33						

* Item is included in the Abbreviated Battery.

Appendix G

Table G-57. Item *p*-Values for TASK 3, Forms A/D, Fall, Grade 12

Item Number	Reading Vocabulary	Reading Comprehension	Mathematics	Language A	Spelling	Science	Social Science	Language D
1	.85*	.62*	.59*	.75	.87*	.68	.49*	.73
2	.73*	.69*	.51	.74*	.69*	.56*	.37*	.60
3	.57*	.57*	.48*	.69*	.86*	.44	.41*	.85
4	.80	.75*	.73	.66*	.50*	.60*	.23	.51
5	.72*	.66*	.49	.50*	.71*	.57*	.38	.73
6	.73*	.65*	.41*	.43	.53*	.54	.41*	.53
7	.38*	.32*	.36*	.45*	.70*	.50*	.27*	.58
8	.66*	.70*	.46*	.40*	.84*	.62*	.26	.79
9	.46	.50*	.51	.61	.65*	.40*	.33	.88
10	.67*	.73*	.46	.61*	.70	.38*	.54	.58
11	.75	.56*	.49*	.34*	.14*	.44*	.25*	.81
12	.33	.43*	.56*	.67*	.81	.22*	.36*	.85
13	.84	.58	.33*	.53	.72*	.62*	.21*	.74
14	.87*	.66	.46	.64*	.46*	.56*	.16*	.79
15	.78*	.72	.35*	.57*	.74	.35*	.31	.54
16	.73	.62*	.25	.64	.70*	.38*	.42*	.44
17	.80*	.69*	.42*	.36*	.57*	.29*	.31*	.78*
18	.65*	.43	.47*	.39*	.57*	.30	.35*	.79*
19	.84*	.67	.42*	.59	.75*	.25*	.25*	.77*
20	.86*	.83	.37	.39	.74	.48*	.31*	.63*
21	.72	.61	.54*	.60*	.62*	.69*	.36	.61*
22	.74*	.74	.20	.52	.66	.40*	.29*	.63*
23	.81	.50	.47	.32	.23	.26*	.34*	.55*
24	.66*	.57	.29*	.62*	.60	.48	.32	.66*
25	.53	.70	.44*	.65*	.44*	.28*	.27*	.77*
26	.80*	.77	.36*	.61*	.58	.49*	.37*	.49*
27	.85	.65	.34*	.51*	.34*	.28*	.39*	.56*
28	.65*	.39	.39*	.54	.50	.45*	.36*	.75*
29	.60*	.52	.40	.46	.79*	.33*	.25*	.80*
30	.85*	.33	.33	.59	.75*	.61*	.28*	.61*
31		.61	.41*	.62	.63*	.39	.38	.50*
32		.30	.39	.56*	.30*	.42	.31*	.35
33		.62	.23*	.57*	.64*	.32	.40*	.71*
34		.67	.33	.40	.41*	.30*	.41*	.37*
35		.67	.29	.52*	.52*	.30*	.43	.55*
36		.70	.49*	.53*	.67*	.51	.38*	.70*
37		.61*	.36*	.49*	.54*	.38	.32*	.53*
38		.50*	.44*	.40*	.70	.36*	.33*	.34*
39		.49*	.43	.52*	.52*	.53*	.24*	.37
40		.54*	.40*	.61	.43*	.23*	.47*	.68*
41		.47*	.43	.58				.53*
42		.29	.32*	.47*				.58*
43		.40*	.30*	.63*				.59*
44		.50*	.26*	.54				.56*
45		.44*	.32	.65				.56*
46		.44*	.27*	.40*				.32*
47		.52*	.25	.56*				.36*
48		.40*	.31*	.48*				.61*
49		.50*	.20*					
50		.37*	.28					
51		.41*						
52		.55*						
53		.50*						
54		.36						

* Item is included in the Abbreviated Battery.

Table G–58. Item p–Values for Primary 1, Forms B/E, Fall, Grade 2

Item Number	Word Study Skills	Word Reading	Sentence Reading	Reading Comprehension	Mathematics Problem Solving	Mathematics Procedures	Language B	Spelling	Environment	Listening	Language E
1	.91*	.93*	.98*	.94*	.97*	.82*	.79*	.96	.99*	.88*	.84*
2	.78*	.74*	.94*	.92*	.77*	.92*	.89*	.87*	.90*	.73*	.89*
3	.80*	.89*	.96	.96*	.91*	.83*	.71	.91	.62*	.73*	.78
4	.90	.89*	.95	.92	.65	.70	.81*	.69	.95	.75*	.87*
5	.96*	.92*	.88*	.76*	.74*	.76	.68	.77*	.95	.53*	.81*
6	.96	.79*	.85	.92*	.53	.70*	.56*	.96*	.80*	.70*	.91*
7	.92	.90*	.92*	.88	.84	.42*	.81*	.96*	.53	.85*	.96*
8	.85*	.88*	.94*	.89*	.91*	.52	.60	.86*	.97*	.74*	.89
9	.95*	.93*	.90*	.83*	.93*	.04	.81*	.77*	.99*	.68*	.83
10	.97*	.87*	.87	.92	.86*	.04*	.40*	.69*	.73*	.65*	.88
11	.95*	.89*	.85*	.95*	.60*	.55	.69*	.86*	.79*	.98*	.83*
12	.95	.79*	.93	.92*	.82	.86*	.64*	.72*	.64*	.90*	.77*
13	.90*	.90*	.94*	.92*	.42*	.85*	.77*	.67*	.83*	.91	.63*
14	.74*	.73*	.83*	.95*	.88	.47*	.61*	.78*	.58	.70	.84*
15	.86*	.92*	.86*	.88*	.71*	.63	.78*	.82*	.80*	.85	.80*
16	.89	.93	.91*	.92*	.62*	.75*	.75	.96*	.80*	.92*	.66*
17	.56*	.76	.83*	.94*	.93*	.56*	.80*	.96*	.93	.50*	.68*
18	.71*	.93	.79*	.92*	.72	.69*	.53*	.84*	.25*	.83*	.88*
19	.65	.90*	.88*	.73*	.77*	.55*	.79	.62*	.92*	.58*	.80*
20	.69*	.83*	.91	.96	.88*	.59	.62*	.86*	.98*	.78*	.67*
21	.89	.78*	.92*	.89	.61*	.77*	.81*	.82	.96	.77*	.48*
22	.68*	.93	.86*	.83	.69*	.80*	.73	.46	.86*	.68*	.73*
23	.70*	.92	.84*	.85*	.84*	.54*	.69*	.70*	.92*	.80	.62*
24	.18*	.85	.85	.86*	.64	.53	.74*	.69*	.56*	.71	.78*
25	.68	.94	.93	.81*	.87	.73*	.69*	.84*	.44*	.78	.19*
26	.75*	.83	.88*	.68*	.75*	.54*	.78	.84*	.90*	.80*	.55*
27	.43	.82	.92*	.85*	.72*	.60	.75*	.96*	.77*	.69*	.43*
28	.69*	.78*	.94*	.67*	.68*	.38*	.50*	.72*	.88*	.50*	.64*
29	.76*	.82*	.92	.54*	.56*	.42*	.74	.60*	.94*	.70*	.70*
30	.29	.83	.93	.65*	.92*	.37	.60	.75*	.76	.67*	.68*
31				.57*	.82		.56*	.79*	.97*	.77*	.87*
32				.63*	.82*		.88*	.48*	.85*	.87*	.73*
33				.73*	.86*		.87*	.79*	.93	.81*	.74*
34				.59*	.92*		.76	.67*	.76*	.49*	.70*
35				.68*	.85*		.51*	.59	.97*	.83*	.63
36				.63*	.89		.84*	.69*	.69	.75*	.79
37				.64	.69*		.71*		.60	.57	.88
38				.61	.76		.88*		.33*	.81	.42
39				.83	.76		.66*		.40*	.66	.83
40				.74	.79*		.73*		.92*	.64	.58
41					.84*						
42					.62*						

* Item is included in the Abbreviated Battery.

Appendix G

Table G–59. Item p–Values for Primary 2, Forms B/E, Fall, Grade 3

Item Number	Word Study Skills	Reading Vocabulary	Reading Compre-hension	Mathematics Problem Solving	Mathematics Procedures	Language B	Spelling	Environment	Listening	Language E
1	.97*	.96	.93*	.94*	.97	.93*	.91	.91*	.85*	.90
2	.96*	.96*	.91*	.95	.91*	.95*	.94*	.90*	.62*	.94*
3	.78*	.88*	.91*	.88	.93*	.81*	.84*	.77*	.82*	.92*
4	.98	.89*	.64*	.92*	.88*	.91	.86*	.77*	.72*	.65*
5	.95*	.92	.62*	.93*	.67*	.82*	.78	.63*	.58*	.89*
6	.96	.94*	.84*	.88*	.70	.87*	.90*	.78	.61*	.84*
7	.97	.88*	.86*	.83*	.80	.77	.94*	.76	.64*	.94
8	.97*	.73	.76*	.77*	.66*	.76*	.91*	.53*	.75*	.67
9	.98*	.77*	.48*	.84	.87*	.81*	.84*	.87*	.62*	.60*
10	.97*	.70	.87*	.66*	.90	.89	.91*	.52*	.72*	.65
11	.95	.91*	.88	.68*	.79*	.90*	.76*	.81	.83*	.83*
12	.97*	.88*	.70	.85	.76*	.86*	.87*	.94*	.90*	.68*
13	.97*	.79*	.85	.76*	.77	.81	.80	.60	.94*	.77
14	.90*	.76*	.67	.57*	.86*	.80*	.82*	.73*	.97*	.75*
15	.68*	.58*	.89	.79	.79*	.59*	.71*	.96*	.69*	.86
16	.82*	.82*	.73	.72*	.65	.77*	.87*	.81*	.99*	.77*
17	.85	.91	.46	.77*	.89*	.85	.92*	.75*	.86*	.79
18	.67	.92	.79*	.89	.92	.84	.87*	.66	.84*	.88*
19	.51*	.75*	.59*	.93*	.88*	.74	.92	.89*	.71*	.69*
20	.69*	.74*	.89*	.70*	.83*	.85*	.83*	.76*	.52*	.77*
21	.62	.76*	.72*	.94*	.89*	.87	.84*	.96	.92	.79*
22	.61	.76*	.76*	.66	.49*	.83*	.87*	.76*	.65	.72*
23	.60*	.62	.77*	.70	.83	.84	.89	.42*	.55	.76*
24	.77	.73	.53*	.39*	.82*	.77*	.89*	.74*	.66	.70*
25	.56*	.95*	.70*	.78*	.35*	.85*	.84*	.66*	.75	.77*
26	.54*	.89*	.77*	.85*	.78	.79*	.93	.55*	.60	.85*
27	.72*	.80	.55*	.83	.80*	.83*	.94*	.39*	.90	.51*
28	.51*	.84*	.89*	.76*	.75*	.85*	.93*	.94*	.60*	.83*
29	.33	.70	.78*	.88*	.66*	.83	.84*	.78*	.62*	.68*
30	.32*	.69*	.84*	.93*	.72	.86*	.85*	.54*	.56*	.83*
31			.48*	.71*		.86*	.78*	.93*	.77*	.85*
32			.65*	.58		.83	.94*	.84*	.62*	.77*
33			.66*	.81		.80	.74*	.43	.82*	.74
34			.62*	.73*		.85*	.88*	.96	.64*	.69*
35			.53*	.94*		.77	.73*	.89*	.41*	.78*
36			.75*	.59*		.75*	.83*	.45*	.85*	.88
37			.78*	.90		.94*		.36	.73*	.63*
38			.72	.67*		.68*		.98	.80	.58*
39			.71	.62*		.82*		.66*	.47	.69
40			.75	.83*		.68		.92*	.72	.83*
41				.67		.80				
42				.45		.81				
43				.29*		.95				
44				.90*		.66*				
45						.64*				
46						.79*				
47						.88				
48						.62*				

* Item is included in the Abbreviated Battery.

Table G–60. Item *p*–Values for Primary 3, Forms B/E, Fall, Grade 4

Item Number	Word Study Skills	Reading Vocabulary	Reading Comprehension	Mathematics Problem Solving	Mathematics Procedures	Language B	Spelling	Science	Social Science	Listening	Language E
1	.96	.94	.76*	.89	.86*	.87*	.86	.93*	.91*	.97*	.60*
2	.94*	.89*	.87*	.90	.79*	.89	.70*	.91*	.89*	.72*	.78*
3	.89*	.85	.56*	.61*	.79	.73	.55*	.51*	.86*	.44*	.77*
4	.94*	.62	.88*	.67*	.70	.75*	.71*	.76*	.78*	.92*	.70*
5	.77*	.91*	.64*	.88*	.78*	.60	.57	.85*	.69	.43*	.56*
6	.74	.79*	.68	.84	.66*	.55	.71*	.83*	.61*	.81*	.64*
7	.69*	.92*	.79	.91	.59*	.47*	.81*	.72*	.81*	.68*	.78*
8	.56*	.73	.85	.86*	.82*	.67*	.48*	.43	.77	.55*	.63*
9	.69	.65*	.71	.84*	.51	.63*	.67*	.42*	.84*	.48*	.69*
10	.69*	.81*	.82	.60	.52*	.33*	.64*	.66*	.88*	.52*	.38*
11	.67	.60*	.69	.77*	.50	.68*	.41*	.40	.65*	.76	.47*
12	.46*	.70*	.85	.64*	.62	.67*	.56*	.50*	.67	.82	.77*
13	.94*	.81	.87	.72*	.59*	.83	.57*	.84	.85	.70	.71*
14	.57*	.57*	.82	.58*	.64*	.60	.72	.72*	.56*	.75	.40*
15	.65*	.68*	.90	.81	.62*	.62*	.47*	.67	.73	.64	.60*
16	.81	.80	.75	.84*	.56	.72*	.71*	.81	.79*	.78	.59*
17	.71	.82*	.47	.72*	.77*	.66*	.79*	.61*	.61	.79	.79*
18	.70*	.72	.85	.77	.47	.57*	.50*	.62*	.47*	.90*	.26*
19	.78*	.70*	.67	.59	.54*	.78	.33	.59*	.60*	.46*	.75*
20	.83	.78*	.61	.77*	.55	.74	.73*	.53*	.79	.56*	.49*
21	.73*	.73*	.79	.64	.40*	.73*	.62*	.63*	.61*	.68*	.43*
22	.80*	.83*	.54	.54*	.62*	.64*	.50*	.43	.81*	.67	.61*
23	.56*	.62*	.69	.78*	.46	.34	.72*	.59	.37*	.59*	.65*
24	.56*	.79*	.63	.76*	.54*	.28*	.70	.71*	.53*	.78*	.42*
25	.65*	.82	.88*	.69*	.53*	.67	.59*	.60	.74*	.62*	.68*
26	.58	.65	.58*	.81*	.46*	.79*	.62*	.60*	.42*	.82*	.74*
27	.57	.76*	.71	.61	.61*	.69*	.79*	.73*	.63*	.60*	.58*
28	.80	.73	.57*	.86*	.61	.69	.76*	.60*	.74*	.83*	.63*
29	.60*	.59*	.72*	.88	.51*	.66*	.48*	.55*	.78*	.75*	.67*
30	.38*	.81*	.49*	.62*	.60*	.71	.69	.71	.64*	.30*	.53*
31			.69*	.80		.62*	.48*	.45*	.75	.36*	.71
32			.64	.76*		.79*	.46*	.48*	.25*	.56*	.46
33			.48*	.70*		.64*	.67*	.73*	.83*	.68*	.50
34			.61*	.70*		.48*	.47	.43*	.76*	.77*	.60
35			.74*	.54*		.60*	.79	.74*	.79*	.57*	.80
36			.67*	.73		.69	.37*	.78	.62*	.76*	.58
37			.77*	.79*		.78*	.52*	.53*	.65*	.77*	.35
38			.45*	.25*		.59	.58*	.45*	.61	.57	.43
39			.65*	.78		.69		.78*	.74	.77	.36
40			.40*	.69*		.33*		.70*	.75*	.74	.52
41			.68	.67		.71					.45
42			.70*	.51*		.69*					.60
43			.66	.72*		.75*					.45
44			.58*	.76*		.66*					.74
45			.37*	.70		.63					.51
46			.75*	.67*		.59*					
47			.57*			.43					
48			.61*			.46*					
49			.59*								
50			.43*								
51			.51*								
52			.44*								
53			.50*								
54			.73								

* Item is included in the Abbreviated Battery.

Appendix G

Table G–61. Item *p*–Values for Intermediate 1, Forms B/E, Fall, Grade 5

Item Number	Word Study Skills	Reading Vocabulary	Reading Comprehension	Mathematics Problem Solving	Mathematics Procedures	Language B	Spelling	Science	Social Science	Listening	Language E
1	.94	.85	.86*	.88	.77*	.94*	.79*	.91*	.87*	.78*	.82*
2	.85*	.80*	.82*	.75*	.78	.52*	.74*	.91*	.90*	.75*	.81*
3	.91*	.67	.71	.79	.84*	.88*	.72	.56	.66	.91*	.80*
4	.91*	.88*	.67*	.82	.54*	.80*	.65	.43	.76*	.48*	.58*
5	.74*	.76*	.77*	.68*	.66	.86	.60*	.82	.70*	.29*	.34*
6	.78	.66*	.53*	.78*	.72*	.48*	.58*	.30*	.70*	.75*	.68*
7	.65	.67*	.86	.67	.70*	.89	.66*	.80*	.81*	.57*	.53
8	.66*	.70*	.66	.79*	.66	.80	.81*	.81*	.63	.46*	.78*
9	.75*	.65*	.50	.84*	.74	.58*	.77*	.63	.70*	.50*	.41*
10	.70	.69	.50	.25*	.78*	.64*	.71*	.29	.77*	.86*	.62*
11	.77*	.77*	.78	.79	.80*	.46*	.40*	.84*	.70*	.67	.71*
12	.40*	.66	.79	.64*	.44*	.66*	.80*	.67*	.49*	.56	.64*
13	.62*	.91*	.86	.79	.63	.79*	.41*	.78*	.84	.74	.72*
14	.67*	.86	.73	.87	.71*	.78*	.75*	.41*	.58	.46*	.53*
15	.63*	.73	.87	.74	.57*	.43	.71*	.71*	.76*	.70*	.77*
16	.52*	.83*	.48	.71*	.61	.81	.48	.74*	.67	.93*	.62*
17	.53	.82*	.58	.74*	.70*	.68	.68*	.76*	.54*	.75*	.73*
18	.61	.81	.75	.67	.66	.73*	.72*	.56	.53*	.43*	.71*
19	.79*	.79*	.66*	.59*	.58*	.63*	.78*	.74*	.53*	.72*	.51*
20	.74	.79*	.71*	.85*	.73	.52*	.47	.60	.59	.91*	.53*
21	.78*	.66*	.57*	.50	.64*	.72	.41	.83*	.71*	.87*	.74*
22	.69	.84	.88*	.69*	.64*	.71*	.78*	.82*	.76*	.84*	.60*
23	.59	.81	.90	.44*	.62*	.67	.59	.65*	.60*	.64*	.55*
24	.55	.81*	.76*	.87*	.74*	.49	.86*	.69	.53*	.52*	.77*
25	.54*	.67*	.67*	.42	.64	.76*	.56*	.45*	.64*	.63	.59*
26	.72*	.82*	.35*	.82*	.70*	.56*	.52	.65*	.67*	.49	.71*
27	.55*	.76*	.57*	.48*	.41*	.68	.51*	.67*	.58*	.47	.61
28	.50*	.79*	.67*	.77	.56*	.59*	.54*	.78*	.67*	.69	.63*
29	.65*	.74	.62*	.70*	.58*	.74	.83	.80*	.83	.41*	.37*
30	.46*	.74*	.72	.73*	.50	.74*	.74	.53*	.45*	.80*	.75*
31			.77*	.63	.62	.64	.90*	.50*	.59	.67*	.86*
32			.66	.66*	.51	.63*	.42*	.44*	.68*	.68*	.71*
33			.43*	.80*		.72	.68	.34	.76	.72*	.53
34			.71*	.67*		.77	.35*	.68*	.71*	.70*	.71
35			.62*	.61		.67*	.70	.86*	.41*	.81*	.77
36			.65*	.73		.67	.50*	.74	.75	.79*	.63
37			.70	.66*		.69*	.70*	.45*	.66*	.72*	.59
38			.62	.67*		.79*	.45*	.73*	.66*	.84	.70
39			.47	.51*		.65*	.74*	.42*	.74*	.37	.42
40			.48	.66		.86*	.48*	.66	.75*	.36	.45
41			.79	.37*		.76*					.53
42			.75	.16*		.68*					.54
43			.44*	.80*		.73*					.47
44			.69*	.64*		.76*					.52
45			.68*	.43		.55*					.44
46			.76*	.61*		.81					.72
47			.41	.61		.73					.53
48			.68*	.62*		.54					.32
49			.51								
50			.54*								
51			.60*								
52			.68*								
53			.59*								
54			.79*								

* Item is included in the Abbreviated Battery.

244

Table G–62. Item p–Values for Intermediate 2, Forms B/E, Fall, Grade 6

Item Number	Reading Vocabulary	Reading Comprehension	Mathematics Problem Solving	Mathematics Procedures	Language B	Spelling	Science	Social Science	Listening	Language E
1	.89	.79	.63*	.81	.89*	.86*	.90	.93*	.91*	.65*
2	.70*	.76*	.76	.78*	.93*	.84*	.91	.57*	.70*	.88*
3	.58*	.81*	.30	.82	.63*	.79*	.76*	.80*	.69*	.91*
4	.74*	.66*	.60*	.52*	.63*	.71	.60*	.68	.44*	.66*
5	.83*	.53*	.51	.70*	.58*	.72*	.72	.62	.65*	.76*
6	.59	.75*	.53	.75*	.77*	.58	.54	.60*	.65*	.58*
7	.54*	.82*	.33*	.81	.77	.79*	.55*	.73*	.58*	.72*
8	.53*	.47*	.85*	.22*	.60*	.65*	.74*	.67	.67*	.64*
9	.70*	.75*	.55*	.68*	.56	.68*	.61*	.64	.69*	.74
10	.74	.58	.48*	.47	.64	.45	.56*	.47*	.82*	.43*
11	.71	.69*	.40	.29*	.22*	.82*	.61	.58	.87	.63*
12	.70*	.58*	.66*	.21	.64	.69*	.68*	.67*	.34	.72*
13	.86*	.43	.82*	.35*	.81*	.58	.63*	.52	.82	.61*
14	.84	.76	.70*	.62*	.65*	.64	.52*	.76	.64*	.40*
15	.72*	.87	.31	.60	.66*	.63	.84*	.49*	.65*	.56*
16	.84*	.60	.65	.64*	.83*	.74*	.55*	.36*	.70*	.68*
17	.86	.54	.60*	.77*	.56	.61*	.54*	.66*	.80*	.86*
18	.59*	.79	.22*	.68*	.74	.81*	.47	.36*	.86*	.64*
19	.77*	.71	.53*	.63	.77	.71*	.64*	.77*	.43*	.69*
20	.88*	.89*	.62	.60	.57	.62*	.61*	.50	.80*	.71*
21	.81	.85*	.57	.47	.50*	.58*	.81*	.63*	.84*	.36*
22	.84	.38*	.54*	.61*	.71*	.81	.66	.44*	.76*	.40*
23	.81*	.63*	.60	.66*	.64	.49*	.71*	.80*	.46*	.70*
24	.82*	.56*	.55*	.66*	.56*	.71*	.62*	.48*	.34*	.33*
25	.73*	.69	.67*	.28*	.66	.64*	.32*	.46*	.50*	.63*
26	.74	.72	.52*	.50*	.63	.48*	.87*	.72*	.59*	.78*
27	.81*	.59	.59	.51	.66	.61*	.54*	.37*	.28*	.54*
28	.68	.55	.60*	.47	.66	.64*	.42	.68	.73*	.50*
29	.69*	.64	.70*	.51	.50*	.87*	.43*	.75	.53*	.49*
30	.74*	.76	.55	.53*	.50*	.76	.48*	.71*	.84*	.71
31		.71*	.79*	.41*	.57	.73	.74*	.53*	.72*	.62*
32		.82*	.75	.44*	.58*	.86*	.35*	.41*	.75*	.55*
33		.66	.65		.65*	.67*	.71	.66*	.77*	.59
34		.75*	.87*		.68*	.47	.37	.47*	.55	.60
35		.77*	.39*		.67*	.77*	.40*	.71*	.78	.69
36		.58*	.68*		.59*	.29*	.71*	.47*	.54	.77
37		.77	.44*		.67*	.55*	.65*	.71*	.47	.56
38		.76	.42		.58	.77	.55*	.41*	.59	.61
39		.45	.54		.63*	.59*	.55*	.63*	.53	.65
40		.62	.45*		.69*	.80*	.77*	.59*	.32	.63
41		.77	.49*		.60					.46
42		.65	.65*		.69					.68
43		.50*	.67*		.65					.53
44		.63*	.27*		.46*					.67
45		.44*	.49*		.59*					.74
46		.61*	.38		.65*					.53
47		.52*	.38*		.47*					.66
48		.63	.66		.45*					.82
49		.67*								
50		.51*								
51		.67*								
52		.54								
53		.39*								
54		.65*								

* Item is included in the Abbreviated Battery.

Appendix G

Table G–63. Item *p*–Values for Intermediate 3, Forms B/E, Fall, Grade 7

Item Number	Reading Vocabulary	Reading Comprehension	Mathematics Problem Solving	Mathematics Procedures	Language B	Spelling	Science	Social Science	Listening	Language E
1	.84*	.80*	.64	.78	.76	.89*	.89*	.52*	.81*	.82*
2	.77*	.69*	.64*	.69*	.81*	.64*	.70*	.79	.85*	.59*
3	.57	.48*	.62*	.53*	.82	.73*	.65	.48*	.59*	.66*
4	.72*	.55*	.76	.69	.66*	.66*	.61*	.60*	.37*	.41*
5	.82*	.87*	.55*	.73	.58	.64	.84	.66	.66*	.47*
6	.46*	.85	.52	.70*	.64	.69*	.70*	.81*	.71*	.31*
7	.66*	.55*	.53*	.49*	.76*	.69*	.57*	.50	.48*	.74*
8	.73	.88*	.43	.72*	.71*	.78*	.59	.57	.90*	.69
9	.50*	.62*	.59*	.51	.61	.59	.61	.68*	.47*	.68*
10	.83	.87*	.65*	.49*	.66*	.64	.73*	.82*	.71*	.70*
11	.76	.38*	.67*	.40*	.65*	.62*	.63*	.59	.70*	.43*
12	.51*	.74	.79*	.80*	.73	.65*	.65*	.44*	.69*	.69*
13	.92	.61	.61	.50	.52	.50*	.76*	.67*	.81*	.67*
14	.87*	.60	.59	.54*	.72*	.82*	.69*	.53*	.81*	.62*
15	.76	.50	.54*	.30*	.80*	.48	.72*	.37*	.48*	.21*
16	.75*	.73	.41*	.31	.77*	.80*	.44*	.63*	.75*	.81*
17	.84	.48	.56*	.71*	.66*	.62*	.57*	.52*	.61*	.71*
18	.63*	.75	.45	.70	.47*	.66*	.79	.35*	.62*	.80*
19	.73*	.87	.60	.70*	.63*	.80*	.47*	.44*	.57*	.53*
20	.81*	.89	.47*	.60	.32*	.79*	.41*	.61*	.62*	.77*
21	.76*	.76	.50*	.54	.79	.66	.72	.56*	.79*	.41*
22	.89	.69	.44*	.53*	.66*	.48*	.82*	.48*	.36*	.66*
23	.79*	.68	.48*	.58*	.56*	.68	.49*	.54*	.49*	.56*
24	.70*	.66	.46*	.63*	.69	.80*	.69*	.66*	.38	.31*
25	.74	.52*	.50	.50	.76	.61*	.31*	.68*	.73	.64*
26	.80	.73*	.59*	.35	.56*	.51	.58	.30*	.82	.49*
27	.80*	.73	.60	.32*	.64	.45*	.42	.53*	.91*	.58*
28	.64*	.31*	.42	.41*	.64*	.63*	.35*	.79*	.71*	.79
29	.75*	.66*	.65*	.24*	.62*	.68	.64	.44	.72*	.70*
30	.64*	.58*	.48*	.35*	.67*	.80*	.43*	.44*	.46*	.67*
31		.45*	.36	.36	.60*	.59*	.75*	.64	.63	.45*
32		.50*	.61*	.42*	.58*	.61	.59	.61*	.77	.49*
33		.67*	.58		.64*	.33*	.39*	.31	.51	.59
34		.56*	.44*		.66	.86	.69*	.51	.63*	.80
35		.57	.48*		.70	.44*	.50*	.74*	.52*	.49
36		.54*	.40		.59	.53*	.51*	.34*	.37*	.71
37		.51*	.39*		.69*	.62*	.66*	.53*	.57	.37
38		.52*	.58*		.72*	.42*	.40*	.65	.58	.64
39		.47*	.65		.52*	.48*	.54*	.46*	.77	.74
40		.66*	.33*		.80*	.55*	.72*	.80*	.54	.47
41		.58*	.60		.79*					.71
42		.73	.75*		.68					.62
43		.57	.39		.72					.30
44		.69	.52*		.58					.57
45		.50	.65*		.50					.74
46		.73	.50		.59*					.52
47		.70	.56*		.60*					.51
48		.54	.50*		.68*					.75
49		.70*								
50		.40								
51		.26*								
52		.54*								
53		.48*								
54		.47*								

* Item is included in the Abbreviated Battery.

246

Table G–64. Item p–Values for Advanced 1, Forms B/E, Fall, Grade 8

Item Number	Reading Vocabulary	Reading Comprehension	Mathematics Problem Solving	Mathematics Procedures	Language B	Spelling	Science	Social Science	Listening	Language E
1	.87*	.88	.68	.54	.76	.87*	.87*	.77*	.82*	.51
2	.68	.87	.53*	.57*	.67*	.70*	.72	.53*	.64*	.78
3	.58*	.86	.59*	.63*	.75	.53	.77*	.48*	.63*	.41
4	.70*	.83	.63*	.67	.72*	.63	.83*	.63*	.95*	.64
5	.79*	.67	.59	.61*	.72	.52*	.83*	.26	.55*	.58
6	.63*	.83	.45*	.50	.64*	.74	.58*	.55*	.75*	.64
7	.53*	.87	.66	.35*	.54*	.79*	.65*	.53*	.46*	.70
8	.75	.72	.57	.59*	.81	.67*	.44*	.57	.52*	.56
9	.53	.86	.58*	.76*	.73*	.67	.70*	.77*	.48*	.56
10	.69*	.78	.62*	.58	.60*	.43*	.83*	.53*	.72*	.73
11	.80	.76	.47	.38*	.73*	.38*	.46	.53	.68*	.74
12	.59*	.64	.74	.49	.75	.56*	.54*	.71*	.75*	.75
13	.88	.63*	.75*	.40*	.64	.55*	.73	.55*	.40*	.46
14	.78*	.64	.38*	.39*	.71	.60*	.78*	.50	.40*	.53
15	.77*	.68*	.61	.49	.59	.37	.59*	.31*	.75*	.71
16	.81*	.73*	.67*	.46*	.76*	.48*	.73*	.46*	.74*	.75
17	.91	.70*	.39*	.63	.72*	.53*	.59	.47*	.76*	.81*
18	.55*	.43*	.74*	.68*	.53*	.72*	.61*	.54*	.72*	.40*
19	.71*	.59	.66*	.52*	.54*	.80*	.46*	.51	.79*	.26*
20	.91	.61*	.49*	.60	.74*	.86*	.80*	.50*	.29*	.39*
21	.55*	.48*	.52	.57	.74*	.30*	.81*	.49	.85*	.38
22	.82*	.38*	.38*	.30	.81*	.72	.61	.27	.76*	.68*
23	.83*	.60*	.77	.60*	.26*	.53	.40	.66*	.72*	.79*
24	.78	.68*	.54*	.63	.63	.67*	.43*	.27*	.81	.64*
25	.56*	.85*	.77*	.43*	.76*	.45*	.66*	.54*	.75	.66*
26	.72	.70*	.40	.46*	.48*	.81	.73	.76*	.66	.65*
27	.78*	.41*	.70*	.52*	.66*	.30*	.53*	.49*	.56	.67*
28	.53*	.42*	.51	.32*	.60	.58*	.55	.58	.67	.84*
29	.53	.78*	.69*	.56*	.63*	.87	.31*	.51*	.49	.43*
30	.67*	.58	.53*	.28*	.62*	.90*	.40*	.67*	.69	.63*
31		.75*	.39*	.59	.59*	.73*	.71	.38	.61	.41*
32		.53*	.58	.45*	.66	.62*	.56*	.46*	.63	.48*
33		.80*	.46		.62	.74*	.57	.32*	.65	.74*
34		.78*	.43		.60	.68*	.41*	.60*	.68*	.66*
35		.68*	.55*		.69*	.40*	.37*	.45*	.67*	.60*
36		.43	.34*		.60*	.75*	.46*	.54	.39*	.58*
37		.64*	.68*		.66*	.42	.40*	.65*	.54*	.53*
38		.67*	.64*		.77	.82*	.27*	.38*	.62*	.44*
39		.49*	.53*		.69*	.55*	.37*	.45*	.80*	.59*
40		.70*	.64		.72	.63*	.58*	.64*	.64*	.63*
41		.74*	.57*		.56					.55*
42		.71	.66		.65*					.63*
43		.45*	.31*		.53*					.69*
44		.69*	.44*		.60					.49*
45		.65*	.45*		.56					.63
46		.56*	.41*		.68*					.72*
47		.64	.54		.62*					.61*
48		.73*	.41		.67*					.55*
49		.29								
50		.60								
51		.53								
52		.59								
53		.56								
54		.62								

* Item is included in the Abbreviated Battery.

Table G–65. Item _p_–Values for Advanced 2, Forms B/E, Fall, Grade 9

Item Number	Reading Vocabulary	Reading Comprehension	Mathematics Problem Solving	Mathematics Procedures	Language B	Spelling	Science	Social Science	Listening	Language E
1	.85	.52*	.56	.56	.85*	.91*	.72*	.69*	.66*	.60
2	.89*	.55*	.66	.59*	.82	.71*	.63*	.59*	.57*	.87
3	.76	.30*	.28*	.48*	.78	.80*	.74	.51*	.83*	.89
4	.77*	.85*	.63	.71	.54*	.36	.69*	.61*	.48*	.88
5	.75*	.67	.43*	.57*	.70*	.57*	.66*	.65*	.73*	.84
6	.67*	.90*	.44*	.45	.73	.59*	.62*	.46*	.60*	.42
7	.64*	.64*	.67	.43*	.68*	.86*	.78	.56*	.47*	.51
8	.58	.70	.39*	.73*	.68*	.77*	.60*	.77	.82*	.78
9	.55*	.75*	.66*	.71*	.74*	.75*	.53	.45*	.28*	.80
10	.76	.83*	.60*	.69*	.71*	.50	.59*	.28*	.77*	.81
11	.62*	.86*	.43*	.46	.79*	.87*	.48	.30*	.71*	.81
12	.42*	.43*	.31	.56*	.60*	.71	.50*	.50*	.73*	.73
13	.96	.74*	.70*	.49	.49*	.55*	.35*	.56	.50*	.73
14	.93	.75*	.66	.48*	.74*	.75*	.48	.41*	.58	.83
15	.67*	.47*	.51*	.64	.34*	.72*	.67*	.40*	.72	.68
16	.85	.52	.54	.65*	.32	.62*	.57	.33*	.71	.85
17	.88*	.60*	.45*	.60	.68	.64	.48*	.63*	.29	.79*
18	.69*	.61*	.55*	.47*	.71*	.87	.76*	.45	.46	.87*
19	.72*	.76	.73*	.71	.66*	.54*	.32*	.45*	.59	.80*
20	.85*	.50*	.56*	.24*	.74	.58*	.65*	.44*	.58	.58*
21	.64*	.76*	.59	.23*	.58	.89*	.32*	.36*	.36*	.63*
22	.83*	.72*	.70*	.73*	.45	.25*	.60*	.51*	.48*	.32*
23	.80*	.85*	.46*	.43*	.39	.57*	.72*	.45	.75*	.67*
24	.76	.48*	.37*	.63	.70*	.52*	.42*	.39*	.59*	.60*
25	.49*	.50*	.44	.43*	.75	.66	.34	.40*	.68*	.49*
26	.79	.74*	.51*	.45*	.59*	.55*	.29*	.51*	.40*	.71*
27	.89*	.67	.50	.47*	.64*	.62*	.70*	.52*	.63*	.50*
28	.61	.71*	.63	.43	.70	.80*	.65*	.38*	.72*	.57*
29	.62*	.73*	.51*	.50	.59*	.84*	.53*	.32*	.55*	.77*
30	.52*	.82*	.38	.32	.55	.88*	.38*	.45	.79*	.66*
31		.50	.64*	.49*	.67*	.83	.37*	.35*	.77*	.57*
32		.83	.62*	.54*	.59	.27*	.53*	.49	.56*	.77
33		.60	.46*		.72	.60*	.37*	.24*	.78	.68*
34		.68	.40		.25*	.74	.52	.38	.61	.53*
35		.39	.39		.71*	.35*	.38*	.46	.69	.75*
36		.53	.36*		.80*	.80*	.29	.44	.64*	.80*
37		.68	.61		.60*	.65	.33*	.48*	.53*	.56*
38		.79	.38*		.62*	.59*	.49*	.40	.64*	.63
39		.49	.50*		.68*	.74*	.51*	.30*	.44*	.44*
40		.77	.41		.65*	.39	.43	.60*	.54*	.56*
41		.56	.53*		.77*					.46*
42		.43	.46*		.63					.63*
43		.60	.39		.40					.71*
44		.68	.56*		.64					.78*
45		.53	.24*		.53					.51*
46		.60	.25		.62*					.75*
47		.64	.35*		.46*					.71*
48		.72	.37*		.57*					.74*
49		.61*								
50		.46*								
51		.48*								
52		.59								
53		.40*								
54		.46*								

* Item is included in the Abbreviated Battery.

Table G–66. Item *p*–Values for TASK 1, Forms B/E, Fall, Grade 9

Item Number	Reading Vocabulary	Reading Comprehension	Mathematics	Language B	Spelling	Science	Social Science	Language E
1	.78	.85*	.61*	.73*	.86	.75*	.58*	.77
2	.60*	.49*	.56	.84*	.71	.82*	.65	.88
3	.57	.77*	.33*	.73	.83*	.64*	.48*	.24
4	.78*	.82*	.83*	.73*	.55*	.66	.38	.38
5	.83*	.56*	.54*	.65*	.71	.54*	.68*	.59
6	.62*	.65*	.74*	.69*	.60*	.24*	.38*	.77
7	.49	.40*	.68	.86*	.67*	.60	.19*	.69
8	.54	.76*	.45	.50*	.79*	.58*	.50*	.77
9	.38*	.84*	.19	.83	.84	.66*	.42*	.39
10	.54*	.93*	.58*	.41*	.55*	.51*	.54*	.56
11	.47*	.53*	.46*	.69*	.41*	.77*	.61	.55
12	.42*	.79*	.68*	.55*	.78*	.47*	.40	.56
13	.96	.78*	.77	.68	.48*	.48*	.28*	.40
14	.89*	.89*	.68	.65*	.64*	.76*	.49	.79
15	.71	.80*	.36	.72*	.63*	.51*	.34	.88
16	.79*	.71*	.36	.56	.61*	.69*	.48*	.62
17	.77*	.69*	.59	.50	.34*	.55*	.49*	.52*
18	.67*	.84	.73*	.13*	.77*	.43	.28*	.53*
19	.87	.64*	.43*	.57*	.83*	.50*	.26	.50*
20	.86*	.72*	.76*	.77	.62*	.39	.65*	.65*
21	.85*	.72*	.37*	.62*	.53	.61	.55	.73*
22	.81*	.75*	.45*	.64	.59*	.69	.22*	.75*
23	.89	.62	.24	.30	.67	.79	.47*	.75*
24	.73	.37*	.35*	.72	.51*	.33*	.35*	.55*
25	.55*	.76*	.29*	.61*	.50*	.48*	.67*	.55*
26	.79*	.72*	.37*	.54	.59*	.40*	.40*	.75*
27	.83*	.60*	.30	.57*	.83*	.57*	.43	.46
28	.63*	.84	.42	.64*	.44	.43*	.38*	.81*
29	.35*	.76	.52*	.47*	.78*	.35*	.29*	.74*
30	.42	.75	.38	.58*	.71*	.54	.59*	.34*
31		.71	.37*	.52*	.57	.27	.29*	.78*
32		.47	.21*	.61*	.37*	.53*	.34*	.47*
33		.88	.42*	.55	.66	.56	.37*	.85*
34		.64	.12*	.55*	.42*	.36*	.54*	.79*
35		.51	.29	.67*	.55	.17*	.28*	.83*
36		.43	.31	.68*	.75*	.53*	.41*	.39*
37		.74	.41*	.66	.38*	.45*	.56	.67*
38		.87	.23	.63	.29*	.21*	.37*	.75*
39		.36	.56*	.65	.48*	.27*	.52*	.56*
40		.75	.43*	.65*	.21*	.39*	.38*	.48
41		.81	.32*	.64*				.52*
42		.58	.29	.56				.55*
43		.18	.30*	.52				.66*
44		.48	.35*	.59				.47*
45		.62	.37*	.51				.68*
46		.53*	.23	.46*				.68*
47		.74	.31	.58*				.77*
48		.50	.27	.43*				.63*
49		.38*	.43*					
50		.61	.31*					
51		.65*						
52		.52*						
53		.46						
54		.40*						

* Item is included in the Abbreviated Battery.

249

Appendix G

Table G–67. Item p–Values for TASK 2, Forms B/E, Fall, Grade 10

Item Number	Reading Vocabulary	Reading Comprehension	Mathematics	Language B	Spelling	Science	Social Science	Language E
1	.66	.80*	.46*	.94*	.68*	.82*	.46*	.53
2	.67*	.74*	.39*	.77*	.76*	.61*	.39*	.72
3	.56*	.43*	.37	.53*	.52*	.46*	.43*	.72
4	.67*	.30*	.42*	.46*	.54	.66	.23*	.34
5	.70*	.46*	.38*	.71	.52*	.54*	.46*	.65
6	.50*	.31*	.32*	.70	.51	.62*	.33*	.67
7	.31	.51*	.66	.71*	.57*	.52*	.25*	.41
8	.45	.40*	.35	.58	.64*	.55*	.26*	.47
9	.45*	.64*	.40	.64*	.67*	.43*	.53	.54
10	.51*	.58	.53	.59*	.57	.62*	.21	.85
11	.58	.62	.62	.49	.33*	.37*	.35*	.51
12	.31*	.62	.27*	.25	.67*	.58	.46*	.21
13	.96	.81	.60*	.71*	.52*	.54	.35*	.23
14	.73*	.54	.24*	.76	.53*	.57*	.37*	.74
15	.67*	.38	.56	.58*	.53*	.34*	.31*	.79
16	.71*	.48	.14*	.66*	.38*	.62*	.42	.55
17	.82	.44	.52*	.27*	.77	.36*	.24*	.85*
18	.49*	.66	.14*	.58	.55*	.54*	.40	.46*
19	.80	.49*	.35	.76*	.58	.42*	.43*	.75*
20	.73*	.60*	.26*	.76	.57	.38	.53*	.77*
21	.64*	.58*	.30*	.67*	.53*	.67*	.52*	.31*
22	.68*	.50*	.27	.37	.40	.40	.29*	.53*
23	.77*	.59	.40	.48*	.71*	.65*	.29*	.54*
24	.47*	.47*	.50*	.41*	.58*	.39*	.22*	.79*
25	.48*	.68*	.65*	.75*	.52*	.36*	.33*	.35*
26	.83	.49*	.47*	.61*	.51*	.59*	.52	.55
27	.72*	.15*	.36	.65*	.53	.48*	.35	.66*
28	.66*	.50	.46	.57*	.61*	.59*	.39*	.51*
29	.58	.39	.28*	.59	.89*	.62	.38*	.63*
30	.44	.68	.27	.72	.51*	.25*	.52	.76*
31		.42	.51*	.60*	.62*	.33	.29	.55*
32		.37	.42*	.68	.25*	.61*	.27*	.30*
33		.41	.29*	.61	.77*	.16*	.20*	.83*
34		.52	.32	.51*	.38*	.29	.39	.83*
35		.49	.33*	.65	.66	.18*	.33*	.33*
36		.58	.40	.65*	.18*	.48	.30	.80*
37		.61*	.46	.55*	.63*	.42*	.30*	.48*
38		.48*	.27*	.74*	.07*	.20*	.33*	.75*
39		.57*	.47	.56*	.43*	.46	.27*	.40*
40		.72*	.45*	.59	.48	.58*	.37*	.68*
41		.62	.22*	.68				.27*
42		.44*	.45*	.53				.56*
43		.74*	.31*	.53				.71
44		.30*	.21*	.56*				.58*
45		.52*	.31	.60*				.71*
46		.35*	.19*	.30*				.59*
47		.53	.22	.64*				.33*
48		.35*	.23	.57*				.55*
49		.59*	.29*					
50		.48*	.25*					
51		.29*						
52		.44						
53		.40						
54		.44						

* Item is included in the Abbreviated Battery.

Table G–68. Item *p*–Values for TASK 3, Forms B/E, Fall, Grade 11

Item Number	Reading Vocabulary	Reading Comprehension	Mathematics	Language B	Spelling	Science	Social Science	Language E
1	.67	.74*	.25	.56	.76*	.41*	.59*	.74
2	.54*	.75*	.41*	.84*	.62*	.75*	.34*	.61
3	.66*	.78*	.37*	.66	.63	.20*	.43*	.86
4	.66*	.57*	.36*	.46*	.32*	.63*	.35*	.24
5	.58*	.70*	.23*	.49	.61	.56*	.33*	.26
6	.46*	.41*	.60	.57	.48	.58*	.31*	.35
7	.28*	.76*	.36*	.71*	.68*	.50*	.27*	.18
8	.49*	.84*	.34	.44*	.64*	.48*	.27*	.24
9	.34*	.38*	.29	.80*	.64*	.46	.53*	.20
10	.66	.59*	.67	.51*	.46*	.52	.49	.66
11	.57	.69	.38	.75	.47*	.38	.39*	.16
12	.41	.72*	.26*	.43	.71*	.40	.17*	.84
13	.80	.45	.12*	.45	.40*	.31*	.36	.28
14	.75*	.64	.48*	.64*	.53*	.60*	.27	.24
15	.67*	.62*	.30	.51*	.75*	.55*	.31	.22
16	.70	.64*	.30	.36*	.74	.73	.35*	.25
17	.78*	.29	.26	.32*	.46*	.53*	.29	.15*
18	.57*	.67*	.46*	.55*	.71*	.24*	.33*	.13*
19	.74*	.72*	.20*	.60*	.71*	.35*	.38*	.22*
20	.74*	.45*	.29*	.74*	.66	.61	.44*	.27*
21	.72*	.65*	.42	.40*	.64	.49*	.56	.22*
22	.73*	.58*	.25*	.35	.51*	.61*	.49	.64
23	.87	.50*	.39*	.21*	.65*	.48	.39*	.18*
24	.58*	.61	.38*	.68	.38*	.41*	.26*	.22*
25	.50*	.36*	.37	.55*	.46*	.26	.24*	.18*
26	.76*	.57*	.51*	.57*	.64*	.57*	.35	.41*
27	.82	.49*	.29	.69	.22*	.32*	.27*	.24*
28	.55	.60	.27	.60	.69*	.53*	.49*	.19*
29	.70	.54	.20*	.46	.74	.23*	.26*	.13*
30	.30*	.49	.41*	.70*	.57*	.50	.27*	.08*
31		.66	.42*	.46*	.68*	.29*	.21	.29*
32		.55	.26*	.74	.15*	.47*	.25	.30*
33		.51	.42	.38*	.65*	.29*	.23*	.23*
34		.59	.17*	.55	.53*	.44*	.20*	.35*
35		.43	.40*	.64*	.46*	.15*	.43*	.47*
36		.39	.35	.57*	.33	.20*	.25*	.29*
37		.59*	.26*	.55*	.65	.49*	.34*	.39*
38		.49	.32	.42*	.47	.47*	.31*	.37*
39		.44*	.56*	.65*	.48*	.20*	.35*	.45*
40		.49*	.23*	.76	.39*	.31	.18*	.23*
41		.45*	.29	.52				.47*
42		.46*	.28*	.48*				.60*
43		.33*	.18*	.65*				.29*
44		.50*	.28	.68				.29*
45		.52*	.21	.63				.29*
46		.61	.26*	.74*				.41
47		.43	.22*	.65*				.25*
48		.38	.32*	.72*				.49*
49		.48	.26					
50		.41	.23*					
51		.33						
52		.55						
53		.34						
54		.33						

* Item is included in the Abbreviated Battery.

Table G–69. Item *p*–Values for TASK 3, Forms B/E, Fall, Grade 12

Item Number	Reading Vocabulary	Reading Comprehension	Mathematics	Language B	Spelling	Science	Social Science	Language E
1	.73	.82*	.35	.54	.85*	.37*	.54*	.49
2	.58*	.80*	.38*	.81*	.67*	.70*	.29*	.76
3	.69*	.82*	.47*	.74	.72	.19*	.45*	.78
4	.67*	.48*	.42*	.51*	.32*	.57*	.37*	.42
5	.73*	.76*	.29*	.44	.61	.49*	.33*	.72
6	.46*	.48*	.63	.54	.54	.57*	.32*	.64
7	.39*	.78*	.38*	.71*	.68*	.44*	.34*	.50
8	.51*	.83*	.37	.37*	.73*	.48*	.26*	.49
9	.37*	.41*	.38	.80*	.65*	.46	.48*	.44
10	.72	.66*	.69	.59*	.57*	.53	.51	.49
11	.68	.65	.46	.77	.57*	.36	.33*	.55
12	.47	.73*	.23*	.51	.74*	.43	.22*	.70
13	.87	.51	.15*	.52	.52*	.34*	.40	.52
14	.85*	.67	.50*	.64*	.58*	.68*	.23	.60
15	.70*	.68*	.23	.53*	.81*	.52*	.31	.33
16	.79	.63*	.31	.40*	.77	.71	.32*	.53
17	.87*	.38	.21	.34*	.60*	.55*	.31	.20*
18	.57*	.73*	.46*	.40*	.77*	.28*	.35*	.40*
19	.84*	.79*	.19*	.65*	.74*	.33*	.36*	.52*
20	.83*	.42*	.30*	.76*	.69	.65	.45*	.54*
21	.80*	.73*	.32	.46*	.73	.46*	.56	.56*
22	.79*	.67*	.25*	.32	.52*	.58*	.46	.55
23	.89	.51*	.39*	.22*	.70*	.46	.42*	.53*
24	.67*	.65	.42*	.67	.42*	.36*	.30*	.50*
25	.63*	.42*	.37	.57*	.50*	.29	.29*	.39*
26	.81*	.64*	.54*	.55*	.65*	.57*	.37	.60*
27	.83	.51*	.29	.68	.25*	.35*	.21*	.57*
28	.61	.66	.25	.58	.71*	.52*	.50*	.48*
29	.79	.57	.23*	.46	.77	.20*	.37*	.19*
30	.34*	.50	.42*	.63*	.67*	.51	.23*	.37*
31		.69	.41*	.49*	.69*	.25*	.23	.43*
32		.56	.33*	.69	.19*	.45*	.24	.54*
33		.52	.46	.41*	.68*	.30*	.29*	.47*
34		.68	.18*	.50	.59*	.44*	.24*	.47*
35		.54	.40*	.66*	.58*	.17*	.39*	.59*
36		.39	.34	.56*	.34	.20*	.34*	.56*
37		.66*	.27*	.54*	.74	.53*	.38*	.56*
38		.54	.32	.42*	.52	.40*	.29*	.44*
39		.48*	.58*	.69*	.48*	.21*	.36*	.46*
40		.58*	.22*	.68	.47*	.30	.21*	.51*
41		.48*	.29	.52				.45*
42		.50*	.24*	.49*				.60*
43		.30*	.21*	.63*				.59*
44		.50*	.29	.70				.56*
45		.54*	.25	.62				.49*
46		.59	.31*	.74*				.50
47		.40	.23*	.60*				.39*
48		.43	.30*	.71*				.38*
49		.52	.21					
50		.42	.25*					
51		.31						
52		.55						
53		.32						
54		.39						

* Item is included in the Abbreviated Battery.

Appendix H:

Mean *p*-Values for the Stanford 10 Standardization Samples

Appendix H

Table H–1. Mean *p*–Values for SESAT 2, Spring, Grade K, Full-Length

Subtest	Number of Items	Form A
Total Reading	100	.69
Sounds and Letters	40	.83
Word Reading	30	.60
Sentence Reading	30	.55
Mathematics	40	.73
Environment	40	.70
Listening to Words and Stories	40	.61

Table H–2. Mean *p*–Values for Primary 1, Spring, Grade 1, Full-Length

Subtest	Number of Items	Form A	Form B
Total Reading	130	.77	.76
Word Study Skills	30	.70	.71
Word Reading	30	.79	.78
Sentence Reading	30	.84	.84
Reading Comprehension	40	.73	.73
Total Mathematics	72	.65	.64
Mathematics Problem Solving	42	.69	.69
Mathematics Procedures	30	.59	.57
Language	40	.63	.62
Spelling	36	.68	.69
Environment	40	.75	.74
Listening	40	.64	.64
Subtest	**Number of Items**	**Form D**	**Form E**
Language	40	.68	.61

Table H–3. Mean *p*–Values for Primary 2, Spring, Grade 2, Full-Length

Subtest	Number of Items	Form A	Form B
Total Reading	100	.71	.71
Word Study Skills	30	.72	.72
Reading Vocabulary	30	.75	.75
Reading Comprehension	40	.65	.66
Total Mathematics	74	.69	.72
Mathematics Problem Solving	44	.68	.71
Mathematics Procedures	30	.71	.73
Language	48	.76	.74
Spelling	36	.79	.81
Environment	40	.74	.70
Listening	40	.64	.69
Subtest	**Number of Items**	**Form D**	**Form E**
Language	40	.70	.72

Table H–4. Mean *p*–Values for Primary 3, Spring, Grade 3, Full-Length

Subtest	Number of Items	Form A	Form B
Total Reading	114	.67	.66
Word Study Skills	30	.68	.67
Reading Vocabulary	30	.67	.71
Reading Comprehension	54	.65	.63
Total Mathematics	76	.64	.66
Mathematics Problem Solving	46	.66	.70
Mathematics Procedures	30	.61	.59
Language	48	.65	.64
Spelling	38	.56	.59
Science	40	.62	.63
Social Science	40	.59	.67
Listening	40	.68	.65
Thinking Skills (Basic)	135	.64	.64
Thinking Skills (Complete)	174	.64	.64
Subtest	**Number of Items**	**Form D**	**Form E**
Language	45	.54	.52
Prewriting	14	.57	.55
Composing	16	.49	.47
Editing	15	.56	.54
Thinking Skills (Basic)	149	.65	.61
Thinking Skills (Complete)	188	.65	.61

Table H–5. Mean *p*–Values for Intermediate 1, Spring, Grade 4, Full-Length

Subtest	Number of Items	Form A	Form B
Total Reading	114	.63	.65
Word Study Skills	30	.64	.64
Reading Vocabulary	30	.69	.71
Reading Comprehension	54	.59	.61
Total Mathematics	80	.59	.62
Mathematics Problem Solving	48	.58	.62
Mathematics Procedures	32	.62	.61
Language	48	.63	.61
Language Mechanics	24	.62	.63
Language Expression	24	.63	.60
Spelling	40	.59	.60
Science	40	.62	.60
Social Science	40	.58	.62
Listening	40	.61	.60
Thinking Skills (Basic)	150	.60	.59
Thinking Skills (Complete)	190	.60	.58
Subtest	**Number of Items**	**Form D**	**Form E**
Language	48	.55	.60
Prewriting	12	.59	.67
Composing	18	.52	.55
Editing	18	.53	.60
Thinking Skills (Basic)	162	.62	.68
Thinking Skills (Complete)	202	.62	.67

Appendix H

Table H–6. Mean *p*–Values for Intermediate 2, Spring, Grade 5, Full-Length

Subtest	Number of Items	Form A	Form B
Total Reading	84	.63	.65
Reading Vocabulary	30	.72	.70
Reading Comprehension	54	.58	.61
Total Mathematics	80	.58	.54
Mathematics Problem Solving	48	.58	.52
Mathematics Procedures	32	.57	.57
Language	48	.66	.61
Language Mechanics	24	.69	.63
Language Expression	24	.63	.58
Spelling	40	.63	.65
Science	40	.54	.56
Social Science	40	.58	.55
Listening	40	.62	.61
Thinking Skills (Basic)	151	.60	.58
Thinking Skills (Complete)	191	.60	.57
Subtest	**Number of Items**	**Form D**	**Form E**
Language	48	.48	.63
Prewriting	12	.57	.74
Composing	18	.44	.58
Editing	18	.46	.61
Thinking Skills (Basic)	163	.53	.64
Thinking Skills (Complete)	203	.53	.63

Table H–7. Mean *p*–Values for Intermediate 3, Spring, Grade 6, Full-Length

Subtest	Number of Items	Form A	Form B
Total Reading	84	.68	.65
Reading Vocabulary	30	.74	.72
Reading Comprehension	54	.65	.61
Total Mathematics	80	.54	.54
Mathematics Problem Solving	48	.53	.53
Mathematics Procedures	32	.55	.54
Language	48	.68	.68
Language Mechanics	24	.68	.69
Language Expression	24	.67	.67
Spelling	40	.66	.63
Science	40	.63	.58
Social Science	40	.57	.55
Listening	40	.64	.63
Thinking Skills (Basic)	151	.62	.62
Thinking Skills (Complete)	192	.61	.62
Subtest	Number of Items	Form D	Form E
Language	48	.61	.52
Prewriting	12	.64	.58
Composing	18	.57	.48
Editing	18	.63	.53
Thinking Skills (Basic)	163	.63	.53
Thinking Skills (Complete)	204	.63	.52

Appendix H

Table H–8. Mean *p*–Values for Advanced 1, Spring, Grade 7, Full-Length

Subtest	Number of Items	Form A	Form B
Total Reading	84	.62	.64
Reading Vocabulary	30	.66	.68
Reading Comprehension	54	.59	.62
Total Mathematics	80	.48	.51
Mathematics Problem Solving	48	.47	.52
Mathematics Procedures	32	.50	.51
Language	48	.65	.62
Language Mechanics	24	.67	.65
Language Expression	24	.63	.60
Spelling	40	.61	.59
Science	40	.57	.57
Social Science	40	.50	.49
Listening	40	.63	.63
Thinking Skills (Basic)	152	.58	.59
Thinking Skills (Complete)	193	.57	.58

Subtest	Number of Items	Form D	Form E
Language	48	.50	.56
Prewriting	12	.53	.54
Composing	18	.47	.57
Editing	18	.50	.56
Thinking Skills (Basic)	164	.55	.62
Thinking Skills (Complete)	205	.54	.61

Table H–9. Mean *p*–Values for Advanced 2, Spring, Grade 8, Full-Length

Subtest	Number of Items	Form A	Form B
Total Reading	84	.65	.63
Reading Vocabulary	30	.70	.69
Reading Comprehension	54	.62	.60
Total Mathematics	80	.46	.47
Mathematics Problem Solving	48	.46	.46
Mathematics Procedures	32	.46	.48
Language	48	.64	.66
Language Mechanics	24	.66	.68
Language Expression	24	.62	.63
Spelling	40	.63	.63
Science	40	.52	.51
Social Science	40	.46	.45
Listening	40	.60	.57
Thinking Skills (Basic)	155	.55	.58
Thinking Skills (Complete)	200	.54	.57
Subtest	**Number of Items**	**Form D**	**Form E**
Language	48	.71	.58
Prewriting	12	.81	.66
Composing	18	.67	.51
Editing	18	.68	.58
Thinking Skills (Basic)	167	.61	.51
Thinking Skills (Complete)	212	.60	.50

Table H–10. Mean *p*–Values for Advanced 2, Spring, Grade 9, Full-Length

Subtest	Number of Items	Form A	Form B
Total Reading	84	.70	.68
Reading Vocabulary	30	.73	.73
Reading Comprehension	54	.69	.64
Total Mathematics	80	.55	.50
Mathematics Problem Solving	48	.54	.49
Mathematics Procedures	32	.54	.50
Language	48	.70	.66
Language Mechanics	24	.70	.67
Language Expression	24	.70	.65
Spelling	40	.67	.64
Science	40	.56	.54
Social Science	40	.51	.48
Listening	40	.65	.61
Thinking Skills (Basic)	155	.63	.59
Thinking Skills (Complete)	200	.61	.58

Subtest	Number of Items	Form D	Form E
Language	48	.64	.62
Prewriting	12	.71	.72
Composing	18	.61	.55
Editing	18	.62	.61
Thinking Skills (Basic)	167	.60	.58
Thinking Skills (Complete)	212	.59	.56

Table H–11. Mean *p*–Values for TASK 1, Spring, Grade 9, Full Length

Subtest	Number of Items	Form A	Form B
Total Reading	84	.65	.68
Reading Vocabulary	30	.72	.69
Reading Comprehension	54	.60	.67
Mathematics	50	.44	.46
Language	48	.63	.65
Language Mechanics	24	.63	.67
Language Expression	24	.63	.64
Spelling	40	.62	.60
Science	40	.52	.51
Social Science	40	.46	.45
Thinking Skills (Basic)	118	.54	.59
Thinking Skills (Complete)	165	.52	.56
Subtest	**Number of Items**	**Form D**	**Form E**
Language	48	.66	.56
Prewriting	12	.66	.61
Composing	18	.69	.56
Editing	18	.63	.54
Thinking Skills (Basic)	130	.59	.59
Thinking Skills (Complete)	177	.57	.56

Table H–12. Mean *p*–Values for TASK 2, Spring, Grade 10, Full-Length

Subtest	Number of Items	Form A	Form B
Total Reading	84	.69	.65
Reading Vocabulary	30	.75	.73
Reading Comprehension	54	.66	.61
Mathematics	50	.45	.41
Language	48	.63	.65
Language Mechanics	24	.60	.64
Language Expression	24	.67	.66
Spelling	40	.64	.64
Science	40	.55	.50
Social Science	40	.43	.40
Thinking Skills (Basic)	118	.60	.56
Thinking Skills (Complete)	166	.56	.54

Subtest	Number of Items	Form D	Form E
Language	48	.66	.58
Prewriting	12	.72	.66
Composing	18	.63	.53
Editing	18	.64	.56
Thinking Skills (Basic)	130	.60	.54
Thinking Skills (Complete)	178	.56	.52

Table H–13. Mean *p*–Values for TASK 3, Spring, Grade 11, Full-Length

Subtest	Number of Items	Form A	Form B
Total Reading	84	.63	.64
Reading Vocabulary	30	.71	.70
Reading Comprehension	54	.59	.60
Mathematics	50	.41	.36
Language	48	.57	.54
Language Mechanics	24	.56	.53
Language Expression	24	.58	.55
Spelling	40	.60	.61
Science	40	.45	.45
Social Science	40	.38	.35
Thinking Skills (Basic)	120	.55	.49
Thinking Skills (Complete)	169	.51	.46

Subtest	Number of Items	Form D	Form E
Language	48	.58	.63
Prewriting	12	.64	.64
Composing	18	.56	.63
Editing	18	.56	.61
Thinking Skills (Basic)	132	.53	.59
Thinking Skills (Complete)	181	.49	.55

Appendix H

Subtest	Number of Items	Form A	Form B
Total Reading	84	.62	.59
Reading Vocabulary	30	.73	.65
Reading Comprehension	54	.56	.55
Mathematics	50	.39	.34
Language	48	.55	.52
Language Mechanics	24	.55	.50
Language Expression	24	.55	.54
Spelling	40	.62	.58
Science	40	.44	.41
Social Science	40	.37	.34
Thinking Skills (Basic)	120	.52	.48
Thinking Skills (Complete)	169	.48	.44
Subtest	**Number of Items**	**Form D**	**Form E**
Language	48	.62	.52
Prewriting	12	.68	.55
Composing	18	.61	.51
Editing	18	.59	.51
Thinking Skills (Basic)	132	.56	.52
Thinking Skills (Complete)	181	.51	.49

Table H–15. Mean *p*–Values for SESAT 1, Fall, Grade K, Full-Length

Subtest	Number of Items	Form A
Total Reading	70	.65
Sounds and Letters	40	.77
Word Reading	30	.48
Mathematics	40	.67
Environment	40	.69
Listening to Words and Stories	40	.70

Table H–16. Mean *p*–Values for SESAT 2, Fall, Grade 1, Full-Length

Subtest	Number of Items	Form A
Total Reading	100	.80
Sounds and Letters	40	.90
Word Reading	30	.72
Sentence Reading	30	.73
Mathematics	40	.82
Environment	40	.76
Listening to Words and Stories	40	.69

Table H–17. Mean *p*–Values for Primary 1, Fall, Grade 2, Full-Length

Subtest	Number of Items	Form A	Form B
Total Reading	130	.83	.83
Word Study Skills	30	.76	.77
Word Reading	30	.87	.86
Sentence Reading	30	.88	.90
Reading Comprehension	40	.80	.81
Total Mathematics	72	.73	.70
Mathematics Problem Solving	42	.76	.77
Mathematics Procedures	30	.68	.60
Language	40	.70	.71
Spelling	36	.78	.78
Environment	40	.78	.78
Listening	40	.72	.74
Subtest	**Number of Items**	**Form D**	**Form E**
Language	40	.72	.73

Table H–18. Mean *p*–Values for Primary 2, Fall, Grade 3, Full-Length

Subtest	Number of Items	Form A	Form B
Total Reading	100	.78	.77
Word Study Skills	30	.78	.77
Reading Vocabulary	30	.83	.81
Reading Comprehension	40	.73	.73
Total Mathematics	74	.75	.77
Mathematics Problem Solving	44	.74	.77
Mathematics Procedures	30	.76	.78
Language	48	.81	.81
Spelling	36	.84	.86
Environment	40	.79	.74
Listening	40	.72	.72
Subtest	**Number of Items**	**Form D**	**Form E**
Language	40	.76	.77

Table H–19. Mean *p*–Values for Primary 3, Fall, Grade 4, Full-Length

Subtest	Number of Items	Form A	Form B
Total Reading	114	.68	.70
Word Study Skills	30	.68	.71
Reading Vocabulary	30	.70	.75
Reading Comprehension	54	.67	.66
Total Mathematics	76	.64	.68
Mathematics Problem Solving	46	.67	.72
Mathematics Procedures	30	.60	.61
Language	48	.68	.64
Spelling	38	.59	.61
Science	40	.66	.64
Social Science	40	.64	.69
Listening	40	.68	.67
Thinking Skills (Basic)	135	.67	.66
Thinking Skills (Complete)	174	.67	.65
Subtest	**Number of Items**	**Form D**	**Form E**
Language	45	.50	.59
Prewriting	14	.54	.63
Composing	16	.45	.52
Editing	15	.51	.61
Thinking Skills (Basic)	149	.59	.68
Thinking Skills (Complete)	188	.60	.68

Appendix H

Table H–20. Mean *p*–Values for Intermediate 1, Fall, Grade 5, Full-Length

Subtest	Number of Items	Form A	Form B
Total Reading	114	.67	.70
Word Study Skills	30	.67	.67
Reading Vocabulary	30	.73	.76
Reading Comprehension	54	.63	.66
Total Mathematics	80	.63	.66
Mathematics Problem Solving	48	.62	.66
Mathematics Procedures	32	.65	.65
Language	48	.66	.69
Language Mechanics	24	.64	.69
Language Expression	24	.68	.70
Spelling	40	.62	.64
Science	40	.66	.65
Social Science	40	.62	.67
Listening	40	.64	.66
Thinking Skills (Basic)	150	.65	.68
Thinking Skills (Complete)	190	.65	.67

Subtest	Number of Items	Form D	Form E
Language	48	.57	.62
Prewriting	12	.62	.68
Composing	18	.56	.58
Editing	18	.54	.61
Thinking Skills (Basic)	162	.64	.66
Thinking Skills (Complete)	202	.64	.66

Table H–21. Mean *p*–Values for Intermediate 2, Fall, Grade 6, Full-Length

Subtest	Number of Items	Form A	Form B
Total Reading	84	.68	.69
Reading Vocabulary	30	.75	.74
Reading Comprehension	54	.64	.65
Total Mathematics	80	.61	.56
Mathematics Problem Solving	48	.62	.56
Mathematics Procedures	32	.58	.56
Language	48	.68	.63
Language Mechanics	24	.70	.66
Language Expression	24	.66	.60
Spelling	40	.66	.68
Science	40	.60	.62
Social Science	40	.64	.60
Listening	40	.64	.64
Thinking Skills (Basic)	151	.63	.61
Thinking Skills (Complete)	191	.63	.60
Subtest	Number of Items	Form D	Form E
Language	48	.58	.63
Prewriting	12	.66	.72
Composing	18	.55	.59
Editing	18	.56	.61
Thinking Skills (Basic)	163	.63	.64
Thinking Skills (Complete)	203	.64	.63

Table H–22. Mean *p*–Values for Intermediate 3, Fall, Grade 7, Full-Length

Subtest	Number of Items	Form A	Form B
Total Reading	84	.66	.66
Reading Vocabulary	30	.71	.73
Reading Comprehension	54	.63	.62
Total Mathematics	80	.52	.54
Mathematics Problem Solving	48	.52	.54
Mathematics Procedures	32	.52	.54
Language	48	.63	.66
Language Mechanics	24	.63	.66
Language Expression	24	.63	.65
Spelling	40	.63	.64
Science	40	.63	.61
Social Science	40	.55	.56
Listening	40	.62	.64
Thinking Skills (Basic)	151	.57	.60
Thinking Skills (Complete)	192	.57	.60
Subtest	**Number of Items**	**Form D**	**Form E**
Language	48	.62	.60
Prewriting	12	.66	.66
Composing	18	.57	.57
Editing	18	.63	.58
Thinking Skills (Basic)	163	.65	.61
Thinking Skills (Complete)	204	.64	.61

Table H–23. Mean *p*–Values for Advanced 1, Fall, Grade 8, Full-Length

Subtest	Number of Items	Form A	Form B
Total Reading	84	.65	.68
Reading Vocabulary	30	.70	.71
Reading Comprehension	54	.62	.65
Total Mathematics	80	.51	.54
Mathematics Problem Solving	48	.51	.56
Mathematics Procedures	32	.51	.52
Language	48	.64	.65
Language Mechanics	24	.65	.67
Language Expression	24	.63	.63
Spelling	40	.63	.62
Science	40	.60	.59
Social Science	40	.53	.52
Listening	40	.65	.65
Thinking Skills (Basic)	152	.57	.62
Thinking Skills (Complete)	193	.57	.60
Subtest	**Number of Items**	**Form D**	**Form E**
Language	48	.64	.60
Prewriting	12	.68	.56
Composing	18	.62	.61
Editing	18	.62	.61
Thinking Skills (Basic)	164	.65	.66
Thinking Skills (Complete)	205	.65	.65

Appendix H

Table H–24. Mean *p*–Values for Advanced 2, Fall, Grade 9, Full-Length

Subtest	Number of Items	Form A	Form B
Total Reading	84	.65	.67
Reading Vocabulary	30	.70	.73
Reading Comprehension	54	.62	.63
Total Mathematics	80	.47	.52
Mathematics Problem Solving	48	.48	.50
Mathematics Procedures	32	.46	.53
Language	48	.63	.63
Language Mechanics	24	.65	.64
Language Expression	24	.61	.61
Spelling	40	.64	.66
Science	40	.51	.52
Social Science	40	.47	.46
Listening	40	.62	.61
Thinking Skills (Basic)	155	.55	.59
Thinking Skills (Complete)	200	.53	.57

Subtest	Number of Items	Form D	Form E
Language	48	.67	.68
Prewriting	12	.74	.75
Composing	18	.64	.60
Editing	18	.66	.70
Thinking Skills (Basic)	167	.62	.64
Thinking Skills (Complete)	212	.61	.62

Table H–25. Mean *p*–Values for TASK 1, Fall, Grade 9, Full-Length

Subtest	Number of Items	Form A	Form B
Total Reading	84	.63	.67
Reading Vocabulary	30	.70	.68
Reading Comprehension	54	.59	.65
Mathematics	50	.41	.43
Language	48	.60	.60
Language Mechanics	24	.60	.63
Language Expression	24	.60	.58
Spelling	40	.61	.60
Science	40	.51	.51
Social Science	40	.45	.44
Thinking Skills (Basic)	118	.52	.55
Thinking Skills (Complete)	165	.50	.52
Subtest	**Number of Items**	**Form D**	**Form E**
Language	48	.67	.63
Prewriting	12	.67	.61
Composing	18	.69	.63
Editing	18	.64	.63
Thinking Skills (Basic)	130	.61	.62
Thinking Skills (Complete)	177	.58	.59

Table H–26. Mean *p*–Values for TASK 2, Fall, Grade 10, Full Length

Subtest	Number of Items	Form A	Form B
Total Reading	84	.59	.55
Reading Vocabulary	30	.62	.62
Reading Comprehension	54	.57	.51
Mathematics	50	.38	.37
Language	48	.54	.60
Language Mechanics	24	.52	.60
Language Expression	24	.57	.60
Spelling	40	.53	.54
Science	40	.49	.48
Social Science	40	.36	.36
Thinking Skills (Basic)	118	.51	.45
Thinking Skills (Complete)	166	.49	.45

Subtest	Number of Items	Form D	Form E
Language	48	.59	.58
Prewriting	12	.62	.66
Composing	18	.57	.54
Editing	18	.59	.55
Thinking Skills (Basic)	130	.51	.55
Thinking Skills (Complete)	178	.49	.52

Table H–27. Mean *p*–Values for TASK 3, Fall, Grade 11, Full-Length

Subtest	Number of Items	Form A	Form B
Total Reading	84	.56	.57
Reading Vocabulary	30	.61	.62
Reading Comprehension	54	.53	.54
Mathematics	50	.35	.33
Language	48	.49	.57
Language Mechanics	24	.48	.54
Language Expression	24	.50	.60
Spelling	40	.54	.56
Science	40	.40	.44
Social Science	40	.32	.34
Thinking Skills (Basic)	120	.52	.50
Thinking Skills (Complete)	169	.47	.47
Subtest	**Number of Items**	**Form D**	**Form E**
Language	48	.54	.33
Prewriting	12	.58	.38
Composing	18	.51	.35
Editing	18	.51	.29
Thinking Skills (Basic)	132	.44	.41
Thinking Skills (Complete)	181	.41	.39

Appendix H

Table H–28. Mean *p*–Values for TASK 3, Fall, Grade 12, Full-Length

Subtest	Number of Items	Form A	Form B
Total Reading	84	.61	.61
Reading Vocabulary	30	.71	.68
Reading Comprehension	54	.56	.57
Mathematics	50	.39	.34
Language	48	.54	.57
Language Mechanics	24	.54	.55
Language Expression	24	.54	.59
Spelling	40	.60	.61
Science	40	.43	.43
Social Science	40	.34	.35
Thinking Skills (Basic)	120	.51	.52
Thinking Skills (Complete)	169	.47	.49

Subtest	Number of Items	Form D	Form E
Language	48	.62	.51
Prewriting	12	.66	.52
Composing	18	.60	.52
Editing	18	.60	.48
Thinking Skills (Basic)	132	.59	.51
Thinking Skills (Complete)	181	.54	.47

Appendix I:

Median Biserial Correlation Coefficients for the Stanford 10 Standardization Samples

Appendix I

Table I–1. Median Biserial Correlation Coefficients for SESAT 2, Spring, Grade K

Subtest	Number of Items	Form A
Sounds and Letters	40	.64
Word Reading	30	.56
Sentence Reading	30	.62
Mathematics	40	.64
Environment	40	.48
Listening to Words and Stories	40	.50

Table I–2. Median Biserial Correlation Coefficients for Primary 1, Spring, Grade 1

Subtest	Number of Items	Form A	Form B
Word Study Skills	30	.63	.59
Word Reading	30	.85	.79
Sentence Reading	30	.82	.84
Reading Comprehension	40	.73	.69
Mathematics Problem Solving	42	.58	.57
Mathematics Procedures	30	.57	.59
Language	40	.53	.56
Spelling	36	.68	.67
Environment	40	.49	.49
Listening	40	.56	.53
Subtest	**Number of Items**	**Form D**	**Form E**
Language	40	.49	.52

Table I–3. Median Biserial Correlation Coefficients for Primary 2, Spring, Grade 2

Subtest	Number of Items	Form A	Form B
Word Study Skills	30	.62	.63
Reading Vocabulary	30	.76	.77
Reading Comprehension	40	.67	.65
Mathematics Problem Solving	44	.57	.61
Mathematics Procedures	30	.62	.65
Language	48	.63	.58
Spelling	36	.67	.73
Environment	40	.45	.45
Listening	40	.51	.52
Subtest	**Number of Items**	**Form D**	**Form E**
Language	40	.48	.53

Table I–4. Median Biserial Correlation Coefficients for Primary 3, Spring, Grade 3

Subtest	Number of Items	Form A	Form B
Word Study Skills	30	.60	.63
Reading Vocabulary	30	.64	.65
Reading Comprehension	54	.62	.60
Mathematics Problem Solving	46	.62	.61
Mathematics Procedures	30	.66	.61
Language	48	.62	.57
Spelling	38	.56	.57
Science	40	.50	.53
Social Science	40	.57	.62
Listening	40	.52	.54
Subtest	**Number of Items**	**Form D**	**Form E**
Language	45	.53	.48

Appendix I

Table I–5. Median Biserial Correlation Coefficients for Intermediate 1, Spring, Grade 4

Subtest	Number of Items	Form A	Form B
Word Study Skills	30	.64	.64
Reading Vocabulary	30	.67	.74
Reading Comprehension	54	.58	.57
Mathematics Problem Solving	48	.57	.61
Mathematics Procedures	32	.65	.67
Language	48	.58	.59
Spelling	40	.55	.58
Science	40	.54	.57
Social Science	40	.54	.60
Listening	40	.47	.53
Subtest	**Number of Items**	**Form D**	**Form E**
Language	48	.51	.54

Table I–6. Median Biserial Correlation Coefficients for Intermediate 2, Spring, Grade 5

Subtest	Number of Items	Form A	Form B
Reading Vocabulary	30	.60	.63
Reading Comprehension	54	.55	.58
Mathematics Problem Solving	48	.56	.53
Mathematics Procedures	32	.60	.62
Language	48	.57	.55
Spelling	40	.58	.55
Science	40	.45	.51
Social Science	40	.56	.54
Listening	40	.49	.49
Subtest	**Number of Items**	**Form D**	**Form E**
Language	48	.53	.57

Table I–7. Median Biserial Correlation Coefficients for Intermediate 3, Spring, Grade 6

Subtest	Number of Items	Form A	Form B
Reading Vocabulary	30	.60	.63
Reading Comprehension	54	.61	.56
Mathematics Problem Solving	48	.56	.53
Mathematics Procedures	32	.62	.63
Language	48	.56	.58
Spelling	40	.55	.55
Science	40	.47	.47
Social Science	40	.52	.53
Listening	40	.47	.54
Subtest	Number of Items	Form D	Form E
Language	48	.50	.55

Table I–8. Median Biserial Correlation Coefficients for Advanced 1, Spring, Grade 7

Subtest	Number of Items	Form A	Form B
Reading Vocabulary	30	.67	.67
Reading Comprehension	54	.59	.58
Mathematics Problem Solving	48	.54	.58
Mathematics Procedures	32	.68	.63
Language	48	.62	.62
Spelling	40	.55	.53
Science	40	.54	.53
Social Science	40	.46	.48
Listening	40	.53	.52
Subtest	Number of Items	Form D	Form E
Language	48	.54	.55

Table I–9. Median Biserial Correlation Coefficients for Advanced 2, Spring, Grade 8

Subtest	Number of Items	Form A	Form B
Reading Vocabulary	30	.62	.62
Reading Comprehension	54	.60	.55
Mathematics Problem Solving	48	.54	.57
Mathematics Procedures	32	.62	.64
Language	48	.60	.65
Spelling	40	.53	.58
Science	40	.47	.48
Social Science	40	.52	.50
Listening	40	.49	.50
Subtest	**Number of Items**	**Form D**	**Form E**
Language	48	.59	.59

Table I–10. Median Biserial Correlation Coefficients for Advanced 2, Spring, Grade 9

Subtest	Number of Items	Form A	Form B
Reading Vocabulary	30	.69	.69
Reading Comprehension	54	.59	.59
Mathematics Problem Solving	48	.61	.57
Mathematics Procedures	32	.61	.65
Language	48	.65	.66
Spelling	40	.57	.56
Science	40	.52	.52
Social Science	40	.56	.53
Listening	40	.48	.51
Subtest	**Number of Items**	**Form D**	**Form E**
Language	48	.65	.59

Table I–11. Median Biserial Correlation Coefficients for TASK 1, Spring, Grade 9

Subtest	Number of Items	Form A	Form B
Reading Vocabulary	30	.69	.66
Reading Comprehension	54	.53	.64
Mathematics	50	.49	.47
Language	48	.67	.70
Spelling	40	.57	.58
Science	40	.48	.52
Social Science	40	.44	.49
Subtest	Number of Items	Form D	Form E
Language	48	.57	.58

Table I–12. Median Biserial Correlation Coefficients for TASK 2, Spring, Grade 10

Subtest	Number of Items	Form A	Form B
Reading Vocabulary	30	.66	.62
Reading Comprehension	54	.60	.57
Mathematics	50	.46	.45
Language	48	.58	.61
Spelling	40	.60	.58
Science	40	.52	.48
Social Science	40	.44	.44
Subtest	Number of Items	Form D	Form E
Language	48	.57	.55

Appendix I

Table I–13. Median Biserial Correlation Coefficients for TASK 3, Spring, Grade 11

Subtest	Number of Items	Form A	Form B
Reading Vocabulary	30	.77	.65
Reading Comprehension	54	.64	.69
Mathematics	50	.48	.45
Language	48	.65	.60
Spelling	40	.64	.62
Science	40	.52	.47
Social Science	40	.44	.39
Subtest	**Number of Items**	**Form D**	**Form E**
Language	48	.66	.59

Table I–14. Median Biserial Correlation Coefficients for TASK 3, Spring, Grade 12

Subtest	Number of Items	Form A	Form B
Reading Vocabulary	30	.72	.71
Reading Comprehension	54	.67	.70
Mathematics	50	.46	.46
Language	48	.66	.64
Spelling	40	.67	.70
Science	40	.47	.50
Social Science	40	.45	.45
Subtest	**Number of Items**	**Form D**	**Form E**
Language	48	.60	.65

Table I–15. Median Biserial Correlation Coefficients for SESAT 1, Fall, Grade K

Subtest	Number of Items	Form A
Sounds and Letters	40	.59
Word Reading	30	.52
Mathematics	40	.60
Environment	40	.52
Listening to Words and Stories	40	.52

Table I–16. Median Biserial Correlation Coefficients for SESAT 2, Fall, Grade 1

Subtest	Number of Items	Form A
Sounds and Letters	40	.67
Word Reading	30	.66
Sentence Reading	30	.78
Mathematics	40	.66
Environment	40	.47
Listening to Words and Stories	40	.49

Table I–17. Median Biserial Correlation Coefficients for Primary 1, Fall, Grade 2

Subtest	Number of Items	Form A	Form B
Word Study Skills	30	.69	.68
Word Reading	30	.94	.88
Sentence Reading	30	.86	1.00
Reading Comprehension	40	.78	.79
Mathematics Problem Solving	42	.59	.65
Mathematics Procedures	30	.60	.62
Language	40	.54	.61
Spelling	36	.78	.72
Environment	40	.48	.53
Listening	40	.61	.58
Subtest	**Number of Items**	**Form D**	**Form E**
Language	40	.54	.52

Appendix I

Table I–18. Median Biserial Correlation Coefficients for Primary 2, Fall, Grade 3

Subtest	Number of Items	Form A	Form B
Word Study Skills	30	.62	.67
Reading Vocabulary	30	.81	.84
Reading Comprehension	40	.70	.67
Mathematics Problem Solving	44	.61	.60
Mathematics Procedures	30	.72	.73
Language	48	.67	.68
Spelling	36	.73	.84
Environment	40	.44	.44
Listening	40	.51	.52
Subtest	**Number of Items**	**Form D**	**Form E**
Language	40	.49	.55

Table I–19. Median Biserial Correlation Coefficients for Primary 3, Fall, Grade 4

Subtest	Number of Items	Form A	Form B
Word Study Skills	30	.65	.64
Reading Vocabulary	30	.73	.72
Reading Comprehension	54	.69	.64
Mathematics Problem Solving	46	.65	.68
Mathematics Procedures	30	.67	.63
Language	48	.64	.62
Spelling	38	.58	.60
Science	40	.61	.56
Social Science	40	.61	.66
Listening	40	.57	.53
Subtest	**Number of Items**	**Form D**	**Form E**
Language	45	.56	.53

Table I–20. Median Biserial Correlation Coefficients for Intermediate 1, Fall, Grade 5

Subtest	Number of Items	Form A	Form B
Word Study Skills	30	.66	.67
Reading Vocabulary	30	.71	.77
Reading Comprehension	54	.62	.61
Mathematics Problem Solving	48	.60	.65
Mathematics Procedures	32	.69	.69
Language	48	.62	.63
Spelling	40	.60	.61
Science	40	.63	.59
Social Science	40	.59	.64
Listening	40	.53	.54
Subtest	Number of Items	Form D	Form E
Language	48	.55	.57

Table I–21. Median Biserial Correlation Coefficients for Intermediate 2, Fall, Grade 6

Subtest	Number of Items	Form A	Form B
Reading Vocabulary	30	.62	.66
Reading Comprehension	54	.57	.62
Mathematics Problem Solving	48	.58	.59
Mathematics Procedures	32	.62	.64
Language	48	.62	.58
Spelling	40	.62	.62
Science	40	.49	.55
Social Science	40	.61	.61
Listening	40	.53	.53
Subtest	Number of Items	Form D	Form E
Language	48	.55	.61

Appendix I

Table I–22. Median Biserial Correlation Coefficients for Intermediate 3, Fall, Grade 7

Subtest	Number of Items	Form A	Form B
Reading Vocabulary	30	.65	.71
Reading Comprehension	54	.61	.61
Mathematics Problem Solving	48	.55	.60
Mathematics Procedures	32	.60	.65
Language	48	.60	.65
Spelling	40	.61	.62
Science	40	.52	.51
Social Science	40	.53	.57
Listening	40	.50	.58
Subtest	**Number of Items**	**Form D**	**Form E**
Language	48	.51	.64

Table I–23. Median Biserial Correlation Coefficients for Advanced 1, Fall, Grade 8

Subtest	Number of Items	Form A	Form B
Reading Vocabulary	30	.70	.72
Reading Comprehension	54	.61	.64
Mathematics Problem Solving	48	.58	.62
Mathematics Procedures	32	.66	.64
Language	48	.63	.64
Spelling	40	.56	.56
Science	40	.60	.53
Social Science	40	.50	.52
Listening	40	.53	.55
Subtest	**Number of Items**	**Form D**	**Form E**
Language	48	.58	.59

Table I–24. Median Biserial Correlation Coefficients for Advanced 2, Fall, Grade 9

Subtest	Number of Items	Form A	Form B
Reading Vocabulary	30	.66	.62
Reading Comprehension	54	.59	.59
Mathematics Problem Solving	48	.58	.58
Mathematics Procedures	32	.59	.72
Language	48	.61	.59
Spelling	40	.55	.59
Science	40	.50	.53
Social Science	40	.55	.52
Listening	40	.53	.51
Subtest	Number of Items	Form D	Form E
Language	48	.62	.61

Table I–25. Median Biserial Correlation Coefficients for TASK 1, Fall, Grade 9

Subtest	Number of Items	Form A	Form B
Reading Vocabulary	30	.67	.59
Reading Comprehension	54	.54	.63
Mathematics	50	.48	.48
Language	48	.60	.60
Spelling	40	.54	.53
Science	40	.49	.47
Social Science	40	.48	.47
Subtest	Number of Items	Form D	Form E
Language	48	.61	.59

Appendix I

Table I–26. Median Biserial Correlation Coefficients for TASK 2, Fall, Grade 10

Subtest	Number of Items	Form A	Form B
Reading Vocabulary	30	.69	.71
Reading Comprehension	54	.63	.65
Mathematics	50	.45	.43
Language	48	.61	.64
Spelling	40	.64	.72
Science	40	.48	.50
Social Science	40	.41	.43
Subtest	**Number of Items**	**Form D**	**Form E**
Language	48	.51	.61

Table I–27. Median Biserial Correlation Coefficients for TASK 3, Fall, Grade 11

Subtest	Number of Items	Form A	Form B
Reading Vocabulary	30	.74	.69
Reading Comprehension	54	.65	.66
Mathematics	50	.43	.38
Language	48	.64	.52
Spelling	40	.59	.62
Science	40	.44	.47
Social Science	40	.40	.37
Subtest	**Number of Items**	**Form D**	**Form E**
Language	48	.56	.72

Table I–28. Median Biserial Correlation Coefficients for TASK 3, Fall, Grade 12

Subtest	Number of Items	Form A	Form B
Reading Vocabulary	30	.71	.59
Reading Comprehension	54	.63	.64
Mathematics	50	.45	.46
Language	48	.65	.58
Spelling	40	.62	.60
Science	40	.50	.50
Social Science	40	.42	.37
Subtest	**Number of Items**	**Form D**	**Form E**
Language	48	.64	.63

Appendix J:

Median Point Biserial Correlation Coefficients for the Stanford 10 Standardization Samples

Appendix J

Table J–1. Median Point Biserial Correlation Coefficients for SESAT 2, Spring, Grade K

Subtest	Number of Items	Form A
Sounds and Letters	40	.41
Word Reading	30	.44
Sentence Reading	30	.48
Mathematics	40	.43
Environment	40	.34
Listening to Words and Stories	40	.36

Table J–2. Median Point Biserial Correlation Coefficients for Primary 1, Spring, Grade 1

Subtest	Number of Items	Form A	Form B
Word Study Skills	30	.45	.37
Word Reading	30	.57	.56
Sentence Reading	30	.50	.54
Reading Comprehension	40	.48	.47
Mathematics Problem Solving	42	.42	.42
Mathematics Procedures	30	.42	.45
Language	40	.40	.44
Spelling	36	.49	.46
Environment	40	.34	.33
Listening	40	.42	.39
Subtest	**Number of Items**	**Form D**	**Form E**
Language	40	.37	.39

Table J–3. Median Point Biserial Correlation Coefficients for Primary 2, Spring, Grade 2

Subtest	Number of Items	Form A	Form B
Word Study Skills	30	.40	.35
Reading Vocabulary	30	.55	.53
Reading Comprehension	40	.49	.49
Mathematics Problem Solving	44	.40	.42
Mathematics Procedures	30	.46	.48
Language	48	.43	.41
Spelling	36	.47	.49
Environment	40	.31	.32
Listening	40	.38	.38
Subtest	**Number of Items**	**Form D**	**Form E**
Language	40	.35	.39

Table J–4. Median Point Biserial Correlation Coefficients for Primary 3, Spring, Grade 3

Subtest	Number of Items	Form A	Form B
Word Study Skills	30	.40	.41
Reading Vocabulary	30	.47	.45
Reading Comprehension	54	.46	.46
Mathematics Problem Solving	46	.47	.44
Mathematics Procedures	30	.50	.49
Language	48	.46	.43
Spelling	38	.43	.44
Science	40	.38	.40
Social Science	40	.44	.46
Listening	40	.39	.37
Subtest	**Number of Items**	**Form D**	**Form E**
Language	45	.42	.37

Table J–5. Median Point Biserial Correlation Coefficients for Intermediate 1, Spring, Grade 4

Subtest	Number of Items	Form A	Form B
Word Study Skills	30	.49	.47
Reading Vocabulary	30	.50	.55
Reading Comprehension	54	.44	.45
Mathematics Problem Solving	48	.44	.46
Mathematics Procedures	32	.49	.52
Language	48	.45	.46
Spelling	40	.42	.44
Science	40	.41	.44
Social Science	40	.42	.47
Listening	40	.36	.40
Subtest	Number of Items	Form D	Form E
Language	48	.40	.41

Table J–6. Median Point Biserial Correlation Coefficients for Intermediate 2, Spring, Grade 5

Subtest	Number of Items	Form A	Form B
Reading Vocabulary	30	.44	.47
Reading Comprehension	54	.43	.45
Mathematics Problem Solving	48	.43	.41
Mathematics Procedures	32	.45	.48
Language	48	.44	.43
Spelling	40	.44	.42
Science	40	.36	.38
Social Science	40	.43	.38
Listening	40	.37	.38
Subtest	Number of Items	Form D	Form E
Language	48	.41	.43

Table J–7. Median Point Biserial Correlation Coefficients for Intermediate 3, Spring, Grade 6

Subtest	Number of Items	Form A	Form B
Reading Vocabulary	30	.43	.46
Reading Comprehension	54	.44	.43
Mathematics Problem Solving	48	.44	.41
Mathematics Procedures	32	.48	.48
Language	48	.43	.44
Spelling	40	.41	.43
Science	40	.36	.37
Social Science	40	.40	.42
Listening	40	.36	.41
Subtest	Number of Items	Form D	Form E
Language	48	.39	.44

Table J–8. Median Point Biserial Correlation Coefficients for Advanced 1, Spring, Grade 7

Subtest	Number of Items	Form A	Form B
Reading Vocabulary	30	.50	.49
Reading Comprehension	54	.45	.45
Mathematics Problem Solving	48	.42	.45
Mathematics Procedures	32	.53	.49
Language	48	.47	.48
Spelling	40	.42	.40
Science	40	.41	.39
Social Science	40	.37	.38
Listening	40	.40	.39
Subtest	Number of Items	Form D	Form E
Language	48	.43	.43

Table J–9. Median Point Biserial Correlation Coefficients for Advanced 2, Spring, Grade 8

Subtest	Number of Items	Form A	Form B
Reading Vocabulary	30	.46	.46
Reading Comprehension	54	.47	.43
Mathematics Problem Solving	48	.43	.45
Mathematics Procedures	32	.49	.51
Language	48	.47	.49
Spelling	40	.41	.43
Science	40	.36	.37
Social Science	40	.41	.39
Listening	40	.37	.39
Subtest	**Number of Items**	**Form D**	**Form E**
Language	48	.42	.45

Table J–10. Median Point Biserial Correlation Coefficients for Advanced 2, Spring, Grade 9

Subtest	Number of Items	Form A	Form B
Reading Vocabulary	30	.52	.49
Reading Comprehension	54	.44	.45
Mathematics Problem Solving	48	.46	.45
Mathematics Procedures	32	.48	.51
Language	48	.47	.51
Spelling	40	.41	.42
Science	40	.40	.41
Social Science	40	.43	.42
Listening	40	.36	.40
Subtest	**Number of Items**	**Form D**	**Form E**
Language	48	.50	.45

Table J–11. Median Point Biserial Correlation Coefficients for TASK 1, Spring, Grade 9

Subtest	Number of Items	Form A	Form B
Reading Vocabulary	30	.47	.45
Reading Comprehension	54	.39	.49
Mathematics	50	.38	.37
Language	48	.50	.54
Spelling	40	.43	.44
Science	40	.37	.42
Social Science	40	.35	.39
Subtest	**Number of Items**	**Form D**	**Form E**
Language	48	.44	.45

Table J–12. Median Point Biserial Correlation Coefficients for TASK 2, Spring, Grade 10

Subtest	Number of Items	Form A	Form B
Reading Vocabulary	30	.44	.44
Reading Comprehension	54	.41	.45
Mathematics	50	.35	.35
Language	48	.44	.43
Spelling	40	.44	.44
Science	40	.40	.37
Social Science	40	.34	.34
Subtest	**Number of Items**	**Form D**	**Form E**
Language	48	.43	.43

Table J–13. Median Point Biserial Correlation Coefficients for TASK 3, Spring, Grade 11

Subtest	Number of Items	Form A	Form B
Reading Vocabulary	30	.57	.51
Reading Comprehension	54	.49	.51
Mathematics	50	.38	.35
Language	48	.50	.47
Spelling	40	.48	.48
Science	40	.40	.37
Social Science	40	.33	.29
Subtest	Number of Items	Form D	Form E
Language	48	.50	.46

Table J–14. Median Point Biserial Correlation Coefficients for TASK 3, Spring, Grade 12

Subtest	Number of Items	Form A	Form B
Reading Vocabulary	30	.53	.57
Reading Comprehension	54	.53	.55
Mathematics	50	.36	.36
Language	48	.51	.51
Spelling	40	.52	.56
Science	40	.38	.40
Social Science	40	.35	.34
Subtest	Number of Items	Form D	Form E
Language	48	.46	.52

Table J–15. Median Point Biserial Correlation Coefficients for SESAT 1, Fall, Grade K

Subtest	Number of Items	Form A
Sounds and Letters	40	.40
Word Reading	30	.41
Mathematics	40	.42
Environment	40	.38
Listening to Words and Stories	40	.37

Table J–16. Median Point Biserial Correlation Coefficients for SESAT 2, Fall, Grade 1

Subtest	Number of Items	Form A
Sounds and Letters	40	.33
Word Reading	30	.48
Sentence Reading	30	.55
Mathematics	40	.37
Environment	40	.31
Listening to Words and Stories	40	.31

Table J–17. Median Point Biserial Correlation Coefficients for Primary 1, Fall, Grade 2

Subtest	Number of Items	Form A	Form B
Word Study Skills	30	.45	.38
Word Reading	30	.56	.53
Sentence Reading	30	.51	.56
Reading Comprehension	40	.50	.50
Mathematics Problem Solving	42	.43	.43
Mathematics Procedures	30	.45	.48
Language	40	.41	.45
Spelling	36	.53	.49
Environment	40	.29	.27
Listening	40	.44	.41
Subtest	**Number of Items**	**Form D**	**Form E**
Language	40	.39	.37

Appendix J

Table J–18. Median Point Biserial Correlation Coefficients for Primary 2, Fall, Grade 3

Subtest	Number of Items	Form A	Form B
Word Study Skills	30	.38	.37
Reading Vocabulary	30	.51	.54
Reading Comprehension	40	.47	.48
Mathematics Problem Solving	44	.42	.40
Mathematics Procedures	30	.51	.49
Language	48	.43	.44
Spelling	36	.47	.52
Environment	40	.28	.27
Listening	40	.36	.37
Subtest	**Number of Items**	**Form D**	**Form E**
Language	40	.34	.39

Table J–19. Median Point Biserial Correlation Coefficients for Primary 3, Fall, Grade 4

Subtest	Number of Items	Form A	Form B
Word Study Skills	30	.45	.45
Reading Vocabulary	30	.54	.50
Reading Comprehension	54	.51	.48
Mathematics Problem Solving	46	.49	.50
Mathematics Procedures	30	.52	.50
Language	48	.49	.48
Spelling	38	.44	.47
Science	40	.46	.42
Social Science	40	.45	.48
Listening	40	.43	.40
Subtest	**Number of Items**	**Form D**	**Form E**
Language	45	.43	.41

Table J–20. Median Point Biserial Correlation Coefficients for Intermediate 1, Fall, Grade 5

Subtest	Number of Items	Form A	Form B
Word Study Skills	30	.49	.52
Reading Vocabulary	30	.52	.55
Reading Comprehension	54	.46	.45
Mathematics Problem Solving	48	.47	.48
Mathematics Procedures	32	.54	.52
Language	48	.46	.47
Spelling	40	.46	.46
Science	40	.47	.45
Social Science	40	.47	.46
Listening	40	.38	.41
Subtest	Number of Items	Form D	Form E
Language	48	.43	.43

Table J–21. Median Point Biserial Correlation Coefficients for Intermediate 2, Fall, Grade 6

Subtest	Number of Items	Form A	Form B
Reading Vocabulary	30	.45	.48
Reading Comprehension	54	.43	.47
Mathematics Problem Solving	48	.45	.45
Mathematics Procedures	32	.48	.49
Language	48	.46	.45
Spelling	40	.47	.45
Science	40	.38	.43
Social Science	40	.45	.46
Listening	40	.39	.41
Subtest	Number of Items	Form D	Form E
Language	48	.43	.47

Appendix J

Table J–22. Median Point Biserial Correlation Coefficients for Intermediate 3, Fall, Grade 7

Subtest	Number of Items	Form A	Form B
Reading Vocabulary	30	.49	.54
Reading Comprehension	54	.47	.47
Mathematics Problem Solving	48	.44	.46
Mathematics Procedures	32	.47	.52
Language	48	.44	.51
Spelling	40	.46	.48
Science	40	.39	.38
Social Science	40	.41	.45
Listening	40	.39	.42
Subtest	**Number of Items**	**Form D**	**Form E**
Language	48	.41	.49

Table J–23. Median Point Biserial Correlation Coefficients for Advanced 1, Fall, Grade 8

Subtest	Number of Items	Form A	Form B
Reading Vocabulary	30	.52	.55
Reading Comprehension	54	.47	.49
Mathematics Problem Solving	48	.46	.49
Mathematics Procedures	32	.52	.50
Language	48	.49	.49
Spelling	40	.43	.42
Science	40	.46	.40
Social Science	40	.40	.41
Listening	40	.40	.42
Subtest	**Number of Items**	**Form D**	**Form E**
Language	48	.43	.45

Table J–24. Median Point Biserial Correlation Coefficients for Advanced 2, Fall, Grade 9

Subtest	Number of Items	Form A	Form B
Reading Vocabulary	30	.49	.46
Reading Comprehension	54	.43	.44
Mathematics Problem Solving	48	.45	.45
Mathematics Procedures	32	.47	.55
Language	48	.47	.45
Spelling	40	.41	.45
Science	40	.38	.41
Social Science	40	.43	.41
Listening	40	.38	.40
Subtest	**Number of Items**	**Form D**	**Form E**
Language	48	.45	.45

Table J–25. Median Point Biserial Correlation Coefficients for TASK 1, Fall, Grade 9

Subtest	Number of Items	Form A	Form B
Reading Vocabulary	30	.47	.43
Reading Comprehension	54	.41	.46
Mathematics	50	.38	.36
Language	48	.46	.47
Spelling	40	.40	.40
Science	40	.37	.37
Social Science	40	.38	.36
Subtest	**Number of Items**	**Form D**	**Form E**
Language	48	.46	.43

Appendix J

Table J–26. Median Point Biserial Correlation Coefficients for TASK 2, Fall, Grade 10

Subtest	Number of Items	Form A	Form B
Reading Vocabulary	30	.54	.57
Reading Comprehension	54	.49	.51
Mathematics	50	.35	.33
Language	48	.48	.50
Spelling	40	.49	.57
Science	40	.38	.38
Social Science	40	.31	.32
Subtest	**Number of Items**	**Form D**	**Form E**
Language	48	.40	.47

Table J–27. Median Point Biserial Correlation Coefficients for TASK 3, Fall, Grade 11

Subtest	Number of Items	Form A	Form B
Reading Vocabulary	30	.59	.54
Reading Comprehension	54	.51	.51
Mathematics	50	.32	.28
Language	48	.48	.42
Spelling	40	.45	.48
Science	40	.34	.37
Social Science	40	.30	.28
Subtest	**Number of Items**	**Form D**	**Form E**
Language	48	.43	.55

Table J–28. Median Point Biserial Correlation Coefficients for TASK 3, Fall, Grade 12

Subtest	Number of Items	Form A	Form B
Reading Vocabulary	30	.54	.47
Reading Comprehension	54	.49	.50
Mathematics	50	.35	.35
Language	48	.51	.44
Spelling	40	.48	.47
Science	40	.38	.39
Social Science	40	.33	.29
Subtest	**Number of Items**	**Form D**	**Form E**
Language	48	.47	.50

Appendix K:

Mean Scaled Scores and Related Summary Data for the Stanford 10 Standardization Samples

Appendix K

Table K–1. Mean Scaled Scores and Related Summary Data for SESAT 1, Grade K

Subtest/Total	Spring		
	N	Mean	SD
Total Reading	1245	493.6	46.1
Sounds and Letters	1253	528.0	40.5
Word Reading	1418	449.9	49.5
Mathematics	1429	491.9	37.5
Environment	1400	557.6	33.2
Listening to Words and Stories	1406	554.5	37.7

Table K–2. Mean Scaled Scores and Related Summary Data for SESAT 2, Grade K

Subtest/Total	Spring		
	N	Mean	SD
Total Reading	2082	476.6	41.7
Sounds and Letters	2176	521.9	43.7
Word Reading	2337	440.3	44.5
Sentence Reading	2356	455.7	49.6
Mathematics	2404	492.3	46.0
Environment	2411	557.3	30.1
Listening to Words and Stories	2402	545.5	34.3

Table K–3. Mean Scaled Scores and Related Summary Data for Primary 1, Grade 1

Subtest/Total	Spring		
	N	Mean	SD
Total Reading	6578	552.8	47.3
Word Study Skills	6936	574.1	49.2
Word Reading	7013	529.9	57.6
Sentence Reading	6878	562.4	52.1
Reading Comprehension	6722	551.4	51.0
Total Mathematics	7025	531.8	36.8
Mathematics Problem Solving	7089	545.0	40.7
Mathematics Procedures	7073	514.3	41.4
Language	5435	550.9	37.6
Spelling	7090	515.6	52.6
Environment	7034	581.0	31.3
Listening	6861	571.4	36.1
Subtest/Total	Spring		
	N	Mean	SD
Language Form D	2594	564.1	36.0

Table K–4. Mean Scaled Scores and Related Summary Data for Primary 2, Grade 2

Subtest/Total	Spring		
	N	**Mean**	**SD**
Total Reading	6415	595.3	42.2
Word Study Skills	6584	609.9	48.6
Reading Vocabulary	6558	583.8	53.7
Reading Comprehension	6615	597.0	45.1
Total Mathematics	6606	576.4	39.3
Mathematics Problem Solving	6643	581.9	41.0
Mathematics Procedures	6661	571.5	47.2
Language	4843	591.4	42.3
Spelling	6664	581.9	49.3
Environment	6572	607.5	28.8
Listening	6549	599.9	35.6

Subtest/Total	Spring		
	N	**Mean**	**SD**
Language Form D	2406	587.7	36.1

Table K–5. Mean Scaled Scores and Related Summary Data for Primary 3, Grade 3

Subtest/Total	Spring		
	N	Mean	SD
Total Reading	4505	621.0	38.1
Word Study Skills	4571	630.3	44.6
Reading Vocabulary	4581	615.5	46.8
Reading Comprehension	4615	621.2	41.8
Total Mathematics	4615	607.6	40.8
Mathematics Problem Solving	4633	612.4	44.0
Mathematics Procedures	4650	602.9	47.2
Language	3867	616.9	39.7
Spelling	4617	605.4	38.8
Science	4603	620.5	36.4
Social Science	4573	610.8	40.8
Listening	4558	614.3	33.0

Subtest/Total	Spring		
	N	Mean	SD
Language Form D	1307	603.3	33.2
Prewriting	1319	597.9	40.4
Composing	1314	602.9	38.4
Editing	1311	608.8	40.0

Appendix K

Table K–6. Mean Scaled Scores and Related Summary Data for Intermediate 1, Grade 4

Subtest/Total	Spring		
	N	Mean	SD
Total Reading	6394	630.4	39.2
Word Study Skills	6514	630.9	48.8
Reading Vocabulary	6486	632.3	50.5
Reading Comprehension	6556	631.1	40.4
Total Mathematics	6514	619.0	39.7
Mathematics Problem Solving	6562	620.5	41.0
Mathematics Procedures	6546	618.7	47.8
Language	4716	621.5	41.0
Language Mechanics	4716	622.6	44.0
Language Expression	4716	621.6	45.4
Spelling	6438	623.4	41.4
Science	6394	623.9	36.1
Social Science	6385	618.1	41.1
Listening	6361	619.8	32.8

Subtest/Total	Spring		
	N	Mean	SD
Language Form D	1552	619.3	34.9
Prewriting	1558	622.6	46.5
Composing	1555	618.7	37.7
Editing	1559	619.6	40.3

Table K–7. Mean Scaled Scores and Related Summary Data for Intermediate 2, Grade 5

Subtest/Total	Spring		
	N	Mean	SD
Total Reading	5759	644.2	36.7
Reading Vocabulary	5794	648.8	43.4
Reading Comprehension	5883	643.1	38.2
Total Mathematics	5847	640.9	35.9
Mathematics Problem Solving	5878	639.0	37.5
Mathematics Procedures	5881	644.4	42.7
Language	4656	637.0	38.6
Language Mechanics	4656	639.2	42.0
Language Expression	4656	636.6	43.0
Spelling	5697	639.9	40.7
Science	5670	636.4	32.8
Social Science	5646	631.1	36.2
Listening	5653	632.8	30.7

Subtest/Total	Spring		
	N	Mean	SD
Language Form D	953	623.0	35.3
Prewriting	956	630.3	45.1
Composing	957	620.4	39.4
Editing	957	621.5	39.9

Appendix K

Table K–8. Mean Scaled Scores and Related Summary Data for Intermediate 3, Grade 6

Subtest/Total	Spring		
	N	Mean	SD
Total Reading	6492	660.8	37.3
Reading Vocabulary	6539	670.0	42.6
Reading Comprehension	6553	657.5	39.4
Total Mathematics	6530	658.4	36.3
Mathematics Problem Solving	6579	656.7	37.0
Mathematics Procedures	6557	661.5	44.2
Language	5088	651.2	38.8
Language Mechanics	5088	653.0	40.3
Language Expression	5088	651.4	44.6
Spelling	6546	654.3	40.7
Science	6538	650.5	30.8
Social Science	6529	646.3	34.9
Listening	6490	647.4	34.6

Subtest/Total	Spring		
	N	Mean	SD
Language Form D	1311	628.1	37.1
Prewriting	1314	629.1	45.1
Composing	1317	628.4	41.7
Editing	1315	627.9	44.7

Table K–9. Mean Scaled Scores and Related Summary Data for Advanced 1, Grade 7

Subtest/Total	Spring		
	N	Mean	SD
Total Reading	5004	668.4	39.1
Reading Vocabulary	5026	680.2	48.8
Reading Comprehension	5105	663.3	39.1
Total Mathematics	5016	668.6	37.9
Mathematics Problem Solving	5099	666.2	37.6
Mathematics Procedures	5055	672.2	48.1
Language	3524	657.6	42.6
Language Mechanics	3523	660.9	44.4
Language Expression	3524	656.0	48.1
Spelling	5085	662.4	36.6
Science	4985	657.9	34.1
Social Science	4956	657.4	31.7
Listening	4831	653.1	33.7

Subtest/Total	Spring		
	N	Mean	SD
Language Form D	1278	638.2	34.8
Prewriting	1280	645.8	41.0
Composing	1292	638.0	39.1
Editing	1292	632.5	44.0

Appendix K

Table K–10. Mean Scaled Scores and Related Summary Data for Advanced 2, Grade 8

Subtest/Total	Spring		
	N	Mean	SD
Total Reading	3676	677.9	37.3
Reading Vocabulary	3706	691.1	44.1
Reading Comprehension	3841	672.5	38.2
Total Mathematics	3672	680.5	36.5
Mathematics Problem Solving	3760	679.4	36.4
Mathematics Procedures	3717	681.2	44.8
Language	2673	666.1	41.2
Language Mechanics	2673	671.4	42.7
Language Expression	2673	662.4	47.5
Spelling	3677	673.2	37.6
Science	3829	668.0	29.9
Social Science	3736	666.2	32.9
Listening	3514	656.8	31.9

Subtest/Total	Spring		
	N	Mean	SD
Language Form D	951	651.2	37.2
Prewriting	951	656.8	46.5
Composing	951	659.2	37.7
Editing	951	642.8	47.2

Table K–11. Mean Scaled Scores and Related Summary Data for Advanced 2, Grade 9

Subtest/Total	Spring		
	N	Mean	SD
Total Reading	3324	688.1	37.5
Reading Vocabulary	3352	701.2	49.0
Reading Comprehension	3368	682.7	38.0
Total Mathematics	3298	689.7	39.3
Mathematics Problem Solving	3356	689.1	40.8
Mathematics Procedures	3329	690.4	47.2
Language	1969	674.9	45.8
Language Mechanics	1969	676.3	47.1
Language Expression	1969	674.7	50.7
Spelling	3380	677.9	38.8
Science	3362	674.4	33.5
Social Science	3350	672.8	36.5
Listening	3232	664.7	33.1
Subtest/Total	**Spring**		
	N	Mean	SD
Language Form D	1388	654.1	38.8
Prewriting	1391	660.4	46.9
Composing	1395	660.7	40.6
Editing	1392	646.0	47.8

Table K–12. Mean Scaled Scores and Related Summary Data for TASK 1, Grade 9

Subtest/Total	Spring		
	N	Mean	SD
Total Reading	2831	688.8	37.2
Reading Vocabulary	2837	703.3	44.3
Reading Comprehension	2842	682.9	39.0
Mathematics	2825	695.2	30.3
Language	1585	677.8	48.3
Language Mechanics	1585	684.1	47.2
Language Expression	1585	672.7	55.8
Spelling	2798	681.0	39.1
Science	2793	679.1	31.2
Social Science	2801	673.5	29.6
Subtest/Total	Spring		
	N	Mean	SD
Language Form D	1193	650.1	36.7
Prewriting	1196	657.6	42.5
Composing	1200	649.0	46.9
Editing	1197	647.9	40.8

Table K–13. Mean Scaled Scores and Related Summary Data for TASK 2, Grade 10

Subtest/Total	Spring		
	N	Mean	SD
Total Reading	5155	699.9	35.6
Reading Vocabulary	5187	715.7	42.5
Reading Comprehension	5187	693.3	37.9
Mathematics	5007	702.4	29.9
Language	3597	687.3	39.6
Language Mechanics	3597	689.5	41.2
Language Expression	3597	687.2	45.6
Spelling	5135	695.8	39.9
Science	4959	688.2	31.8
Social Science	4951	682.2	26.5
Subtest/Total	**Spring**		
	N	Mean	SD
Language Form D	1447	664.6	35.2
Prewriting	1451	680.9	42.9
Composing	1448	662.3	38.0
Editing	1452	658.5	44.2

Appendix K

Table K–14. Mean Scaled Scores and Related Summary Data for TASK 3, Grade 11

Subtest/Total	Spring		
	N	Mean	SD
Total Reading	3642	707.6	43.0
Reading Vocabulary	3659	725.6	50.2
Reading Comprehension	3647	699.4	44.0
Mathematics	3582	707.8	28.2
Language	2278	687.7	41.8
Language Mechanics	2278	690.3	41.4
Language Expression	2275	686.6	50.2
Spelling	3527	695.2	44.7
Science	3516	692.0	30.0
Social Science	3531	686.4	26.7
Subtest/Total	Spring		
	N	Mean	SD
Language Form D	1150	662.7	40.7
Prewriting	1154	677.5	48.4
Composing	1157	657.4	46.7
Editing	1153	659.9	45.2

Table K–15. Mean Scaled Scores and Related Summary Data for TASK 3, Grade 12

Subtest/Total	Spring		
	N	Mean	SD
Total Reading	3326	702.8	45.3
Reading Vocabulary	3353	723.7	52.8
Reading Comprehension	3348	693.2	46.8
Mathematics	3327	704.2	28.7
Language	2157	684.9	45.2
Language Mechanics	2156	688.0	44.5
Language Expression	2157	682.9	53.3
Spelling	3278	695.5	48.8
Science	3224	688.3	31.9
Social Science	3211	684.2	27.6
Subtest/Total	Spring		
	N	Mean	SD
Language Form D	1000	656.7	42.2
Prewriting	1012	672.3	49.2
Composing	1008	650.7	46.9
Editing	1005	652.9	49.5

Appendix K

Table K–16. Mean Scaled Scores and Related Summary Data for SESAT 1, Grade K

Subtest/Total	Fall		
	N	Mean	SD
Total Reading	925	426.80	36.96
Sounds and Letters	937	474.93	43.33
Word Reading	1111	381.17	38.30
Mathematics	1096	459.95	38.49
Environment	1096	542.83	33.77
Listening to Words and Stories	1085	535.73	34.52

Table K–17. Mean Scaled Scores and Related Summary Data for SESAT 2, Grade 1

Subtest/Total	Fall		
	N	Mean	SD
Total Reading	1220	512.20	56.17
Sounds and Letters	1241	547.91	43.02
Word Reading	1349	469.77	52.81
Sentence Reading	1220	502.53	59.81
Mathematics	1359	516.48	41.49
Environment	1359	568.94	27.45
Listening to Words and Stories	1349	562.07	28.60

Table K–18. Mean Scaled Scores and Related Summary Data for Primary 1, Grade 2

Subtest/Total	Fall		
	N	Mean	SD
Total Reading	3268	573.43	48.70
Word Study Skills	3369	591.83	53.17
Word Reading	3403	552.02	52.10
Sentence Reading	3364	580.22	49.19
Reading Comprehension	3308	573.17	52.47
Total Mathematics	3375	545.89	38.02
Mathematics Problem Solving	3390	563.45	43.09
Mathematics Procedures	3392	525.56	42.54
Language	2370	568.68	41.36
Spelling	3386	538.45	54.70
Environment	3387	590.05	28.55
Listening	3294	590.74	40.37
Subtest/Total	Fall		
	N	Mean	SD
Language Form D	1553	580.89	37.09

Appendix K

Table K–19. Mean Scaled Scores and Related Summary Data for Primary 2, Grade 3

Subtest/Total	Fall		
	N	Mean	SD
Total Reading	2522	613.10	43.15
Word Study Skills	2540	629.77	51.53
Reading Vocabulary	2543	603.17	51.70
Reading Comprehension	2558	613.12	44.34
Total Mathematics	2541	590.46	41.04
Mathematics Problem Solving	2552	596.06	41.08
Mathematics Procedures	2551	586.02	50.18
Language	1677	609.16	44.80
Spelling	2550	597.55	48.76
Environment	2517	618.57	27.83
Listening	2531	613.15	35.44
Subtest/Total	Fall		
	N	Mean	SD
Language Form D	1266	601.28	36.12

Table K–20. Mean Scaled Scores and Related Summary Data for Primary 3, Grade 4

Subtest/Total	Fall		
	N	Mean	SD
Total Reading	7169	626.67	41.57
Word Study Skills	7240	633.62	47.80
Reading Vocabulary	7228	624.69	51.17
Reading Comprehension	7263	627.03	45.53
Total Mathematics	7217	609.70	43.52
Mathematics Problem Solving	7251	615.27	47.51
Mathematics Procedures	7252	604.46	48.95
Language	5058	620.75	43.70
Spelling	7237	610.98	42.86
Science	7226	625.42	40.97
Social Science	7234	617.59	42.71
Listening	7217	616.23	37.01

Subtest/Total	Fall		
	N	Mean	SD
Language Form D	2066	606.80	35.21
Prewriting	2082	605.10	45.40
Composing	2073	604.78	38.34
Editing	2080	612.17	44.83

Table K–21. Mean Scaled Scores and Related Summary Data for Intermediate 1, Grade 5

Subtest/Total	Fall		
	N	Mean	SD
Total Reading	7348	639.28	41.54
Word Study Skills	7507	637.32	50.29
Reading Vocabulary	7409	644.24	52.18
Reading Comprehension	7538	640.74	43.01
Total Mathematics	7534	626.42	41.98
Mathematics Problem Solving	7553	628.45	42.81
Mathematics Procedures	7567	625.80	50.18
Language	4763	632.15	42.28
Language Mechanics	4763	630.05	43.78
Language Expression	4763	636.30	48.28
Spelling	7525	630.55	43.35
Science	7470	633.06	38.12
Social Science	7461	627.45	42.81
Listening	7450	628.19	35.39
Subtest/Total	Fall		
	N	Mean	SD
Language Form D	2597	622.04	37.39
Prewriting	2614	624.33	48.37
Composing	2630	623.23	41.79
Editing	2620	619.99	42.89

Table K–22. Mean Scaled Scores and Related Summary Data for Intermediate 2, Grade 6

Subtest/Total	Fall		
	N	Mean	SD
Total Reading	8776	635.09	38.86
Reading Vocabulary	8796	657.58	44.49
Reading Comprehension	8880	652.28	40.19
Total Mathematics	8859	645.64	37.61
Mathematics Problem Solving	8887	646.91	39.79
Mathematics Procedures	8879	644.21	42.45
Language	6075	641.91	41.12
Language Mechanics	6075	643.84	45.03
Language Expression	6075	642.00	45.10
Spelling	8866	646.01	44.72
Science	8851	646.39	34.10
Social Science	8825	641.87	39.37
Listening	8726	637.31	32.15

Subtest/Total	Fall		
	N	Mean	SD
Language Form D	2584	635.05	36.42
Prewriting	2585	641.09	46.70
Composing	2595	633.95	40.16
Editing	2594	633.75	41.45

Appendix K

Table K–23. Mean Scaled Scores and Related Summary Data for Intermediate 3, Grade 7

Subtest/Total	Fall		
	N	Mean	SD
Total Reading	8236	660.66	41.18
Reading Vocabulary	8267	669.93	47.94
Reading Comprehension	8372	656.98	42.50
Total Mathematics	8354	657.15	37.44
Mathematics Problem Solving	8386	657.42	39.88
Mathematics Procedures	8389	657.84	44.83
Language	5489	645.42	43.07
Language Mechanics	5495	646.09	44.82
Language Expression	5489	646.25	48.21
Spelling	8378	653.45	44.02
Science	8341	653.21	34.28
Social Science	8317	645.81	36.26
Listening	8315	646.49	37.37

Subtest/Total	Fall		
	N	Mean	SD
Language Form D	2569	639.91	37.43
Prewriting	2572	644.92	46.21
Composing	2580	641.50	40.45
Editing	2579	636.79	46.04

Table K–24. Mean Scaled Scores and Related Summary Data for Advanced 1, Grade 8

Subtest/Total	Fall		
	N	Mean	SD
Total Reading	8835	675.23	42.37
Reading Vocabulary	8856	687.45	53.16
Reading Comprehension	8938	670.09	42.11
Total Mathematics	8692	673.80	41.08
Mathematics Problem Solving	8832	673.43	42.17
Mathematics Procedures	8757	674.18	48.88
Language	6071	659.02	45.32
Language Mechanics	6071	660.95	46.80
Language Expression	6072	658.61	51.33
Spelling	8850	667.29	39.77
Science	8707	661.58	37.42
Social Science	8660	662.25	34.28
Listening	8663	658.21	35.93

Subtest/Total	Fall		
	N	Mean	SD
Language Form D	2517	654.26	37.83
Prewriting	2530	662.16	44.16
Composing	2527	654.69	43.41
Editing	2528	650.16	43.83

Appendix K

Table K–25. Mean Scaled Scores and Related Summary Data for Advanced 2, Grade 9

Subtest/Total	Fall		
	N	Mean	SD
Total Reading	2462	681.42	36.07
Reading Vocabulary	2492	694.99	45.73
Reading Comprehension	2492	675.70	37.48
Total Mathematics	2265	684.68	35.74
Mathematics Problem Solving	2451	683.39	38.40
Mathematics Procedures	2288	683.85	43.88
Language	1502	662.57	39.54
Language Mechanics	1502	666.10	39.84
Language Expression	1502	660.43	47.76
Spelling	2421	667.09	38.29
Science	2370	669.38	31.00
Social Science	2321	667.69	35.75
Listening	2150	662.35	33.24

Subtest/Total	Fall		
	N	Mean	SD
Language Form D	817	663.36	37.57
Prewriting	820	666.86	45.55
Composing	826	668.04	41.33
Editing	823	658.76	45.42

Table K–26. Mean Scaled Scores and Related Summary Data for TASK 1, Grade 9

Subtest/Total	Fall		
	N	Mean	SD
Total Reading	2900	685.77	36.33
Reading Vocabulary	2927	698.66	42.86
Reading Comprehension	2941	679.56	39.45
Mathematics	2920	690.90	30.19
Language	1586	668.74	41.59
Language Mechanics	1586	675.60	42.18
Language Expression	1586	663.76	50.26
Spelling	2897	680.09	36.66
Science	2876	678.28	30.68
Social Science	2870	670.73	29.58

Subtest/Total	Fall		
	N	Mean	SD
Language Form D	1224	654.69	35.74
Prewriting	1225	658.70	42.16
Composing	1230	654.60	45.77
Editing	1227	654.96	40.47

Appendix K

Table K–27. Mean Scaled Scores and Related Summary Data for TASK 2, Grade 10

Subtest/Total	Fall		
	N	Mean	SD
Total Reading	2789	680.80	42.14
Reading Vocabulary	2900	689.95	49.36
Reading Comprehension	2851	675.98	41.91
Mathematics	2907	692.74	26.51
Language	2148	672.05	40.50
Language Mechanics	2148	675.08	40.88
Language Expression	2148	670.36	48.29
Spelling	2953	674.70	46.33
Science	2903	680.83	31.46
Social Science	2900	672.66	25.86

Subtest/Total	Fall		
	N	Mean	SD
Language Form D	726	660.44	36.07
Prewriting	730	674.76	43.20
Composing	733	660.95	39.30
Editing	731	650.74	45.13

Table K–28. Mean Scaled Scores and Related Summary Data for TASK 3, Grade 11

Subtest/Total	Fall		
	N	Mean	SD
Total Reading	2668	693.65	42.35
Reading Vocabulary	2679	705.44	47.53
Reading Comprehension	2694	687.88	42.66
Mathematics	2678	699.63	24.40
Language	2113	682.37	37.38
Language Mechanics	2113	683.24	37.01
Language Expression	2113	682.62	45.87
Spelling	2654	683.74	41.66
Science	2733	686.90	27.74
Social Science	2712	679.71	23.31
Subtest/Total	Fall		
	N	Mean	SD
Language Form D	591	636.56	39.28
Prewriting	598	650.70	44.32
Composing	605	631.23	42.99
Editing	600	630.04	48.02

Appendix K

Table K–29. Mean Scaled Scores and Related Summary Data for TASK 3, Grade 12

Subtest/Total	Fall		
	N	Mean	SD
Total Reading	1988	702.25	39.73
Reading Vocabulary	1999	722.37	45.65
Reading Comprehension	2005	692.88	42.24
Mathematics	1937	703.73	27.27
Language	1353	687.10	40.77
Language Mechanics	1352	690.37	38.96
Language Expression	1352	685.42	50.27
Spelling	1873	694.43	41.44
Science	1857	688.88	30.33
Social Science	1858	682.62	24.36
Subtest/Total	**Fall**		
	N	Mean	SD
Language Form D	456	654.35	42.00
Prewriting	459	666.55	46.85
Composing	463	649.99	47.81
Editing	462	649.76	48.92

Appendix L:

Pearson Product-Moment Coefficients of Correlation (r) for Corresponding Subtests and Totals for Stanford 10 and Stanford 9

Appendix L

Table L–1. Pearson Product-Moment Coefficients of Correlation (r) for Corresponding Stanford 10 and Stanford 9 Subtests and Totals for SESAT 1

Subtest	N	SAT10 – Form A			SAT9 – Form S			r
		Number of Items	Mean	SD	Number of Items	Mean	SD	
Total Reading	376	70	45.5	11.4	78	51.4	13.4	0.85
Sounds and Letters	534	40	28.9	7.3	48	33.5	9.0	0.83
Word Reading	739	30	13.4	5.9	30	14.9	6.3	0.75
Mathematics	712	40	23.4	6.9	40	22.0	7.0	0.74
Environment	648	40	24.1	6.8	40	22.2	6.8	0.78
Listening to Words and Stories	645	40	25.4	6.9	40	22.6	6.9	0.80

Table L–2. Pearson Product-Moment Coefficients of Correlation (r) for Corresponding Stanford 10 and Stanford 9 Subtests and Totals for SESAT 2

Subtest	N	SAT10 – Form A			SAT9 – Form S			r
		Number of Items	Mean	SD	Number of Items	Mean	SD	
Total Reading	827	100	72.8	17.5	110	74.5	21.7	0.92
Sounds and Letters	999	40	33.9	5.6	40	30.0	7.7	0.82
Word Reading	1047	30	19.3	6.6	40	26.8	9.1	0.85
Sentence Reading	1059	30	18.5	7.5	30	15.8	7.4	0.78
Mathematics	905	40	30.2	6.9	40	26.4	7.6	0.78
Environment	869	40	28.6	4.9	40	26.1	6.4	0.68
Listening to Words and Stories	922	40	24.9	6.1	40	26.1	6.8	0.78

Table L–3. Pearson Product-Moment Coefficients of Correlation (r) for Corresponding Stanford 10 and Stanford 9 Subtests and Totals for Primary 1

Subtest	N	SAT10 – Form A			SAT9 – Form S			r
		Number of Items	Mean	SD	Number of Items	Mean	SD	
Total Reading	1138	130	106.2	19.6	106	82.5	18.4	0.91
Word Study Skills	1628	30	22.6	5.1	36	27.2	6.7	0.82
Word Reading	1664	30	25.6	5.5	30	23.3	6.0	0.80
Sentence Reading	1462	30	26.2	4.4	(40)*	(31.5)	(7.5)	0.74
Reading Comprehension	1447	40	31.7	7.0	40	31.5	7.5	0.84
Total Mathematics	696	72	49.3	12.5	69	47.7	11.5	0.87
Mathematics Problem Solving	909	42	30.2	7.3	44	30.3	8.1	0.79
Mathematics Procedures	886	30	19.5	6.0	25	16.3	5.3	0.77
Language	528	40	28.6	7.1	44	30.5	8.2	0.84
Spelling	909	36	28.1	7.1	30	22.8	5.6	0.87
Environment	703	40	30.3	5.1	40	24.8	5.3	0.71
Listening	655	40	26.8	7.3	40	26.9	6.7	0.83
Subtest	N	SAT10 – Form D			SAT9 – Form SA			r
		Number of Items	Mean	SD	Number of Items	Mean	SD	
Language	214	40	27.5	6.9	46	29.3	8.5	0.83

* Numbers in parentheses represent a subtest, content total, or test of similar content used as a stand-in when there is no corresponding test.

Appendix L

Table L–4. Pearson Product-Moment Coefficients of Correlation (r) for Corresponding Stanford 10 and Stanford 9 Subtests and Totals for Primary 2

Subtest	N	SAT10 – Form A			SAT9 – Form S			r
		Number of Items	Mean	SD	Number of Items	Mean	SD	
Total Reading	1526	100	74.0	16.3	118	81.9	20.4	0.90
Word Study Skills	1740	30	22.5	4.6	48	36.1	8.5	0.80
Reading Vocabulary	1715	30	23.5	6.1	30	20.7	5.5	0.83
Reading Comprehension	1770	40	27.6	7.9	40	24.9	8.5	0.80
Total Mathematics	618	74	52.9	11.8	74	54.4	11.5	0.87
Mathematics Problem Solving	781	44	29.2	8.6	46	33.3	7.6	0.82
Mathematics Procedures	733	30	22.3	5.9	28	20.5	5.5	0.76
Language	786	48	38.0	7.0	44	30.7	6.6	0.81
Spelling	1253	36	29.3	6.3	30	21.1	6.3	0.80
Environment	620	40	29.9	4.5	40	26.7	5.1	0.63
Listening	642	40	27.0	6.1	40	27.2	6.1	0.76

Subtest	N	SAT10 – Form D			SAT9 – Form SA			r
		Number of Items	Mean	SD	Number of Items	Mean	SD	
Language	123	40	27.3	5.7	46	29.1	6.7	0.73

Table L–5. Pearson Product-Moment Coefficients of Correlation (r) for Corresponding Stanford 10 and Stanford 9 Subtests and Totals for Primary 3

Subtest	N	SAT10 – Form A			SAT9 – Form S			r
		Number of Items	Mean	SD	Number of Items	Mean	SD	
Total Reading	1432	114	76.4	21.3	84	57.0	15.6	0.90
Word Study Skills	1566	30	20.1	5.6	(30)*	(21.7)	(5.5)	0.70
Reading Vocabulary	1710	30	20.9	6.4	30	21.7	5.5	0.82
Reading Comprehension	1726	54	35.4	11.3	54	35.2	10.9	0.85
Total Mathematics	609	76	49.4	15.2	76	54.0	13.2	0.87
Mathematics Problem Solving	798	46	30.5	9.7	46	33.4	8.2	0.78
Mathematics Procedures	748	30	18.6	7.1	30	20.3	6.3	0.82
Language	648	48	32.7	9.6	48	30.2	9.3	0.86
Spelling	985	38	22.0	7.7	30	19.7	5.9	0.83
Science	608	40	25.0	7.2	40	26.5	7.3	0.80
Social Science	610	40	24.3	8.5	40	21.5	7.5	0.83
Listening	640	40	28.5	6.4	40	26.0	5.7	0.73
Subtest	N	SAT10 – Form D			SAT9 – Form SA			r
		Number of Items	Mean	SD	Number of Items	Mean	SD	
Language	469	45	23.6	8.7	54	29.4	11.1	0.82
Prewriting	677	14	7.9	3.0	15	9.1	3.7	0.64
Composing	671	16	7.6	3.4	15	7.5	3.7	0.64
Editing	673	15	8.2	3.3	24	12.7	5.3	0.70

* Numbers in parentheses represent a subtest, content total, or test of similar content used as a stand-in when there is no corresponding test.

Appendix L

Table L–6. Pearson Product-Moment Coefficients of Correlation (r) for Corresponding Stanford 10 and Stanford 9 Subtests and Totals for Intermediate 1

Subtest	N	SAT10 – Form A			SAT9 – Form S			r
		Number of Items	Mean	SD	Number of Items	Mean	SD	
Total Reading	1251	114	77.1	21.9	84	59.4	15.9	0.89
Word Study Skills	1436	30	20.9	6.1	(30)*	(22.8)	(5.3)	0.67
Reading Vocabulary	1653	30	22.4	6.4	30	22.8	5.3	0.79
Reading Comprehension	1676	54	33.6	11.4	54	36.4	11.5	0.85
Total Mathematics	480	80	47.6	16.1	78	49.7	15.0	0.86
Mathematics Problem Solving	628	48	28.3	9.8	48	31.8	9.0	0.78
Mathematics Procedures	609	32	20.4	7.7	30	19.1	7.0	0.79
Language	435	48	33.6	9.6	48	32.3	9.8	0.83
Language Mechanics	573	24	16.4	5.1	24	16.1	5.0	0.73
Language Expression	573	24	16.8	5.3	24	15.5	5.7	0.78
Spelling	1235	40	24.5	8.0	30	19.6	5.4	0.82
Science	476	40	26.3	7.4	40	26.9	7.7	0.78
Social Science	461	40	24.8	8.3	40	21.7	7.8	0.80
Listening	476	40	25.9	5.9	40	28.6	6.1	0.73

Subtest	N	SAT10 – Form D			SAT9 – Form SA			r
		Number of Items	Mean	SD	Number of Items	Mean	SD	
Language	220	48	29.7	8.9	54	36.0	10.3	0.81
Prewriting	1094	12	7.3	2.6	15	9.9	3.2	0.69
Composing	1094	18	10.1	4.0	15	9.5	3.4	0.62
Editing	1094	18	9.5	3.9	24	14.8	5.4	0.73

* Numbers in parentheses represent a subtest, content total, or test of similar content used as a stand-in when there is no corresponding test.

Table L–7. Pearson Product-Moment Coefficients of Correlation (r) for Corresponding Stanford 10 and Stanford 9 Subtests and Totals for Intermediate 2

Subtest	N	SAT10 – Form A			SAT9 – Form S			r
		Number of Items	Mean	SD	Number of Items	Mean	SD	
Total Reading	1467	84	56.2	15.1	84	56.7	14.8	0.85
Reading Vocabulary	1632	30	22.6	5.6	30	21.8	5.3	0.78
Reading Comprehension	1651	54	33.2	10.6	54	34.3	10.8	0.80
Total Mathematics	593	80	47.4	16.0	78	51.8	15.1	0.85
Mathematics Problem Solving	741	48	27.8	10.0	48	31.7	9.3	0.82
Mathematics Procedures	735	32	18.7	7.3	30	19.8	6.7	0.74
Language	580	48	33.4	9.5	48	31.5	9.4	0.85
Language Mechanics	687	24	16.7	5.1	24	15.9	4.7	0.74
Language Expression	687	24	16.1	5.4	24	15.6	5.2	0.80
Spelling	1061	40	26.5	7.9	30	20.4	5.4	0.83
Science	593	40	23.3	7.1	40	25.3	7.1	0.70
Social Science	597	40	24.5	8.3	40	21.7	7.9	0.77
Listening	602	40	25.4	6.4	40	27.7	6.4	0.77
Subtest	N	SAT10 – Form D			SAT9 – Form SA			r
		Number of Items	Mean	SD	Number of Items	Mean	SD	
Language	347	48	26.5	8.7	54	33.4	10.8	0.85
Prewriting	908	12	7.6	2.7	15	9.7	3.6	0.72
Composing	909	18	9.3	3.7	15	8.3	3.7	0.72
Editing	909	18	9.9	3.5	24	14.1	5.0	0.70

Appendix L

Table L–8. Pearson Product-Moment Coefficients of Correlation (r) for Corresponding Stanford 10 and Stanford 9 Subtests and Totals for Intermediate 3

Subtest	N	SAT10 – Form A			SAT9 – Form S			r
		Number of Items	Mean	SD	Number of Items	Mean	SD	
Total Reading	949	84	61.4	13.7	84	61.2	13.8	0.83
Reading Vocabulary	1038	30	23.7	4.7	30	22.4	5.0	0.70
Reading Comprehension	1041	54	37.1	10.2	54	38.4	10.3	0.79
Total Mathematics	687	80	45.0	16.0	78	51.9	15.1	0.88
Mathematics Problem Solving	789	48	26.7	9.7	48	31.7	9.1	0.83
Mathematics Procedures	789	32	18.4	7.2	30	20.2	6.7	0.80
Language	465	48	34.8	8.0	48	31.7	8.7	0.78
Language Mechanics	513	24	17.5	4.1	24	16.4	4.4	0.65
Language Expression	513	24	17.5	4.6	24	15.7	4.9	0.70
Spelling	911	40	26.7	7.5	30	21.5	5.2	0.77
Science	689	40	26.6	6.6	40	26.3	6.3	0.77
Social Science	688	40	23.9	7.5	40	24.2	7.2	0.81
Listening	550	40	26.5	6.7	40	28.6	6.1	0.76

Subtest	N	SAT10 – Form D			SAT9 – Form SA			r
		Number of Items	Mean	SD	Number of Items	Mean	SD	
Language	219	48	29.8	8.7	54	35.5	9.7	0.80
Prewriting	535	12	7.8	2.3	15	10.4	3.0	0.57
Composing	535	18	10.6	3.9	15	9.9	3.2	0.68
Editing	535	18	11.5	3.7	24	14.8	4.9	0.65

Table L–9. Pearson Product-Moment Coefficients of Correlation (r) for Corresponding Stanford 10 and Stanford 9 Subtests and Totals for Advanced 1

Subtest	N	SAT10 – Form A			SAT9 – Form S			r
		Number of Items	Mean	SD	Number of Items	Mean	SD	
Total Reading	1274	84	58.0	14.3	84	60.6	13.2	0.84
Reading Vocabulary	1429	30	22.8	5.5	30	23.1	4.5	0.78
Reading Comprehension	1433	54	35.6	9.7	54	37.7	9.8	0.76
Total Mathematics	516	80	44.5	16.0	80	48.8	16.5	0.90
Mathematics Problem Solving	633	48	25.5	9.3	50	31.1	10.0	0.85
Mathematics Procedures	633	32	19.0	7.8	30	18.4	7.3	0.84
Language	329	48	34.1	9.7	48	31.7	9.1	0.80
Language Mechanics	378	24	17.2	4.7	24	15.9	4.6	0.72
Language Expression	378	24	17.1	5.5	24	15.3	5.0	0.72
Spelling	1117	40	25.4	7.6	30	20.1	5.5	0.77
Science	514	40	25.2	7.0	40	24.9	6.2	0.78
Social Science	511	40	22.3	7.1	40	21.7	7.0	0.73
Listening	481	40	26.7	6.2	40	27.8	6.1	0.78

Subtest	N	SAT10 – Form D			SAT9 – Form SA			r
		Number of Items	Mean	SD	Number of Items	Mean	SD	
Language	324	48	29.3	8.7	54	35.1	9.6	0.77
Prewriting	1070	12	7.5	2.6	15	9.7	3.2	0.59
Composing	1072	18	10.6	3.8	15	10.3	3.3	0.69
Editing	1072	18	10.6	4.1	24	15.8	4.4	0.66

Appendix L

Table L–10. Pearson Product-Moment Coefficients of Correlation (r) for Corresponding Stanford 10 and Stanford 9 Subtests and Totals for Advanced 2

Subtest	N	SAT10 – Form A			SAT9 – Form S			r
		Number of Items	Mean	SD	Number of Items	Mean	SD	
Total Reading	737	84	60.2	13.3	84	61.5	13.3	0.80
Reading Vocabulary	841	30	23.1	5.2	30	22.5	5.2	0.70
Reading Comprehension	839	54	36.9	9.3	54	38.2	9.9	0.73
Total Mathematics	565	80	44.0	15.5	82	49.8	15.9	0.85
Mathematics Problem Solving	693	48	25.7	9.5	52	31.5	10.2	0.78
Mathematics Procedures	693	32	17.9	7.4	30	17.7	6.9	0.79
Language	329	48	35.2	8.8	48	33.3	8.6	0.82
Language Mechanics	360	24	17.7	4.4	24	16.8	4.9	0.75
Language Expression	360	24	17.3	5.3	24	15.5	4.8	0.73
Spelling	682	40	27.0	6.2	30	21.1	5.1	0.73
Science	573	40	21.7	6.6	40	25.1	7.3	0.74
Social Science	564	40	20.7	7.5	40	22.0	7.3	0.79
Listening	479	40	26.3	6.1	40	27.9	6.3	0.71

Subtest	N	SAT10 – Form D			SAT9 – Form SA			r
		Number of Items	Mean	SD	Number of Items	Mean	SD	
Language	247	48	30.7	8.8	54	35.0	8.7	0.80
Prewriting	472	12	8.9	2.5	15	10.7	2.7	0.51
Composing	475	18	11.0	3.6	15	10.2	3.0	0.68
Editing	474	18	11.4	3.7	24	14.5	4.4	0.66

Table L–11. Pearson Product-Moment Coefficients of Correlation (r) for Corresponding Stanford 10 and Stanford 9 Subtests and Totals for TASK 1

Subtest	N	SAT10 – Form A			SAT9 – Form S			r
		Number of Items	Mean	SD	Number of Items	Mean	SD	
Total Reading	651	84	60.5	11.8	84	61.0	13.0	0.82
Reading Vocabulary	674	30	24.4	4.4	30	21.5	4.7	0.76
Reading Comprehension	673	54	36.3	8.3	54	39.5	8.9	0.75
Mathematics	252	50	24.6	6.9	48	25.8	8.6	0.77
Language	207	48	35.3	9.3	48	31.3	10.1	0.82
Language Mechanics	232	24	17.3	4.7	24	15.7	4.9	0.77
Language Expression	232	24	18.0	5.2	24	16.3	4.8	0.75
Spelling	390	40	26.3	7.4	30	19.7	5.0	0.75
Science	243	40	21.6	6.0	40	22.7	6.1	0.68
Social Science	235	40	19.9	6.1	40	18.9	6.4	0.73
Subtest	N	SAT10 – Form D			SAT9 – Form SA			r
		Number of Items	Mean	SD	Number of Items	Mean	SD	
Language	260	48	35.6	6.6	54	35.6	7.6	0.72
Prewriting	500	12	8.5	2.8	15	10.1	2.4	0.46
Composing	484	18	12.7	4.3	15	9.8	2.5	0.67
Editing	444	18	12.7	3.1	24	15.8	3.9	0.47

Appendix L

Table L–12. Pearson Product-Moment Coefficients of Correlation (r) for Corresponding Stanford 10 and Stanford 9 Subtests and Totals for TASK 2

Subtest	N	SAT10 – Form A			SAT9 – Form S			r
		Number of Items	Mean	SD	Number of Items	Mean	SD	
Total Reading	606	84	62.9	11.0	84	59.5	13.8	0.82
Reading Vocabulary	630	30	24.2	4.2	30	21.7	4.8	0.77
Reading Comprehension	632	54	38.5	7.9	54	37.6	10.1	0.76
Mathematics	175	50	23.7	8.3	48	24.7	9.2	0.78
Language	187	48	34.0	9.4	48	32.6	9.5	0.87
Language Mechanics	199	24	16.1	5.0	24	15.2	5.2	0.77
Language Expression	199	24	17.6	5.2	24	16.4	5.1	0.79
Spelling	307	40	27.1	7.2	30	19.6	5.6	0.79
Science	170	40	22.6	7.1	40	21.1	6.6	0.79
Social Science	169	40	18.3	6.1	40	18.6	6.0	0.72
Subtest	N	SAT10 – Form D			SAT9 – Form SA			r
		Number of Items	Mean	SD	Number of Items	Mean	SD	
Language	312	48	33.4	7.2	54	36.4	9.8	0.79
Prewriting	428	12	8.6	2.4	15	9.5	3.0	0.63
Composing	428	18	11.7	3.3	15	10.7	3.2	0.65
Editing	428	18	12.1	3.5	24	16.0	4.7	0.73

Table L–13. Pearson Product-Moment Coefficients of Correlation (r) for Corresponding Stanford 10 and Stanford 9 Subtests and Totals for TASK 3

| Subtest | N | SAT10 – Form A | | | SAT9 – Form S | | | r |
		Number of Items	Mean	SD	Number of Items	Mean	SD	
Total Reading	478	84	60.8	14.7	84	58.6	13.7	0.81
Reading Vocabulary	512	30	23.6	5.4	30	22.4	4.8	0.75
Reading Comprehension	508	54	36.7	10.6	54	35.7	10.1	0.74
Mathematics	129	50	23.0	9.3	48	24.3	9.0	0.73
Language	143	48	33.2	10.2	48	31.2	9.0	0.75
Language Mechanics	156	24	15.9	5.2	24	15.6	4.8	0.68
Language Expression	156	24	17.3	5.5	24	14.8	4.9	0.69
Spelling	350	40	26.6	7.6	30	18.5	5.6	0.80
Science	127	40	19.5	6.8	40	19.1	6.5	0.69
Social Science	125	40	17.9	6.6	40	19.3	7.0	0.74
Subtest	N	SAT10 – Form D			SAT9 – Form SA			r
		Number of Items	Mean	SD	Number of Items	Mean	SD	
Language	140	48	32.1	7.9	54	38.4	9.0	0.87
Prewriting	348	12	8.5	2.3	15	11.1	2.9	0.62
Composing	348	18	11.6	3.6	15	10.5	2.7	0.65
Editing	348	18	11.9	3.6	24	16.8	4.4	0.85

Appendix M:

Intercorrelations Among Stanford 10 Subtests and *Otis-Lennon School Ability Test*®, Eighth Edition (OLSAT® 8)

Table M–1.   Intercorrelations Among Stanford 10 SESAT 1 and *Otis-Lennon School Ability Test*®, Eighth Edition, Spring, Grade K (N=1213)

Total/Subtest	Variable	2	3	4	5	6	7	8	9
Total Reading	1	.88	.94	.72	.59	.59	.63	.60	.56
Sounds and Letters	2		.67	.68	.57	.57	.59	.58	.51
Word Reading	3			.64	.51	.52	.57	.53	.52
Mathematics	4				.72	.68	.73	.71	.63
Environment	5					.74	.70	.70	.60
Listening to Words and Stories	6						.70	.72	.58
OLSAT Total	7							.92	.92
OLSAT Verbal	8								.70
OLSAT Nonverbal	9								

Table M–2. Intercorrelations Among Stanford 10 SESAT 2 and *Otis-Lennon School Ability Test*®, Eighth Edition, Spring, Grade K (N=2023)

Total/Subtest	Variable	2	3	4	5	6	7	8	9	10
Total Reading	1	.84	.92	.90	.71	.56	.54	.68	.64	.61
Sounds and Letters	2		.69	.59	.70	.56	.54	.63	.61	.55
Word Reading	3			.77	.63	.47	.48	.60	.56	.54
Sentence Reading	4				.57	.47	.43	.59	.55	.54
Mathematics	5					.65	.64	.73	.72	.62
Environment	6						.68	.68	.69	.56
Listening to Words and Stories	7							.68	.71	.54
OLSAT Total	8								.92	.92
OLSAT Verbal	9									.69
OLSAT Nonverbal	10									

355

Table M–3. Intercorrelations Among Stanford 10 Primary 1 Form A and *Otis-Lennon School Ability Test®*, Eighth Edition, Spring, Grade 1 (N=2189)

Total/Subtest	Variable	2	3	4	5	6	7	8	9	10	11	12	13	14	15
Total Reading	1	.84	.92	.95	.88	.70	.66	.62	.75	.83	.53	.55	.61	.59	.54
Word Study Skills	2		.69	.74	.65	.68	.65	.58	.71	.69	.54	.54	.62	.61	.53
Word Reading	3			.83	.75	.58	.53	.52	.62	.77	.42	.42	.49	.46	.45
Reading Comprehension	4				.79	.67	.63	.59	.72	.82	.49	.52	.57	.55	.50
Sentence Reading	5					.60	.57	.52	.63	.68	.49	.50	.54	.52	.47
Total Mathematics	6						.93	.90	.73	.59	.61	.63	.71	.68	.63
Mathematics Problem Solving	7							.68	.72	.53	.63	.67	.71	.69	.62
Mathematics Procedures	8								.60	.56	.47	.47	.58	.54	.53
Language	9									.66	.62	.69	.68	.67	.58
Spelling	10										.37	.41	.47	.45	.42
Environment	11											.70	.61	.62	.51
Listening	12												.64	.66	.51
OLSAT Total	13													.91	.92
OLSAT Verbal	14														.69
OLSAT Nonverbal	15														

Table M–4. Intercorrelations Among Stanford 10 Primary 2 Form A and *Otis-Lennon School Ability Test*®, Eighth Edition, Spring, Grade 2 (N=2550)

Total/Subtest	Variable	2	3	4	5	6	7	8	9	10	11	12	13	14
Total Reading	1	.82	.92	.94	.73	.70	.65	.79	.73	.57	.66	.67	.64	.60
Word Study Skills	2		.67	.65	.65	.62	.58	.65	.60	.46	.52	.61	.56	.56
Reading Vocabulary	3			.79	.64	.62	.57	.71	.73	.51	.57	.57	.54	.52
Reading Comprehension	4				.68	.65	.60	.75	.65	.54	.65	.64	.61	.56
Total Mathematics	5					.95	.91	.75	.57	.62	.63	.77	.74	.67
Mathematics Problem Solving	6						.72	.73	.52	.63	.66	.77	.76	.66
Mathematics Procedures	7							.67	.55	.49	.50	.64	.61	.57
Language	8								.62	.62	.67	.71	.66	.64
Spelling	9									.38	.41	.48	.44	.44
Environment	10										.68	.65	.64	.54
Listening	11											.66	.68	.54
OLSAT Total	12												.93	.91
OLSAT Verbal	13													.69
OLSAT Nonverbal	14													

Table M–5. Intercorrelations Among Stanford 10 Primary 3 Form A and *Otis-Lennon School Ability Test* ®, Eighth Edition, Spring, Grade 3 (N=1680)

Total/Subtest	Variable	2	3	4	5	6	7	8	9	10	11	12	13	14	15
Total Reading	1	.83	.91	.95	.76	.76	.66	.84	.70	.83	.82	.71	.74	.76	.58
Word Study Skills	2		.69	.67	.63	.62	.55	.68	.59	.63	.61	.53	.64	.62	.55
Reading Vocabulary	3			.80	.68	.69	.58	.77	.68	.77	.77	.66	.68	.72	.52
Reading Comprehension	4				.74	.73	.65	.81	.63	.81	.80	.69	.68	.72	.52
Total Mathematics	5					.96	.92	.76	.58	.72	.71	.57	.73	.69	.65
Mathematics Problem Solving	6						.76	.74	.57	.73	.71	.59	.75	.71	.66
Mathematics Procedures	7							.68	.52	.60	.60	.47	.61	.57	.54
Language	8								.67	.76	.78	.65	.70	.71	.56
Spelling	9									.59	.59	.44	.52	.55	.38
Science	10										.81	.71	.71	.74	.55
Social Science	11											.70	.69	.73	.52
Listening	12												.58	.60	.45
OLSAT Total	13													.92	.91
OLSAT Verbal	14														.66
OLSAT Nonverbal	15														

Table M–6. *Intercorrelations Among Stanford 10 Intermediate 1 Form A and Otis-Lennon School Ability Test*®, *Eighth Edition, Spring, Grade 4 (N=2336)*

Total/Subtest	Variable	2	3	4	5	6	7	8	9	10	11	12	13	14	15	16	17
Total Reading	1	.86	.92	.94	.79	.79	.68	.84	.76	.81	.72	.79	.80	.74	.71	.73	.60
Word Study Skills	2		.73	.68	.66	.68	.56	.71	.67	.65	.67	.63	.62	.57	.61	.62	.52
Reading Vocabulary	3			.80	.70	.72	.60	.76	.67	.74	.66	.74	.74	.71	.65	.67	.55
Reading Comprehension	4				.76	.76	.66	.76	.67	.74	.65	.76	.78	.73	.67	.68	.57
Total Mathematics	5					.96	.93	.81	.72	.79	.62	.76	.77	.66	.69	.65	.65
Mathematics Problem Solving	6						.78	.79	.73	.74	.60	.74	.76	.66	.68	.65	.63
Mathematics Procedures	7							.71	.71	.72	.56	.63	.68	.58	.62	.57	.59
Language	8								.66	.66	.72	.73	.79	.67	.64	.64	.57
Language Mechanics	9									.93	.66	.65	.70	.61	.59	.57	.54
Language Expression	10										.74	.71	.77	.65	.61	.62	.53
Spelling	11											.68	.57	.53	.53	.55	.44
Science	12												.63	.71	.61	.55	.52
Social Science	13													.69	.65	.66	.57
Listening	14														.55	.58	.46
OLSAT Total	15															.94	.94
OLSAT Verbal	16																.76
OLSAT Nonverbal	17																

Table M–7. Intercorrelations Among Stanford 10 Intermediate 2 Form A and *Otis-Lennon School Ability Test*®, Eighth Edition, Spring, Grade 5 (N=2224)

Total/Subtest	Variable	2	3	4	5	6	7	8	9	10	11	12	13	14	15	16
Total Reading	1	.90	.97	.74	.75	.60	.81	.72	.79	.67	.76	.79	.72	.67	.70	.57
Reading Vocabulary	2		.76	.64	.65	.51	.73	.66	.70	.60	.69	.68	.67	.58	.60	.49
Reading Comprehension	3			.73	.74	.60	.79	.69	.77	.65	.73	.78	.68	.67	.69	.56
Total Mathematics	4				.95	.91	.77	.69	.74	.62	.65	.69	.58	.71	.64	.68
Mathematics Problem Solving	5					.73	.74	.66	.71	.61	.66	.69	.60	.70	.65	.67
Mathematics Procedures	6						.69	.62	.66	.55	.53	.58	.46	.59	.53	.59
Language	7							.92	.94	.70	.71	.77	.64	.65	.64	.59
Language Mechanics	8								.73	.66	.61	.69	.56	.60	.57	.55
Language Expression	9									.65	.71	.75	.62	.62	.62	.54
Spelling	10										.51	.60	.50	.52	.53	.45
Science	11											.74	.67	.63	.63	.55
Social Science	12												.67	.62	.65	.52
Listening	13													.53	.55	.45
OLSAT Total	14														.94	.94
OLSAT Verbal	15															.77
OLSAT Nonverbal	16															

Table M–8. Intercorrelations Among Stanford 10 Intermediate 3 Form A and *Otis-Lennon School Ability Test*®, Eighth Edition, Spring, Grade 6 (N=2510)

Total/Subtest	Variable	2	3	4	5	6	7	8	9	10	11	12	13	14	15	16
Total Reading	1	.88	.97	.74	.73	.65	.77	.66	.77	.66	.75	.78	.74	.69	.67	.62
Reading Vocabulary	2		.75	.65	.65	.56	.68	.58	.68	.60	.68	.71	.68	.61	.60	.54
Reading Comprehension	3			.72	.70	.64	.75	.65	.75	.63	.72	.75	.71	.67	.64	.60
Total Mathematics	4				.96	.92	.72	.64	.70	.61	.69	.73	.63	.73	.63	.73
Mathematics Problem Solving	5					.78	.70	.62	.68	.60	.70	.72	.64	.73	.62	.72
Mathematics Procedures	6						.65	.58	.63	.55	.58	.65	.54	.64	.55	.64
Language	7							.92	.94	.67	.67	.70	.60	.66	.61	.61
Language Mechanics	8								.74	.63	.59	.61	.52	.59	.54	.56
Language Expression	9									.63	.66	.70	.60	.63	.59	.59
Spelling	10										.55	.60	.46	.55	.53	.49
Science	11											.74	.68	.66	.62	.60
Social Science	12												.71	.66	.62	.60
Listening	13													.61	.59	.55
OLSAT Total	14														.92	.94
OLSAT Verbal	15															.72
OLSAT Nonverbal	16															

Appendix M

Table M–9.   Intercorrelations Among Stanford 10 Advanced 1 Form A and *Otis-Lennon School Ability Test*®, Eighth Edition, Spring, Grade 7 (N=1468)

Total/Subtest	Variable	2	3	4	5	6	7	8	9	10	11	12	13	14	15	16
Total Reading	1	.91	.97	.71	.69	.64	.82	.75	.79	.73	.76	.76	.76	.73	.75	.61
Reading Vocabulary	2		.78	.65	.64	.57	.76	.71	.72	.71	.73	.71	.72	.68	.71	.56
Reading Comprehension	3			.68	.66	.62	.78	.71	.76	.68	.71	.73	.72	.69	.70	.59
Total Mathematics	4				.95	.93	.70	.65	.67	.61	.69	.68	.57	.78	.72	.73
Mathematics Problem Solving	5					.77	.66	.61	.65	.59	.70	.67	.57	.75	.70	.70
Mathematics Procedures	6						.65	.62	.62	.55	.59	.61	.50	.71	.65	.67
Language	7							.94	.96	.71	.67	.71	.66	.70	.68	.62
Language Mechanics	8								.79	.67	.63	.66	.61	.65	.63	.59
Language Expression	9									.68	.64	.68	.65	.67	.65	.59
Spelling	10										.57	.64	.59	.61	.63	.50
Science	11											.73	.69	.65	.66	.56
Social Science	12												.69	.69	.70	.59
Listening	13													.62	.66	.50
OLSAT Total	14														.92	.93
OLSAT Verbal	15															.72
OLSAT Nonverbal	16															

Table M–10. Intercorrelations Among Stanford 10 Advanced 2 Form A and *Otis-Lennon School Ability Test*®, Eighth Edition, Spring, Grade 8 (N=1507)

Total/Subtest	Variable	2	3	4	5	6	7	8	9	10	11	12	13	14	15	16
Total Reading	1	.90	.97	.68	.69	.59	.82	.72	.81	.70	.75	.78	.72	.73	.72	.65
Reading Vocabulary	2		.78	.61	.62	.53	.74	.68	.70	.66	.68	.71	.69	.64	.65	.56
Reading Comprehension	3			.67	.67	.58	.80	.69	.80	.67	.73	.76	.68	.72	.71	.65
Total Mathematics	4				.96	.93	.70	.62	.68	.61	.68	.70	.57	.74	.67	.72
Mathematics Problem Solving	5					.79	.70	.61	.67	.60	.69	.71	.58	.73	.66	.70
Mathematics Procedures	6						.63	.56	.61	.54	.59	.62	.48	.66	.59	.64
Language	7							.92	.95	.73	.68	.74	.66	.71	.69	.65
Language Mechanics	8								.76	.69	.59	.63	.60	.63	.62	.58
Language Expression	9									.68	.67	.74	.64	.69	.67	.63
Spelling	10										.59	.60	.56	.61	.60	.55
Science	11											.75	.63	.64	.62	.59
Social Science	12												.68	.68	.68	.61
Listening	13													.55	.58	.46
OLSAT Total	14														.93	.95
OLSAT Verbal	15															.78
OLSAT Nonverbal	16															

Table M–11. Intercorrelations Among Stanford 10 Advanced 2 Form A and *Otis-Lennon School Ability Test*®, Eighth Edition, Spring, Grade 9 (N=774)

Total/Subtest	Variable	2	3	4	5	6	7	8	9	10	11	12	13	14	15	16
Total Reading	1	.91	.97	.66	.70	.51	.75	.73	.69	.69	.69	.74	.70	.65	.66	.59
Reading Vocabulary	2		.79	.61	.66	.45	.65	.66	.57	.63	.64	.67	.65	.62	.64	.55
Reading Comprehension	3			.64	.67	.51	.74	.71	.70	.67	.67	.72	.68	.62	.63	.56
Total Mathematics	4				.95	.91	.67	.64	.62	.61	.72	.72	.60	.69	.65	.67
Mathematics Problem Solving	5					.74	.67	.65	.62	.63	.74	.73	.60	.71	.67	.70
Mathematics Procedures	6						.57	.53	.54	.48	.59	.60	.50	.55	.52	.54
Language	7							.93	.96	.67	.68	.73	.68	.66	.65	.62
Language Mechanics	8								.78	.67	.70	.70	.65	.66	.65	.61
Language Expression	9									.62	.68	.67	.63	.60	.58	.56
Spelling	10										.59	.60	.52	.57	.55	.53
Science	11											.78	.66	.69	.66	.66
Social Science	12												.67	.73	.71	.68
Listening	13													.56	.57	.49
OLSAT Total	14														.95	.96
OLSAT Verbal	15															.83
OLSAT Nonverbal	16															

Table M–12. Intercorrelations Among Stanford 10 TASK 1 Form A and *Otis-Lennon School Ability Test*®, Eighth Edition, Spring, Grade 9 (N=859)

Total/Subtest	Variable	2	3	4	5	6	7	8	9	10	11	12	13
Total Reading	1	.90	.97	.69	.81	.75	.79	.73	.75	.74	.74	.73	.67
Reading Vocabulary	2		.76	.61	.73	.69	.69	.71	.70	.67	.66	.65	.59
Reading Comprehension	3			.67	.79	.73	.77	.68	.71	.71	.72	.71	.65
Mathematics	4				.68	.65	.65	.62	.66	.64	.70	.62	.69
Language	5					.94	.96	.80	.68	.67	.71	.66	.66
Language Mechanics	6						.80	.76	.64	.62	.67	.62	.63
Language Expression	7							.75	.64	.65	.67	.64	.62
Spelling	8								.61	.61	.63	.61	.56
Science	9									.66	.66	.63	.60
Social Science	10										.65	.63	.58
OLSAT Total	11											.93	.94
OLSAT Verbal	12												.75
OLSAT Nonverbal	13												

Table M–13. Intercorrelations Among Stanford 10 TASK 2 Form A and *Otis-Lennon School Ability Test*®, Eighth Edition, Spring, Grade 10 (N=1575)

Total/Subtest	Variable	2	3	4	5	6	7	8	9	10	11	12	13
Total Reading	1	.87	.97	.64	.75	.69	.73	.71	.71	.66	.62	.63	.53
Reading Vocabulary	2		.71	.54	.65	.62	.60	.65	.63	.58	.54	.56	.47
Reading Comprehension	3			.62	.73	.65	.72	.66	.68	.63	.59	.61	.51
Mathematics	4				.63	.60	.58	.53	.70	.63	.59	.55	.56
Language	5					.93	.95	.73	.67	.63	.60	.59	.53
Language Mechanics	6						.76	.69	.63	.58	.56	.57	.50
Language Expression	7							.68	.63	.60	.56	.55	.50
Spelling	8								.59	.58	.55	.57	.47
Science	9									.73	.59	.57	.53
Social Science	10										.60	.59	.53
OLSAT Total	11											.94	.95
OLSAT Verbal	12												.77
OLSAT Nonverbal	13												

Table M–14. Intercorrelations Among Stanford 10 TASK 3 Form A and *Otis-Lennon School Ability Test®*, Eighth Edition, Spring, Grade 11 (N=1365)

Total/Subtest	Variable	2	3	4	5	6	7	8	9	10	11	12	13
Total Reading	1	.92	.97	.65	.82	.77	.78	.76	.73	.67	.52	.54	.45
Reading Vocabulary	2		.81	.55	.72	.70	.67	.74	.65	.59	.45	.48	.39
Reading Comprehension	3			.66	.82	.75	.79	.72	.72	.67	.52	.53	.46
Mathematics	4				.68	.65	.65	.56	.74	.67	.62	.59	.59
Language	5					.94	.96	.77	.74	.69	.55	.56	.49
Language Mechanics	6						.80	.74	.72	.66	.54	.54	.49
Language Expression	7							.72	.69	.66	.51	.52	.45
Spelling	8								.60	.54	.47	.49	.41
Science	9									.74	.56	.55	.52
Social Science	10										.53	.54	.48
OLSAT Total	11											.95	.96
OLSAT Verbal	12												.81
OLSAT Nonverbal	13												

Table M–15. Intercorrelations Among Stanford 10 TASK 3 Form A and *Otis-Lennon School Ability Test®*, Eighth Edition, Spring, Grade 12 (N=1375)

Total/Subtest	Variable	2	3	4	5	6	7	8	9	10	11	12	13
Total Reading	1	.88	.97	.61	.78	.73	.75	.69	.65	.64	.59	.60	.52
Reading Vocabulary	2		.72	.51	.64	.61	.60	.64	.56	.51	.53	.55	.47
Reading Comprehension	3			.59	.78	.72	.76	.66	.63	.65	.56	.57	.50
Mathematics	4				.64	.63	.60	.56	.64	.56	.56	.51	.54
Language	5					.94	.96	.72	.69	.64	.62	.63	.55
Language Mechanics	6						.81	.71	.64	.58	.59	.59	.53
Language Expression	7							.66	.66	.64	.60	.61	.53
Spelling	8								.55	.50	.56	.58	.49
Science	9									.65	.57	.55	.54
Social Science	10										.54	.53	.50
OLSAT Total	11											.94	.95
OLSAT Verbal	12												.79
OLSAT Nonverbal	13												

Table M–16. Intercorrelations Among Stanford 10 Primary 1 Form D and *Otis-Lennon School Ability Test*®, Eighth Edition, Spring, Grade 1 (N=1233)

Total/Subtest	Variable	2	3	4	5	6	7	8	9	10	11	12	13	14	15
Total Reading	1	.85	.92	.94	.86	.74	.71	.66	.80	.55	.60	.70	.63	.60	.57
Word Study Skills	2		.72	.74	.62	.75	.73	.65	.69	.56	.62	.68	.68	.64	.61
Word Reading	3			.82	.77	.63	.59	.57	.74	.42	.44	.55	.50	.47	.46
Reading Comprehension	4				.76	.71	.68	.63	.78	.55	.60	.69	.58	.56	.52
Sentence Reading	5					.57	.54	.51	.65	.46	.47	.55	.51	.49	.46
Total Mathematics	6						.94	.90	.61	.70	.72	.78	.77	.75	.69
Mathematics Problem Solving	7							.90	.56	.73	.76	.78	.77	.76	.67
Mathematics Procedures	8								.57	.54	.55	.65	.64	.60	.58
Spelling	9									.35	.41	.56	.49	.46	.46
Environment	10										.77	.72	.68	.68	.58
Listening	11											.77	.70	.71	.59
Language	12												.71	.72	.61
OLSAT Total	13													.92	.93
OLSAT Verbal	14														.72
OLSAT Nonverbal	15														

Appendix M

Table M–17. Intercorrelations Among Stanford 10 Primary 2 Form D and *Otis-Lennon School Ability Test*®, Eighth Edition, Spring, Grade 2 (N=958)

Total/Subtest	Variable	2	3	4	5	6	7	8	9	10	11	12	13	14
Total Reading	1	.79	.92	.94	.75	.74	.64	.70	.61	.69	.74	.70	.65	.64
Word Study Skills	2		.61	.62	.65	.64	.57	.52	.51	.50	.61	.63	.60	.56
Reading Vocabulary	3			.80	.66	.65	.56	.70	.54	.63	.65	.59	.54	.56
Reading Comprehension	4				.69	.69	.59	.63	.58	.68	.70	.65	.61	.60
Total Mathematics	5					.95	.91	.52	.66	.65	.73	.77	.75	.68
Mathematics Problem Solving	6						.74	.48	.68	.69	.74	.80	.77	.70
Mathematics Procedures	7							.48	.53	.49	.60	.62	.59	.56
Spelling	8								.37	.40	.52	.42	.36	.42
Environment	9									.75	.71	.69	.67	.60
Listening	10										.72	.69	.69	.59
Language	11											.71	.68	.62
OLSAT Total	12												.93	.92
OLSAT Verbal	13													.71
OLSAT Nonverbal	14													

Table M–18. Intercorrelations Among Stanford 10 Primary 3 Form D and *Otis-Lennon School Ability Test®*, Eighth Edition, Spring, Grade 3 (N=600)

Total/Subtest	Variable	2	3	4	5	6	7	8	9	10	11	12	13	14	15	16	17	18
Total Reading	1	.83	.93	.96	.81	.82	.67	.68	.85	.87	.77	.83	.74	.72	.74	.79	.80	.66
Word Study Skills	2		.70	.69	.70	.69	.60	.64	.65	.68	.58	.69	.62	.59	.62	.68	.65	.59
Reading Vocabulary	3			.85	.74	.76	.60	.62	.80	.81	.73	.76	.69	.66	.67	.74	.75	.60
Reading Comprehension	4				.77	.79	.63	.63	.84	.85	.75	.80	.72	.70	.71	.76	.77	.61
Total Mathematics	5					.95	.91	.57	.77	.77	.67	.78	.67	.67	.73	.80	.76	.70
Mathematics Problem Solving	6						.73	.55	.79	.80	.69	.76	.66	.65	.71	.82	.79	.72
Mathematics Procedures	7							.50	.61	.60	.54	.69	.58	.60	.65	.63	.60	.56
Spelling	8								.58	.59	.46	.60	.51	.53	.55	.53	.56	.41
Science	9									.85	.77	.76	.71	.66	.66	.77	.78	.62
Social Science	10										.77	.79	.74	.67	.69	.76	.79	.61
Listening	11											.68	.66	.60	.56	.68	.69	.55
Language	12												.87	.91	.88	.72	.72	.59
Prewriting	13													.70	.65	.62	.65	.49
Composing	14														.68	.62	.62	.52
Editing	15															.67	.66	.56
OLSAT Total	16																.92	.91
OLSAT Verbal	17																	.68
OLSAT Nonverbal	18																	

Table M–19. Intercorrelations Among Stanford 10 Intermediate 1 Form D and *Otis-Lennon School Ability Test*®, Eighth Edition, Spring, Grade 4 (N=308)

Total/Subtest	Variable	2	3	4	5	6	7	8	9	10	11	12	13	14	15	16	17	18
Total Reading	1	.86	.91	.93	.78	.82	.62	.74	.76	.80	.74	.79	.64	.73	.67	.69	.71	.60
Word Study Skills	2		.75	.66	.65	.70	.49	.66	.60	.61	.57	.64	.48	.60	.54	.57	.59	.49
Reading Vocabulary	3			.75	.69	.72	.55	.65	.72	.71	.72	.69	.57	.61	.59	.60	.63	.50
Reading Comprehension	4				.76	.78	.61	.70	.72	.80	.70	.78	.64	.71	.66	.68	.68	.59
Total Mathematics	5					.94	.91	.61	.69	.68	.62	.77	.56	.69	.72	.72	.69	.67
Mathematics Problem Solving	6						.72	.61	.70	.69	.66	.78	.60	.69	.70	.74	.71	.68
Mathematics Procedures	7							.51	.56	.56	.48	.65	.43	.59	.62	.59	.57	.54
Spelling	8								.54	.62	.51	.67	.56	.61	.55	.53	.55	.45
Science	9									.77	.67	.75	.64	.64	.66	.61	.60	.55
Social Science	10										.66	.73	.59	.66	.63	.64	.65	.56
Listening	11											.67	.60	.58	.57	.54	.54	.47
Language	12												.78	.71	.87	.68	.68	.60
Prewriting	13													.61	.53	.51	.52	.45
Composing	14														.67	.59	.59	.52
Editing	15															.63	.62	.57
OLSAT Total	16																.94	.94
OLSAT Verbal	17																	.77
OLSAT Nonverbal	18																	

Table M–20. Intercorrelations Among Stanford 10 Intermediate 2 Form D and *Otis-Lennon School Ability Test*, Eighth Edition, Spring, Grade 5 (N=576)

Total/Subtest	Variable	2	3	4	5	6	7	8	9	10	11	12	13	14	15	16	17
Total Reading	1	.91	.97	.73	.75	.57	.68	.79	.79	.73	.81	.75	.75	.69	.69	.73	.59
Reading Vocabulary	2		.78	.63	.65	.49	.61	.72	.72	.68	.70	.65	.64	.60	.61	.66	.50
Reading Comprehension	3			.73	.74	.58	.66	.76	.76	.70	.82	.75	.76	.69	.69	.71	.60
Total Mathematics	4				.95	.90	.58	.68	.76	.64	.75	.63	.70	.68	.78	.73	.75
Mathematics Problem Solving	5					.72	.58	.70	.73	.66	.76	.64	.71	.68	.76	.72	.73
Mathematics Procedures	6						.49	.53	.57	.50	.62	.51	.57	.58	.66	.62	.65
Spelling	7							.59	.62	.55	.69	.62	.64	.61	.56	.58	.49
Science	8								.78	.70	.70	.66	.64	.58	.63	.64	.55
Social Science	9									.70	.74	.71	.67	.62	.67	.70	.58
Listening	10										.67	.61	.63	.55	.63	.66	.55
Language	11											.87	.91	.90	.68	.69	.61
Prewriting	12												.71	.70	.58	.60	.51
Composing	13													.71	.65	.66	.59
Editing	14														.59	.59	.54
OLSAT Total	15															.95	.95
OLSAT Verbal	16																.82
OLSAT Nonverbal	17																

Table M–21. Intercorrelations Among Stanford 10 Intermediate 3 Form D and *Otis-Lennon School Ability Test*, Eighth Edition, Spring, Grade 6 (N=316)

Total/Subtest	Variable	2	3	4	5	6	7	8	9	10	11	12	13	14	15	16	17
Total Reading	1	.92	.98	.72	.72	.65	.76	.81	.84	.76	.83	.69	.74	.73	.74	.72	.67
Reading Vocabulary	2		.83	.63	.63	.55	.67	.77	.79	.71	.73	.61	.65	.64	.69	.68	.62
Reading Comprehension	3			.73	.72	.66	.77	.79	.82	.74	.83	.69	.75	.73	.72	.70	.66
Total Mathematics	4				.97	.94	.72	.67	.72	.67	.73	.56	.66	.67	.76	.68	.75
Mathematics Problem Solving	5					.81	.70	.69	.72	.66	.70	.55	.63	.63	.76	.68	.75
Mathematics Procedures	6						.67	.56	.64	.61	.70	.52	.63	.63	.68	.61	.68
Spelling	7							.66	.72	.60	.75	.57	.68	.68	.68	.68	.64
Science	8								.74	.66	.75	.59	.69	.65	.69	.64	.65
Social Science	9									.68	.78	.61	.67	.73	.68	.64	.62
Listening	10										.68	.51	.63	.61	.64	.64	.57
Language	11											.80	.90	.89	.70	.66	.66
Prewriting	12												.63	.58	.54	.51	.51
Composing	13													.67	.62	.60	.57
Editing	14														.65	.60	.63
OLSAT Total	15															.94	.95
OLSAT Verbal	16																.78
OLSAT Nonverbal	17																

Table M–22.   Intercorrelations Among Stanford 10 Advanced 1 Form D and *Otis-Lennon School Ability Test*®, Eighth Edition, Spring, Grade 7 (N=553)

Total/Subtest	Variable	2	3	4	5	6	7	8	9	10	11	12	13	14	15	16	17
Total Reading	1	.92	.97	.74	.74	.66	.69	.81	.75	.83	.82	.75	.75	.73	.71	.74	.60
Reading Vocabulary	2		.80	.70	.72	.61	.68	.81	.72	.81	.74	.68	.68	.66	.71	.72	.62
Reading Comprehension	3			.70	.70	.63	.64	.75	.71	.78	.80	.73	.73	.72	.65	.69	.54
Total Mathematics	4				.97	.94	.59	.78	.74	.68	.74	.69	.64	.69	.83	.80	.76
Mathematics Problem Solving	5					.82	.59	.79	.73	.68	.71	.66	.62	.66	.81	.80	.72
Mathematics Procedures	6						.53	.69	.67	.62	.71	.66	.61	.65	.77	.72	.72
Spelling	7							.66	.61	.60	.68	.62	.63	.60	.59	.61	.50
Science	8								.74	.74	.70	.64	.65	.61	.71	.72	.61
Social Science	9									.74	.75	.70	.68	.67	.71	.71	.62
Listening	10										.72	.69	.65	.63	.70	.70	.60
Language	11											.87	.92	.92	.75	.73	.68
Prewriting	12												.72	.71	.68	.67	.61
Composing	13													.74	.67	.67	.59
Editing	14														.69	.65	.65
OLSAT Total	15															.93	.94
OLSAT Verbal	16																.75
OLSAT Nonverbal	17																

Table M–23. Intercorrelations Among Stanford 10 Advanced 2 Form D and *Otis-Lennon School Ability Test*, Eighth Edition, Spring, Grade 8 (N=181)

Total/Subtest	Variable	2	3	4	5	6	7	8	9	10	11	12	13	14	15	16	17
Total Reading	1	.88	.97	.72	.74	.60	.62	.79	.87	.77	.79	.75	.72	.66	.75	.76	.68
Reading Vocabulary	2		.73	.63	.64	.55	.59	.73	.81	.74	.71	.69	.69	.54	.68	.70	.61
Reading Comprehension	3			.69	.72	.58	.58	.75	.82	.71	.76	.71	.67	.66	.71	.72	.65
Total Mathematics	4				.96	.93	.59	.73	.74	.69	.72	.63	.63	.68	.79	.74	.77
Mathematics Problem Solving	5					.80	.56	.73	.76	.65	.70	.62	.60	.66	.78	.73	.75
Mathematics Procedures	6						.56	.64	.63	.66	.67	.56	.60	.63	.73	.68	.71
Spelling	7							.59	.60	.50	.70	.62	.65	.59	.55	.57	.49
Science	8								.85	.71	.68	.65	.62	.57	.66	.66	.61
Social Science	9									.77	.79	.71	.72	.70	.75	.75	.68
Listening	10										.70	.60	.67	.61	.74	.74	.68
Language	11											.87	.92	.89	.75	.74	.70
Prewriting	12												.72	.68	.61	.62	.56
Composing	13													.70	.70	.67	.66
Editing	14														.68	.68	.63
OLSAT Total	15															.95	.96
OLSAT Verbal	16																.83
OLSAT Nonverbal	17																

Table M–24. Intercorrelations Among Stanford 10 Advanced 2 Form D and *Otis-Lennon School Ability Test*®, Eighth Edition, Spring, Grade 9 (N=581)

Total/Subtest	Variable	2	3	4	5	6	7	8	9	10	11	12	13	14	15	16	17
Total Reading	1	.80	.89	.58	.57	.50	.64	.47	.60	.52	.57	.58	.50	.50	.65	.66	.59
Reading Vocabulary	2		.79	.61	.66	.45	.65	.66	.57	.63	.64	.67	.65	.63	.62	.64	.55
Reading Comprehension	3			.67	.65	.57	.65	.57	.71	.60	.67	.64	.60	.62	.55	.52	.50
Total Mathematics	4				.95	.90	.57	.69	.66	.54	.68	.56	.64	.65	.68	.56	.70
Mathematics Problem Solving	5					.71	.55	.68	.65	.55	.64	.50	.61	.61	.65	.52	.68
Mathematics Procedures	6						.49	.59	.56	.44	.61	.53	.56	.58	.61	.52	.60
Spelling	7							.50	.57	.49	.55	.50	.48	.51	.48	.45	.44
Science	8								.79	.66	.67	.60	.63	.60	.66	.60	.63
Social Science	9									.68	.74	.65	.68	.69	.64	.61	.58
Listening	10										.59	.52	.57	.52	.54	.53	.48
Language	11											.89	.93	.92	.58	.53	.55
Prewriting	12												.75	.74	.49	.43	.48
Composing	13													.76	.51	.46	.50
Editing	14														.58	.55	.52
OLSAT Total	15															.92	.94
OLSAT Verbal	16																.74
OLSAT Nonverbal	17																

Table M–25. Intercorrelations Among Stanford 10 TASK 1 Form D and *Otis-Lennon School Ability Test*®, Eighth Edition, Spring, Grade 9 (N=746)

Total/Subtest	Variable	2	3	4	5	6	7	8	9	10	11	12	13	14
Total Reading	1	.86	.96	.63	.66	.70	.75	.74	.64	.64	.69	.68	.66	.62
Reading Vocabulary	2		.68	.55	.59	.66	.63	.59	.52	.49	.56	.59	.57	.55
Reading Comprehension	3			.60	.62	.64	.72	.74	.63	.64	.68	.64	.63	.59
Mathematics	4				.48	.61	.64	.49	.49	.38	.45	.68	.58	.69
Spelling	5					.52	.51	.64	.56	.53	.61	.49	.49	.45
Science	6						.70	.54	.50	.44	.51	.59	.54	.57
Social Science	7							.62	.55	.55	.56	.64	.60	.61
Language	8								.85	.91	.88	.55	.50	.53
Prewriting	9									.69	.65	.47	.41	.47
Composing	10										.68	.44	.40	.43
Editing	11											.54	.51	.51
OLSAT Total	12												.94	.95
OLSAT Verbal	13													.78
OLSAT Nonverbal	14													

Table M–26. Intercorrelations Among Stanford 10 TASK 2 Form D and *Otis-Lennon School Ability Test*®, Eighth Edition, Spring, Grade 10 (N=527)

Total/Subtest	Variable	2	3	4	5	6	7	8	9	10	11	12	13	14
Total Reading	1	.84	.97	.67	.77	.77	.74	.75	.61	.67	.72	.74	.74	.64
Reading Vocabulary	2		.68	.53	.68	.66	.60	.60	.50	.53	.58	.63	.66	.52
Reading Comprehension	3			.67	.73	.75	.72	.74	.59	.67	.71	.71	.70	.63
Mathematics	4				.61	.70	.75	.70	.58	.66	.64	.74	.69	.69
Spelling	5					.64	.66	.77	.68	.66	.74	.63	.64	.53
Science	6						.76	.67	.59	.57	.64	.69	.69	.59
Social Science	7							.69	.61	.59	.66	.67	.69	.57
Language	8								.86	.91	.92	.65	.61	.60
Prewriting	9									.68	.73	.51	.47	.48
Composing	10										.74	.59	.53	.56
Editing	11											.64	.63	.57
OLSAT Total	12												.93	.93
OLSAT Verbal	13													.74
OLSAT Nonverbal	14													

Table M–27. Intercorrelations Among Stanford 10 TASK 3 Form D and *Otis-Lennon School Ability Test*®, Eighth Edition, Spring, Grade 11 (N=499)

Total/Subtest	Variable	2	3	4	5	6	7	8	9	10	11	12	13	14
Total Reading	1	.87	.97	.56	.74	.66	.62	.73	.71	.68	.64	.49	.44	.46
Reading Vocabulary	2		.72	.52	.70	.57	.56	.64	.63	.58	.57	.49	.47	.42
Reading Comprehension	3			.52	.69	.64	.59	.71	.68	.66	.62	.45	.38	.43
Mathematics	4				.53	.62	.62	.60	.56	.59	.53	.59	.48	.59
Spelling	5					.48	.54	.71	.65	.62	.69	.50	.46	.45
Science	6						.72	.51	.47	.48	.45	.54	.49	.50
Social Science	7							.51	.47	.47	.47	.58	.53	.52
Language	8								.90	.94	.92	.45	.35	.47
Prewriting	9									.81	.74	.36	.29	.37
Composing	10										.79	.43	.32	.45
Editing	11											.45	.36	.45
OLSAT Total	12												.90	.92
OLSAT Verbal	13													.66
OLSAT Nonverbal	14													

Table M–28.   Intercorrelations Among Stanford 10 TASK 3 Form D and *Otis-Lennon School Ability Test*, Eighth Edition, Spring, Grade 12 (N=517)

Total/Subtest	Variable	2	3	4	5	6	7	8	9	10	11	12	13	14
Total Reading	1	.84	.97	.62	.65	.75	.66	.78	.67	.73	.72	.59	.58	.51
Reading Vocabulary	2		.70	.53	.67	.64	.56	.74	.66	.68	.69	.54	.52	.48
Reading Comprehension	3			.59	.59	.72	.65	.72	.61	.69	.67	.55	.55	.48
Mathematics	4				.47	.69	.58	.61	.43	.59	.61	.61	.55	.59
Spelling	5					.50	.50	.64	.54	.58	.63	.49	.48	.44
Science	6						.72	.70	.60	.68	.62	.58	.54	.53
Social Science	7							.63	.52	.61	.58	.52	.47	.49
Language	8								.87	.95	.92	.51	.48	.47
Prewriting	9									.77	.69	.37	.34	.35
Composition	10										.80	.50	.47	.46
Editing	11											.50	.48	.46
OLSAT Total	12												.92	.94
OLSAT Verbal	13													.73
OLSAT Nonverbal	14													

Appendix N:

Pearson Product-Moment Coefficients of Correlation (r) for Corresponding Subtests and Totals for the Equating of Levels Samples

Appendix N

Table N–1. Pearson Product-Moment Coefficients of Correlation (r) for Corresponding Subtests and Totals for the Equating of Levels Sample - SESAT 1 and SESAT 2

Total/Subtest	N	SESAT 1 – Form A			SESAT 2 – Form A			r
		Number of Items	Mean	SD	Number of Items	Mean	SD	
Total Reading	633	70	63.3	6.9	100	79.6	15.3	0.75
Sounds and Letters	633	40	37.5	3.0	40	35.6	4.5	0.67
Word Reading	633	30	25.9	4.7	30	21.7	6.0	0.69
Sentence Reading	633	(30)*	(25.5)	(5.0)	30	21.9	6.6	0.66
Mathematics	747	40	33.5	4.8	40	32.0	6.3	0.78
Environment	754	40	31.9	5.1	40	29.6	4.8	0.72
Listening to Words and Stories	756	40	32.5	5.4	40	27.7	6.2	0.78

* Numbers in parentheses represent a subtest, content total, or test of similar content used as a stand-in when there is no corresponding test.

Table N–2. Pearson Product-Moment Coefficients of Correlation (r) for Corresponding Subtests and Totals for the Equating of Levels Sample - SESAT 2 and Primary 1

Total/Subtest	N	SESAT 2 - Form A			Primary 1 – Form A			r
		Number of Items	Mean	SD	Number of Items	Mean	SD	
Total Reading	1296	100	93.9	9.4	130	104.3	19.9	0.79
Word Study Skills	1296	40	38.1	2.9	30	22.1	5.0	0.54
Word Reading	1296	30	27.9	3.8	30	25.2	5.7	0.73
Sentence Reading	1296	30	27.9	4.0	30	25.6	4.8	0.71
Reading Comprehension	1296	(30)	(27.9)	(4.0)	40	31.0	7.3	0.63
Total Mathematics	1451	40	35.7	5.0	72	48.9	12.8	0.68
Mathematics Problem Solving	1451	(40)	(35.7)	(5.0)	42	30.1	7.5	0.71
Mathematics Procedures	1451	(40)	(35.7)	(5.0)	30	18.7	6.3	0.52
Language	1179	(30)	(27.8)	(4.1)	40	26.7	7.5	0.55
Spelling	1179	(40)	(37.9)	(3.2)	36	26.9	7.6	0.47
Environment	1436	40	31.7	4.7	40	30.3	5.3	0.73
Listening	1432	40	30.1	6.1	40	27.3	7.3	0.77

Total/Subtest	N	SESAT 2 - Form D			Primary 1 – Form D			r
		Number of Items	Mean	SD	Number of Items	Mean	SD	
Language	240	(30)	(28.6)	(2.0)	40	26.8	5.6	0.36

* Numbers in parentheses represent a subtest, content total, or test of similar content used as a stand-in when there is no corresponding test.

Table N–3. Pearson Product-Moment Coefficients of Correlation (r) for Corresponding Subtests and Totals for the Equating of Levels Sample - Primary 1 and Primary 2

| Total/Subtest | N | Primary 1 – Form A | | | Primary 2 – Form A | | | r |
		Number of Items	Mean	SD	Number of Items	Mean	SD	
Total Reading	1256	130	116.4	13.8	100	72.1	17.7	0.81
Word Study Skills	1256	30	25.0	4.5	30	22.4	4.7	0.78
Reading Vocabulary	1256	30	28.1	3.3	30	22.9	6.4	0.63
Sentence Reading	1316	30	27.7	3.3	(40)*	(26.8)	(8.3)	0.51
Reading Comprehension	1438	40	35.5	5.2	40	26.6	8.3	0.69
Total Mathematics	950	72	60.9	9.4	74	52.7	12.5	0.84
Mathematics Problem Solving	950	42	35.4	5.7	44	30.2	7.7	0.79
Mathematics Procedures	950	30	25.5	4.7	30	22.5	5.8	0.72
Language	859	40	32.7	6.4	48	37.3	7.9	0.82
Spelling	859	36	33.0	4.6	36	29.4	6.5	0.74
Environment	939	40	33.4	4.1	40	29.4	5.2	0.66
Listening	931	40	31.6	6.1	40	26.5	6.6	0.76
Total/Subtest	N	Primary 1 – Form D			Primary 2 – Form D			r
		Number of Items	Mean	SD	Number of Items	Mean	SD	
Language	554	40	31.6	5.1	40	28.5	5.5	0.74

* Numbers in parentheses represent a subtest, content total, or test of similar content used as a stand-in when there is no corresponding test.

Appendix N

Table N–4. Pearson Product-Moment Coefficients of Correlation (r) for Corresponding Subtests and Totals for the Equating of Levels Sample - Primary 2 and Primary 3

Total/Subtest	N	Primary 2 – Form A			Primary 3 – Form A			r
		Number of Items	Mean	SD	Number of Items	Mean	SD	
Total Reading	1039	100	81.0	13.8	114	77.3	20.3	0.88
Word Study Skills	1039	30	23.7	4.3	30	20.7	5.4	0.80
Reading Vocabulary	1039	30	26.4	4.4	30	21.0	6.1	0.76
Reading Comprehension	1039	40	30.9	6.8	54	35.7	10.7	0.80
Total Mathematics	563	74	60.3	10.1	76	49.2	15.3	0.78
Mathematics Problem Solving	563	44	34.9	6.4	46	30.9	9.2	0.73
Mathematics Procedures	563	30	25.5	4.6	30	18.3	7.1	0.68
Language	200	48	39.4	7.1	48	30.3	10.4	0.80
Spelling	200	36	31.8	5.3	38	21.8	7.8	0.60
Environment	515	40	32.2	4.3	(80)*	(50.9)	(14.3)	0.59
Listening	601	40	29.3	6.1	40	27.9	5.9	0.71

Total/Subtest	N	Primary 2 – Form D			Primary 3 – Form D			r
		Number of Items	Mean	SD	Number of Items	Mean	SD	
Language	1056	40	30.7	5.4	45	23.8	8.1	0.73
Prewriting	1056	(40)	(30.7)	(5.4)	14	7.9	2.8	0.65
Composing	1056	(40)	(30.7)	(5.4)	16	7.5	3.4	0.63
Editing	1056	(40)	(30.7)	(5.4)	15	8.3	3.1	0.64

* Numbers in parentheses represent a subtest, content total, or test of similar content used as a stand-in when there is no corresponding test.

Table N–5. Pearson Product-Moment Coefficients of Correlation (r) for Corresponding Subtests and Totals for the Equating of Levels Sample – Primary 3 and Intermediate 1

Total/Subtest	N	Primary 3 – Form A			Intermediate 1 – Form A			r
		Number of Items	Mean	SD	Number of Items	Mean	SD	
Total Reading	945	114	83.2	21.2	114	75.7	21.7	0.88
Word Study Skills	945	30	21.9	5.4	30	20.9	6.3	0.80
Reading Vocabulary	945	30	23.3	6.2	30	21.8	6.4	0.81
Reading Comprehension	945	54	38.0	11.6	54	32.9	11.0	0.80
Total Mathematics	734	76	54.9	16.0	80	49.4	15.6	0.85
Mathematics Problem Solving	734	46	33.1	9.8	48	28.2	9.3	0.81
Mathematics Procedures	734	30	21.7	7.1	32	21.2	7.4	0.75
Language	571	48	35.4	9.7	48	32.8	9.7	0.84
Language Mechanics	571	(48)*	(35.4)	(9.7)	24	16.3	4.9	0.77
Language Expression	571	(48)	(35.4)	(9.7)	24	16.5	5.4	0.81
Spelling	571	38	26.2	7.7	40	25.1	8.1	0.82
Science	623	40	27.6	8.3	40	26.0	7.6	0.76
Social Science	623	40	27.0	9.2	40	24.6	8.6	0.81
Listening	747	40	29.5	6.4	40	25.2	6.3	0.76
Total/Subtest	N	Primary 3 – Form D			Intermediate 1 – Form D			r
		Number of Items	Mean	SD	Number of Items	Mean	SD	
Language	327	45	25.6	9.5	48	25.1	9.5	0.82
Prewriting	327	14	8.8	3.1	12	6.9	2.7	0.64
Composing	327	16	8.1	3.8	18	9.2	4.1	0.74
Editing	327	15	8.6	3.7	18	9.0	3.7	0.68

* Numbers in parentheses represent a subtest, content total, or test of similar content used as a stand-in when there is no corresponding test.

Appendix N

Table N–6. Pearson Product-Moment Coefficients of Correlation (r) for Corresponding Subtests and Totals for the Equating of Levels Sample - Intermediate 1 and Intermediate 2

Total/Subtest	N	Intermediate 1 – Form A			Intermediate 2 – Form A			r
		Number of Items	Mean	SD	Number of Items	Mean	SD	
Total Reading	1366	114	82.8	20.5	84	54.9	16.1	0.88
Word Study Skills	1366	30	21.7	6.0	(30)*	(22.0)	(6.2)	0.70
Reading Vocabulary	1366	30	24.1	5.8	30	22.0	22.0	0.82
Reading Comprehension	1366	54	36.9	10.5	54	32.8	10.7	0.83
Total Mathematics	1159	80	58.7	14.1	80	48.6	15.3	0.84
Mathematics Problem Solving	1159	48	33.7	8.7	48	28.8	9.3	0.80
Mathematics Procedures	1159	32	25.0	6.3	32	19.8	6.8	0.76
Language	887	48	36.0	9.3	48	33.0	9.6	0.86
Language Mechanics	887	24	18.0	4.6	24	17.0	4.7	0.79
Language Expression	887	24	18.0	5.3	24	16.0	5.4	0.80
Spelling	887	40	28.1	7.4	40	26.3	8.2	0.87
Science	1337	40	29.5	6.5	40	23.8	6.8	0.75
Social Science	1339	40	28.3	7.7	40	25.2	8.0	0.79
Listening	1375	40	28.0	5.8	40	23.6	6.2	0.75

Total/Subtest	N	Intermediate 1 – Form D			Intermediate 2 – Form D			r
		Number of Items	Mean	SD	Number of Items	Mean	SD	
Language	481	48	29.3	9.5	48	26.3	9.0	0.86
Prewriting	481	12	8.1	2.7	12	7.5	2.7	0.65
Composing	481	18	10.8	4.1	18	9.2	3.9	0.73
Editing	481	18	10.4	3.9	18	9.5	3.7	0.75

* Numbers in parentheses represent a subtest, content total, or test of similar content used as a stand-in when there is no corresponding test.

Table N–7. Pearson Product-Moment Coefficients of Correlation (r) for Corresponding Subtests and Totals for the Equating of Levels Sample – Intermediate 2 and Intermediate 3

Total/Subtest	N	Intermediate 2 - Form A			Intermediate 3 – Form A			r
		Number of Items	Mean	SD	Number of Items	Mean	SD	
Total Reading	1511	84	61.1	14.4	84	57.9	15.6	0.87
Reading Vocabulary	1511	30	24.4	5.0	30	22.4	5.6	0.79
Reading Comprehension	1511	54	36.7	10.1	54	35.4	10.8	0.83
Total Mathematics	858	80	57.1	14.2	80	47.7	14.9	0.87
Mathematics Problem Solving	858	48	33.3	8.9	48	27.4	9.0	0.81
Mathematics Procedures	858	32	23.8	6.3	32	20.3	7.1	0.78
Language	968	48	34.7	9.2	48	33.2	9.0	0.85
Language Mechanics	968	24	17.7	4.7	24	16.7	4.4	0.77
Language Expression	968	24	17.0	5.2	24	16.5	5.1	0.79
Spelling	968	40	28.9	7.8	40	26.5	8.0	0.86
Science	1026	40	26.4	6.6	40	26.1	6.4	0.76
Social Science	1020	40	28.2	7.2	40	24.0	7.1	0.75
Listening	1055	40	28.1	6.0	40	26.3	6.8	0.76
Total/Subtest	N	Intermediate 2 - Form D			Intermediate 3 – Form D			r
		Number of Items	Mean	SD	Number of Items	Mean	SD	
Language	528	48	30.8	9.0	48	29.7	9.2	0.83
Prewriting	528	12	8.4	2.6	12	7.7	2.5	0.56
Composition	528	18	11.0	4.0	18	10.4	4.0	0.74
Editing	528	18	11.4	3.5	18	11.7	3.8	0.70

Appendix N

Table N–8. Pearson Product-Moment Coefficients of Correlation (r) for Corresponding Subtests and Totals for the Equating of Levels Sample - Intermediate 3 and Advanced 1

Total/Subtest	N	Intermediate 3 - Form A			Advanced 1 – Form A			r
		Number of Items	Mean	SD	Number of Items	Mean	SD	
Total Reading	1113	84	61.4	15.0	84	55.4	15.6	0.87
Reading Vocabulary	1113	30	24.2	5.0	30	21.7	6.0	0.77
Reading Comprehension	1113	54	37.1	10.9	54	33.7	10.4	0.82
Total Mathematics	648	80	53.1	15.0	80	44.6	15.2	0.88
Mathematics Problem Solving	648	48	30.9	9.2	48	25.5	8.9	0.82
Mathematics Procedures	648	32	22.2	6.9	32	19.1	7.3	0.80
Language	593	48	36.3	8.6	48	34.2	9.7	0.86
Language Mechanics	593	24	18.3	4.2	24	17.2	4.8	0.77
Language Expression	593	24	18.0	4.9	24	17.0	5.5	0.80
Spelling	593	40	29.5	7.3	40	26.7	7.3	0.84
Science	668	40	28.0	5.9	40	25.5	6.5	0.73
Social Science	668	40	26.4	6.6	40	22.3	6.6	0.77
Listening	849	40	27.3	6.7	40	26.1	6.5	0.78
Total/Subtest	N	Intermediate 3 - Form D			Advanced 1 – Form D			r
		Number of Items	Mean	SD	Number of Items	Mean	SD	
Language	429	48	29.4	9.3	48	26.9	9.7	0.81
Prewriting	429	12	7.6	2.4	12	6.9	2.7	0.53
Composition	429	18	10.5	4.1	18	9.8	4.0	0.72
Editing	429	18	11.3	3.9	18	10.2	4.1	0.72

Table N–9. Pearson Product-Moment Coefficients of Correlation (r) for Corresponding Subtests and Totals for the Equating of Levels Sample – Advanced 1 and Advanced 2

Total/Subtest	N	Advanced 1 – Form A			Advanced 2 – Form A			r
		Number of Items	Mean	SD	Number of Items	Mean	SD	
Total Reading	1055	84	61.1	14.7	84	58.4	14.8	0.84
Reading Vocabulary	1055	30	24.0	5.4	30	22.8	5.6	0.79
Reading Comprehension	1055	54	37.1	9.9	54	35.5	10.1	0.79
Total Mathematics	643	80	49.7	16.2	80	43.2	16.2	0.90
Mathematics Problem Solving	643	48	28.9	9.4	48	25.1	9.4	0.86
Mathematics Procedures	643	32	20.8	7.7	32	18.2	7.8	0.83
Language	591	48	36.2	9.0	48	34.3	9.6	0.85
Language Mechanics	591	24	18.2	4.4	24	17.2	4.7	0.75
Language Expression	591	24	18.0	5.3	24	17.1	5.5	0.80
Spelling	591	40	28.7	7.4	40	27.3	7.1	0.84
Science	664	40	27.2	7.0	40	21.5	6.4	0.72
Social Science	664	40	24.7	7.2	40	20.8	7.5	0.78
Listening	907	40	28.3	6.4	40	26.0	6.3	0.75
Total/Subtest	N	Advanced 1 – Form D			Advanced 2 – Form D			r
		Number of Items	Mean	SD	Number of Items	Mean	SD	
Language	359	48	31.8	9.1	48	30.9	9.3	0.92
Prewriting	359	12	8.3	2.5	12	8.6	2.6	0.66
Composition	359	18	11.7	3.8	18	10.9	3.7	0.79
Editing	359	18	11.8	3.9	18	11.4	4.0	0.80

Appendix N

Table N–10.   Pearson Product-Moment Coefficients of Correlation (r) for Corresponding Subtests and Totals for the Equating of Levels Sample - Advanced 2 and TASK 1

Total/Subtest	N	Advanced 2 – Form A			TASK 1 – Form A			r
		Number of Items	Mean	SD	Number of Items	Mean	SD	
Total Reading	558	84	60.8	14.5	84	56.6	15.0	0.78
Reading Vocabulary	558	30	23.9	5.4	30	23.1	6.0	0.72
Reading Comprehension	558	54	36.9	10.1	54	33.5	10.1	0.71
Total Mathematics	279	80	45.6	16.2	50	23.1	8.2	0.81
Mathematics Problem Solving	279	48	26.4	9.4	(50)*	(23.1)	(8.2)	0.78
Mathematics Procedures	279	32	19.2	8.1	(50)	(23.1)	(8.2)	0.72
Language	333	48	33.6	10.4	48	30.9	11.5	0.77
Language Mechanics	333	24	17.1	5.1	24	15.4	5.8	0.73
Language Expression	333	24	16.6	6.0	24	15.4	6.2	0.72
Spelling	333	40	26.9	7.2	40	24.7	8.0	0.70
Science	377	40	21.5	6.9	40	20.1	6.8	0.58
Social Science	377	40	19.9	7.9	40	18.0	6.8	0.68
Total/Subtest	N	Advanced 2 – Form D			TASK 1 – Form D			r
		Number of Items	Mean	SD	Number of Items	Mean	SD	
Language	160	48	34.0	8.7	48	33.9	9.1	0.93
Prewriting	160	12	9.4	2.3	12	8.6	2.4	0.68
Composing	160	18	11.9	3.6	18	13.1	4.0	0.81
Editing	160	18	12.7	3.7	18	12.3	3.7	0.79

* Numbers in parentheses represent a subtest, content total, or test of similar content used as a stand-in when there is no corresponding test.

Table N–11. Pearson Product-Moment Coefficients of Correlation (r) for Corresponding Subtests and Totals for the Equating of Levels Sample – TASK 1 and TASK 2

Total/Subtest	N	TASK 1 – Form A			TASK 2 – Form A			r
		Number of Items	Mean	SD	Number of Items	Mean	SD	
Total Reading	401	84	60.8	14.0	84	60.1	13.5	0.82
Reading Vocabulary	401	30	24.9	5.1	30	24.2	4.7	0.75
Reading Comprehension	401	54	36.0	9.9	54	35.9	9.7	0.77
Mathematics	272	50	24.5	8.9	50	22.1	8.3	0.83
Language	217	48	35.7	9.5	48	33.3	9.7	0.84
Language Mechanics	217	24	17.8	4.6	24	16.2	4.8	0.75
Language Expression	217	24	17.9	5.7	24	17.2	5.6	0.79
Spelling	217	40	28.6	7.3	40	27.6	7.6	0.78
Science	231	40	23.5	6.5	40	22.7	7.0	0.73
Social Science	231	40	21.6	6.7	40	18.6	6.3	0.75
Total/Subtest	N	TASK 1 – Form D			TASK 2 – Form D			r
		Number of Items	Mean	SD	Number of Items	Mean	SD	
Language	135	48	36.2	8.8	48	33.2	9.4	0.84
Prewriting	135	12	9.3	2.4	12	8.5	2.7	0.61
Composing	135	18	14.0	3.7	18	12.2	3.8	0.73
Editing	135	18	13.0	3.6	18	12.4	4.0	0.78

Appendix N

Table N–12. Pearson Product-Moment Coefficients of Correlation (r) for Corresponding Subtests and Totals for the Equating of Levels Sample - TASK 2 and TASK 3

Total/Subtest	N	TASK 2 – Form A			TASK 3 – Form A			r
		Number of Items	Mean	SD	Number of Items	Mean	SD	
Total Reading	756	84	59.7	14.9	84	52.8	17.6	0.78
Reading Vocabulary	756	30	23.4	5.3	30	21.7	6.5	0.71
Reading Comprehension	756	54	36.3	10.6	54	31.1	12.2	0.73
Mathematics	379	50	25.8	9.9	50	22.8	10.0	0.84
Language	498	48	31.3	10.2	48	27.3	11.4	0.81
Language Mechanics	498	24	15.0	5.0	24	13.3	5.5	0.76
Language Expression	498	24	16.4	5.9	24	14.0	6.5	0.74
Spelling	498	40	26.5	8.3	40	24.8	9.0	0.80
Science	462	40	24.0	7.9	40	19.3	7.4	0.77
Social Science	459	40	19.2	7.0	40	16.3	6.6	0.75
Total/Subtest	N	TASK 2 – Form D			TASK 3 – Form D			r
		Number of Items	Mean	SD	Number of Items	Mean	SD	
Language	212	48	32.0	9.4	48	31.6	9.4	0.80
Prewriting	212	12	8.6	2.7	12	8.5	2.5	0.62
Composing	212	18	11.6	3.8	18	11.7	4.1	0.72
Editing	212	18	11.9	3.9	18	11.4	3.7	0.69

Appendix O:

Scaled Scores at Key Percentile Ranks

Appendix O

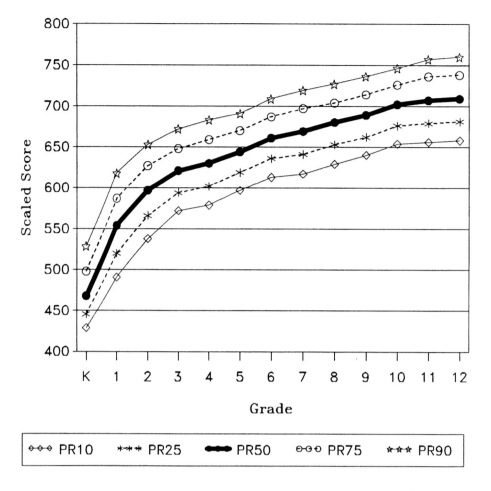

Total Reading – Spring Norms

PR	Grade												
	K	1	2	3	4	5	6	7	8	9	10	11	12
90	529	618	653	672	683	691	709	719	727	736	746	757	760
75	498	587	627	648	659	670	687	697	704	714	726	736	738
50	468	554	597	621	630	644	661	669	680	689	702	707	709
25	446	520	566	594	602	619	636	641	653	662	676	679	681
10	429	491	538	572	579	597	613	617	629	640	654	656	658

Figure O–1.   Scaled Scores at Key Percentile Ranks by Grade
Total Reading – Spring Norms

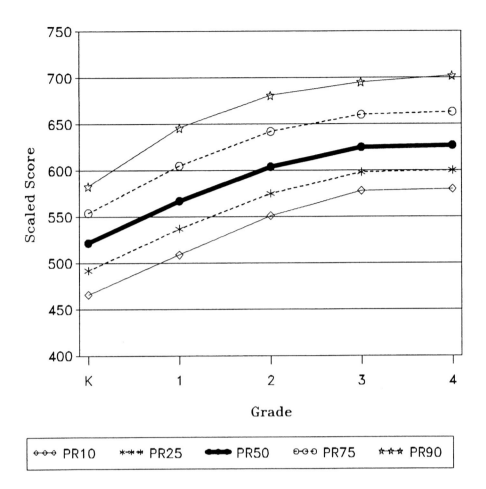

Sounds and Letters/Word Study Skills – Spring Norms

PR	Grade												
	K	1	2	3	4	5	6	7	8	9	10	11	12
90	583	646	681	695	702	-----	-----	-----	-----	-----	-----	-----	-----
75	554	605	642	660	663	-----	-----	-----	-----	-----	-----	-----	-----
50	522	567	604	625	627	-----	-----	-----	-----	-----	-----	-----	-----
25	492	537	575	598	600	-----	-----	-----	-----	-----	-----	-----	-----
10	466	509	551	578	580	-----	-----	-----	-----	-----	-----	-----	-----

Figure O–2. **Scaled Scores at Key Percentile Ranks by Grade**
Sounds and Letters/Word Study Skills – Spring Norms

Appendix O

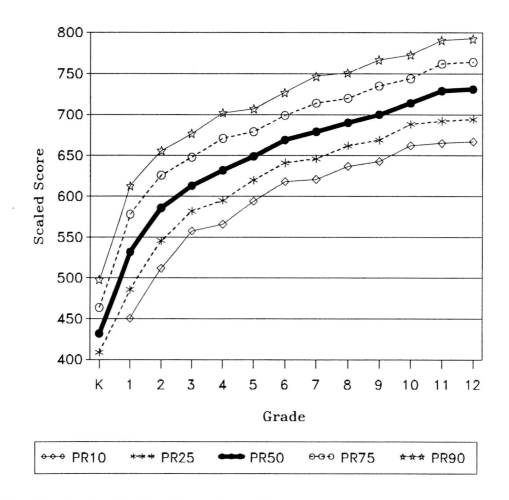

Word Reading/Reading Vocabulary – Spring Norms

PR	Grade												
	K	1	2	3	4	5	6	7	8	9	10	11	12
90	498	613	656	677	702	707	727	747	751	767	773	791	793
75	464	578	626	648	671	679	699	714	720	735	744	762	764
50	432	532	586	613	632	649	669	679	690	700	714	729	731
25	409	486	546	582	595	620	641	646	662	669	688	692	694
10	391	451	512	558	566	594	618	621	637	643	662	665	667

Figure O–3.   Scaled Scores at Key Percentile Ranks by Grade
Word Reading/Reading Vocabulary – Spring Norms

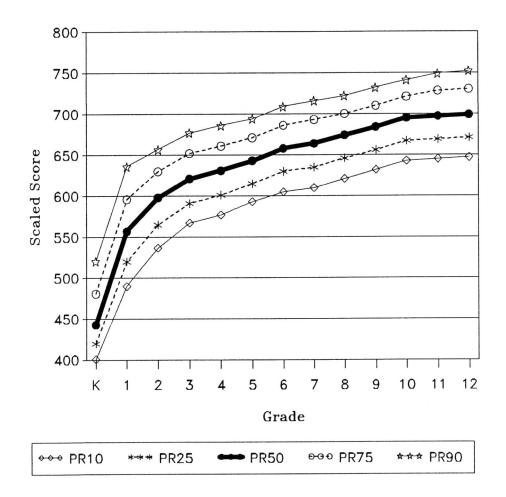

Sentence Reading/Reading Comprehension – Spring Norms

PR	Grade												
	K	**1**	**2**	**3**	**4**	**5**	**6**	**7**	**8**	**9**	**10**	**11**	**12**
90	521	636	657	677	686	694	709	716	722	732	741	749	752
75	481	596	630	652	661	671	686	693	700	710	721	728	730
50	443	557	598	621	631	643	658	664	674	684	695	697	699
25	420	520	565	591	601	615	630	635	646	656	667	669	671
10	401	490	537	567	577	593	605	610	621	632	643	645	647

Figure O–4.   Scaled Scores at Key Percentile Ranks by Grade
              Sentence Reading/Reading Comprehension – Spring Norms

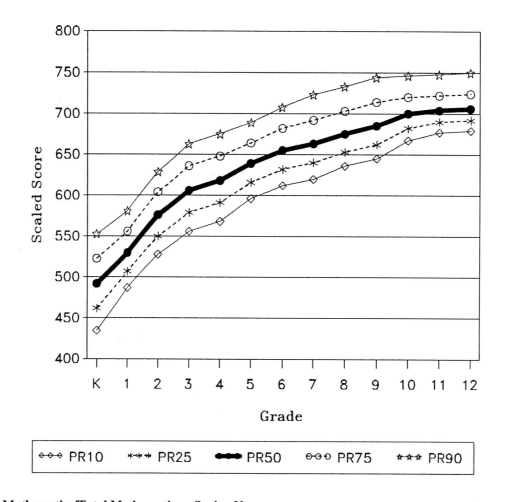

Mathematics/Total Mathematics – Spring Norms

PR	K	1	2	3	4	5	6	7	8	9	10	11	12
							Grade						
90	553	581	629	663	675	689	708	723	733	744	746	748	750
75	523	556	604	636	648	664	682	692	703	714	720	722	724
50	492	530	576	606	618	639	655	663	675	685	700	704	706
25	462	507	550	579	591	616	632	640	653	662	682	690	692
10	435	487	528	556	568	596	612	620	636	645	667	677	679

Figure O–5. **Scaled Scores at Key Percentile Ranks by Grade**
Mathematics/Total Mathematics – Spring Norms

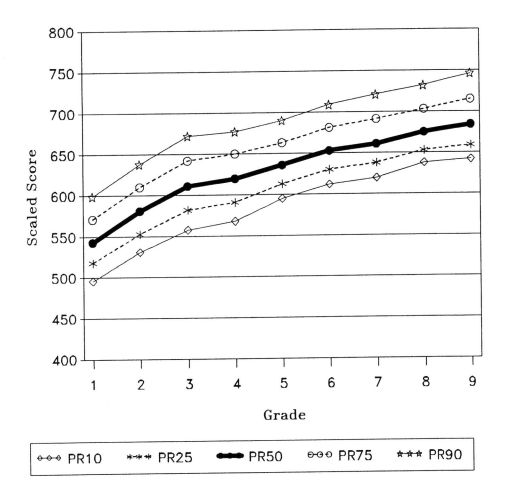

| ⋄-⋄ PR10 | ＊-＊ PR25 | ●●● PR50 | ⊖-⊖ PR75 | ★★★ PR90 |

Mathematics Problem Solving – Spring Norms

PR							Grade						
	K	**1**	**2**	**3**	**4**	**5**	**6**	**7**	**8**	**9**	**10**	**11**	**12**
90	-----	599	638	672	677	690	709	721	732	746	-----	-----	-----
75	-----	571	610	642	650	663	681	691	703	715	-----	-----	-----
50	-----	543	581	611	620	636	653	661	675	684	-----	-----	-----
25	-----	518	553	582	591	613	630	638	653	659	-----	-----	-----
10	-----	496	531	558	568	595	612	620	638	642	-----	-----	-----

Figure O–6.   **Scaled Scores at Key Percentile Ranks by Grade**
Mathematics Problem Solving – Spring Norms

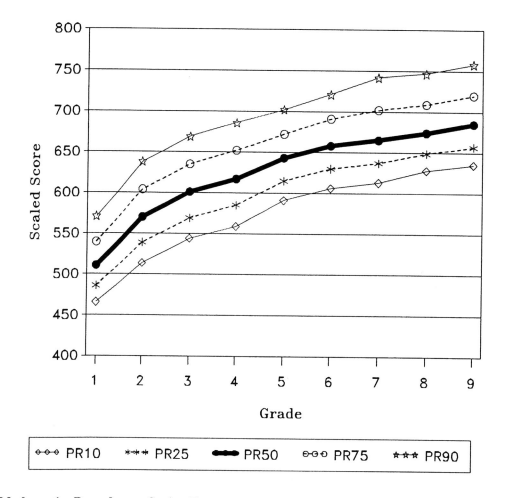

Mathematics Procedures – Spring Norms

PR	K	1	2	3	4	5	6	7	8	9	10	11	12
						Grade							
90	-----	571	638	669	686	702	721	742	747	758	-----	-----	-----
75	-----	540	604	635	652	672	691	702	709	720	-----	-----	-----
50	-----	511	570	601	617	643	658	665	674	685	-----	-----	-----
25	-----	486	539	569	585	615	630	637	649	657	-----	-----	-----
10	-----	466	514	544	559	591	606	613	628	635	-----	-----	-----

Figure O–7.   Scaled Scores at Key Percentile Ranks by Grade
Mathematics Procedures – Spring Norms

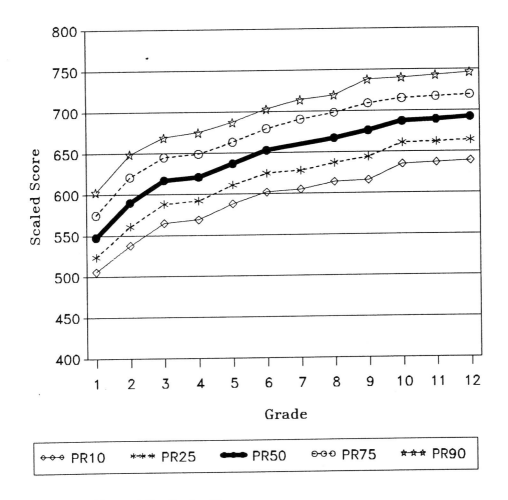

Total Language Forms A/B – Spring Norms

PR	K	| Grade											
		1	**2**	**3**	**4**	**5**	**6**	**7**	**8**	**9**	**10**	**11**	**12**
90	-----	603	649	669	675	687	703	714	719	738	740	743	746
75	-----	575	621	645	649	663	679	690	698	709	715	717	719
50	-----	548	590	617	621	637	653	-----	667	676	687	689	692
25	-----	524	561	588	592	611	625	628	637	644	661	662	664
10	-----	506	538	565	569	588	602	605	614	616	635	637	639

Figure O–8. Scaled Scores at Key Percentile Ranks by Grade
Total Language Forms A/B – Spring Norms

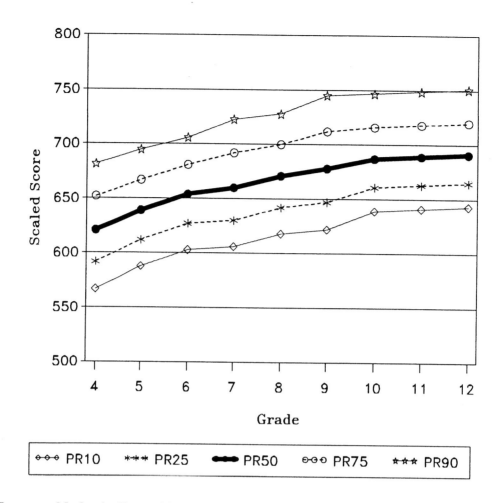

Language Mechanics Forms A/B – Spring Norms

	Grade												
PR	**K**	**1**	**2**	**3**	**4**	**5**	**6**	**7**	**8**	**9**	**10**	**11**	**12**
90	-----	-----	-----	-----	682	695	706	723	728	745	747	749	751
75	-----	-----	-----	-----	652	667	681	692	700	712	716	718	720
50	-----	-----	-----	-----	621	639	654	660	671	678	687	689	691
25	-----	-----	-----	-----	592	612	627	630	642	647	661	663	665
10	-----	-----	-----	-----	567	588	603	606	618	622	639	641	643

Figure O–9. **Scaled Scores at Key Percentile Ranks by Grade**
Language Mechanics Forms A/B – Spring Norms

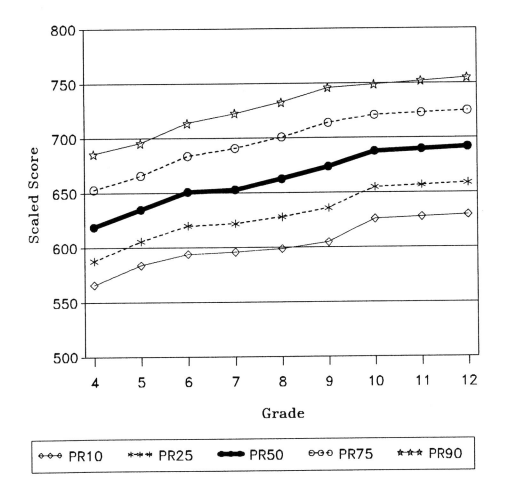

Language Expression Forms A/B – Spring Norms

PR	Grade												
	K	1	2	3	4	5	6	7	8	9	10	11	12
90	-----	-----	-----	-----	686	696	714	723	733	746	749	752	755
75	-----	-----	-----	-----	653	666	684	691	701	714	721	723	725
50	-----	-----	-----	-----	619	635	651	653	663	674	688	690	692
25	-----	-----	-----	-----	588	606	620	622	628	636	655	657	659
10	-----	-----	-----	-----	566	584	594	596	599	605	626	628	630

Figure O–10. Scaled Scores at Key Percentile Ranks by Grade
Language Expression Forms A/B – Spring Norms

Appendix O

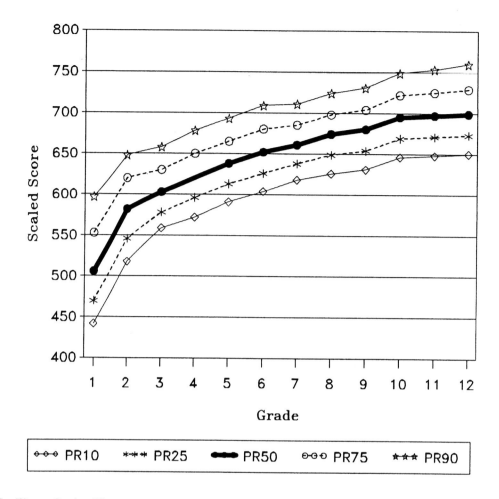

Spelling – Spring Norms

PR	K	1	2	3	4	5	6	7	8	9	10	11	12
						Grade							
90	-----	597	648	658	678	693	709	711	724	731	749	753	760
75	-----	553	620	630	650	665	680	685	698	704	722	725	729
50	-----	506	582	603	-----	638	652	661	674	680	695	697	699
25	-----	470	546	578	596	613	626	638	649	654	669	671	673
10	-----	442	518	559	572	591	604	618	626	631	646	648	650

Figure O–11. Scaled Scores at Key Percentile Ranks by Grade
Spelling – Spring Norms

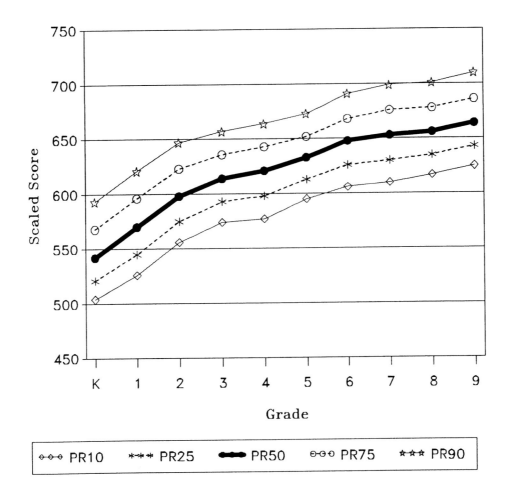

Listening To Words and Stories/Listening – Spring Norms

PR	Grade												
	K	1	2	3	4	5	6	7	8	9	10	11	12
90	593	621	647	657	664	673	691	699	701	710	-----	-----	-----
75	568	596	623	636	643	652	668	676	678	686	-----	-----	-----
50	542	570	598	614	621	633	648	653	656	664	-----	-----	-----
25	521	545	575	593	598	613	626	630	635	643	-----	-----	-----
10	504	526	556	574	577	595	606	610	617	625	-----	-----	-----

**Figure O–12. Scaled Scores at Key Percentile Ranks by Grade
Listening to Words and Stories/Listening – Spring Norms**

Appendix O

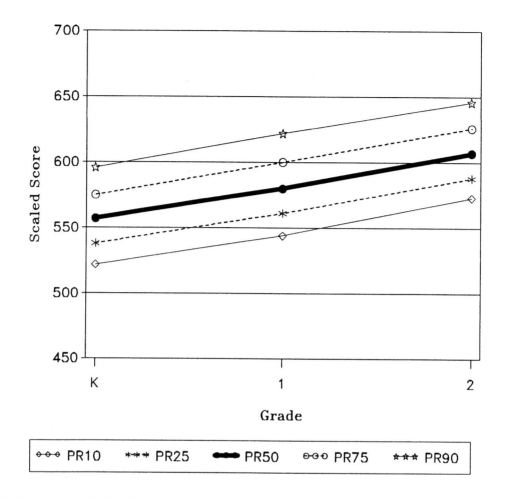

Environment – Spring Norms

	Grade												
PR	**K**	**1**	**2**	**3**	**4**	**5**	**6**	**7**	**8**	**9**	**10**	**11**	**12**
90	596	622	646	-----	-----	-----	-----	-----	-----	-----	-----	-----	-----
75	575	600	626	-----	-----	-----	-----	-----	-----	-----	-----	-----	-----
50	557	580	607	-----	-----	-----	-----	-----	-----	-----	-----	-----	-----
25	538	561	588	-----	-----	-----	-----	-----	-----	-----	-----	-----	-----
10	522	544	573	-----	-----	-----	-----	-----	-----	-----	-----	-----	-----

Figure O–13. Scaled Scores at Key Percentile Ranks by Grade
           Environment – Spring Norms

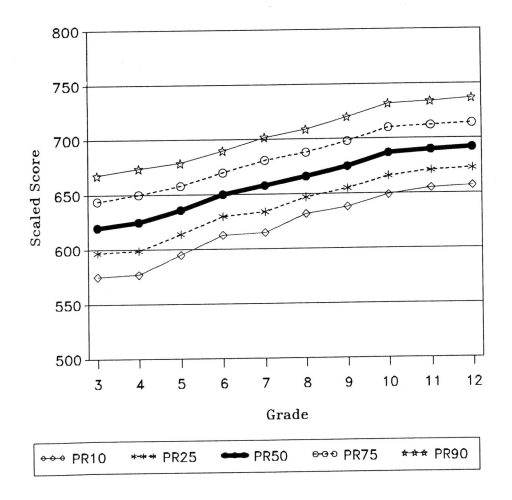

Science – Spring Norms

	Grade												
PR	**K**	**1**	**2**	**3**	**4**	**5**	**6**	**7**	**8**	**9**	**10**	**11**	**12**
90	-----	-----	-----	668	674	679	690	702	709	720	732	734	737
75	-----	-----	-----	644	650	658	670	681	688	698	710	712	714
50	-----	-----	-----	620	625	636	650	658	666	675	687	690	692
25	-----	-----	-----	597	599	614	630	634	647	655	666	671	673
10	-----	-----	-----	575	577	595	613	615	632	638	649	655	657

Figure O–14. Scaled Scores at Key Percentile Ranks by Grade
          Science – Spring Norms

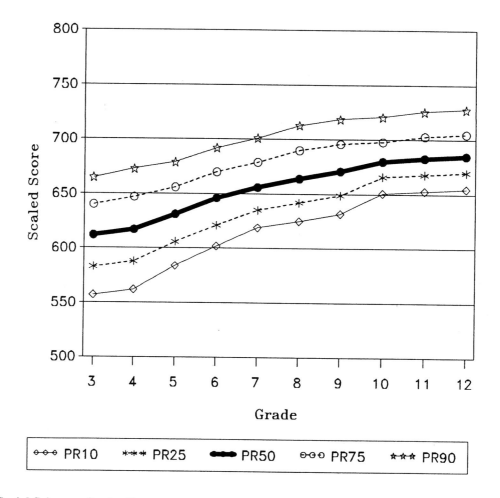

Social Science – Spring Norms

	Grade												
PR	**K**	**1**	**2**	**3**	**4**	**5**	**6**	**7**	**8**	**9**	**10**	**11**	**12**
90	-----	-----	-----	665	673	679	692	701	713	719	721	726	728
75	-----	-----	-----	640	647	656	670	679	690	696	698	703	705
50	-----	-----	-----	612	617	631	646	656	664	671	680	683	685
25	-----	-----	-----	583	588	606	621	635	642	649	666	668	670
10	-----	-----	-----	557	562	584	602	619	625	632	651	653	655

Figure O–15. Scaled Scores at Key Percentile Ranks by Grade Social Science – Spring Norms

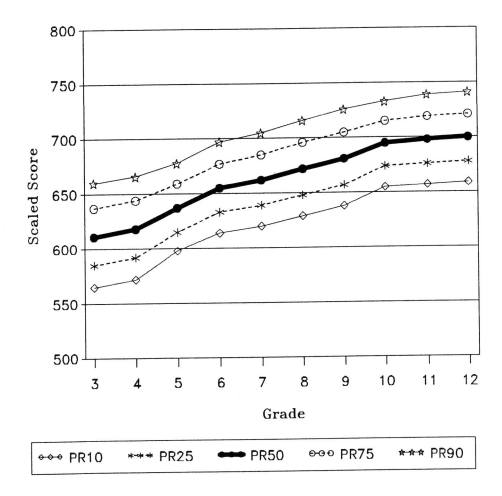

Thinking Skills (Basic Battery) – Spring Norms

PR							Grade						
	K	1	2	3	4	5	6	7	8	9	10	11	12
90	-----	-----	-----	660	666	678	697	705	716	726	733	739	741
75	-----	-----	-----	637	644	659	677	685	696	705	715	719	721
50	-----	-----	-----	611	618	637	655	662	672	681	695	698	700
25	-----	-----	-----	585	592	615	633	639	648	657	674	676	678
10	-----	-----	-----	565	572	598	614	620	629	638	655	657	659

**Figure O–16. Scaled Scores at Key Percentile Ranks by Grade
Thinking Skills (Basic Battery) – Spring Norms**

411

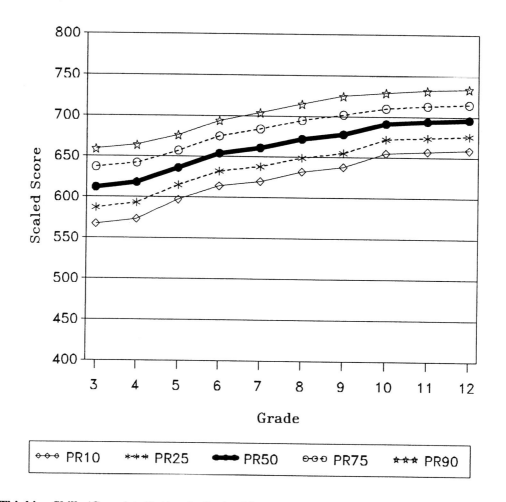

Thinking Skills (Complete Battery) - Spring Norms

PR	K	1	2	3	4	5	6	7	8	9	10	11	12
								Grade					
90	-----	-----	-----	659	664	676	694	704	715	725	729	732	734
75	-----	-----	-----	637	642	657	675	684	695	702	710	713	715
50	-----	-----	-----	612	618	636	654	661	672	678	691	694	696
25	-----	-----	-----	587	593	615	632	638	649	655	672	674	676
10	-----	-----	-----	567	573	597	614	620	632	638	655	657	659

Figure O–17. Scaled Scores at Key Percentile Ranks by Grade
Thinking Skills (Complete Battery) – Spring Norms

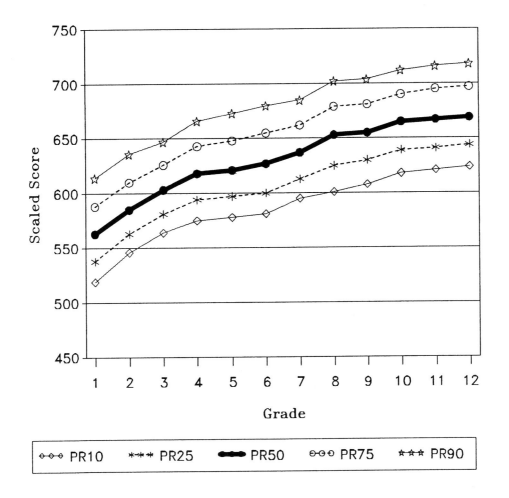

Comprehensive Language Forms D/E - Spring Norms

PR	K	\multicolumn{12}{c}{Grade}											
		1	2	3	4	5	6	7	8	9	10	11	12
90	-----	614	636	647	666	673	680	685	702	704	712	716	718
75	-----	588	610	626	643	648	655	662	679	681	690	695	697
50	-----	563	585	603	618	621	627	637	653	655	665	667	669
25	-----	538	563	581	594	597	600	613	625	630	639	641	644
10	-----	519	546	564	575	578	581	595	601	608	618	621	624

Figure O–18. Scaled Scores at Key Percentile Ranks by Grade
Comprehensive Language Forms D/E – Spring Norms

Appendix O

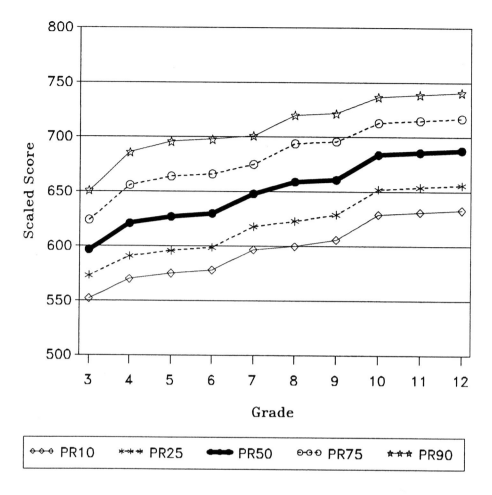

Prewriting Forms D/E - Spring Norms

PR	Grade											
	K	1	2	3	4	5	6	7	8	9	10	11	12
90	-----	-----	-----	651	686	696	698	701	720	722	737	739	741
75	-----	-----	-----	624	656	664	666	675	694	696	713	715	717
50	-----	-----	-----	597	621	627	630	648	659	661	684	686	688
25	-----	-----	-----	573	591	596	599	618	623	629	652	654	656
10	-----	-----	-----	552	570	575	578	597	600	606	629	631	633

**Figure O–19. Scaled Scores at Key Percentile Ranks by Grade
Prewriting Forms D/E – Spring Norms**

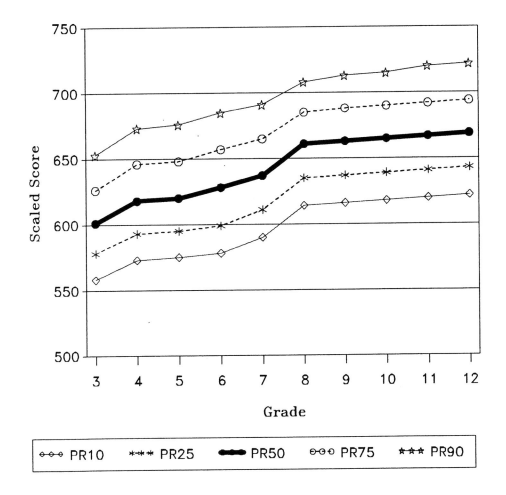

Composing Forms D/E - Spring Norms

PR	K	1	2	3	4	5	6	7	8	9	10	11	12
							Grade						
90	-----	-----	-----	653	673	676	685	691	708	713	715	720	722
75	-----	-----	-----	626	646	648	657	665	685	688	690	692	694
50	-----	-----	-----	601	618	620	628	637	661	663	665	667	669
25	-----	-----	-----	578	593	595	599	611	635	637	639	641	643
10	-----	-----	-----	558	573	575	578	590	614	616	618	620	622

**Figure O–20. Scaled Scores at Key Percentile Ranks by Grade
Composing Forms D/E – Spring Norms**

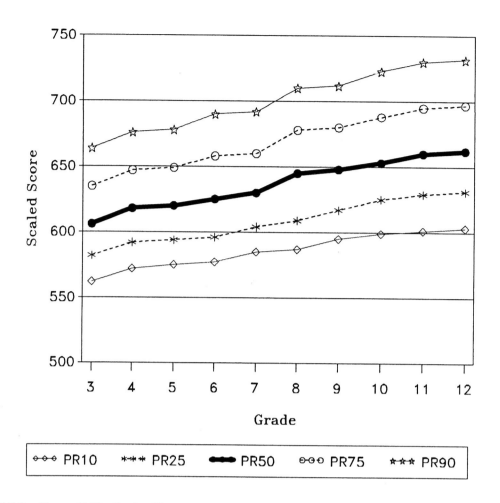

Editing Forms D/E – Spring Norms

PR	K	1	2	3	4	5	6	7	8	9	10	11	12
								Grade					
90	-----	-----	-----	664	676	678	690	692	710	712	723	730	732
75	-----	-----	-----	635	647	649	658	660	678	680	688	695	697
50	-----	-----	-----	606	618	620	625	630	645	648	653	660	662
25	-----	-----	-----	582	592	594	596	604	609	617	625	629	631
10	-----	-----	-----	562	572	575	577	585	587	595	599	601	603

Figure O–21. Scaled Scores at Key Percentile Ranks by Grade
Editing Forms D/E – Spring Norms

Appendix P:

Performance Levels Corresponding to Scaled Scores

Appendix P

Table P–1. Performance Levels Corresponding to Scaled Scores at Primary 1

Subtest	Level 1	Level 2	Level 3	Level 4
Total Reading	Below 488	488–541	542–593	Above 593
Word Study Skills	Below 506	506–561	562–613	Above 613
Word Reading	Below 469	469–517	518–572	Above 572
Sentence Reading	Below 497	497–560	561–605	Above 605
Reading Comprehension	Below 479	479–534	535–580	Above 580
Total Mathematics	Below 485	485–536	537–585	Above 585
Mathematics Problem Solving	Below 494	494–540	541–593	Above 593
Mathematics Procedures	Below 469	469–530	531–568	Above 568
Language (Forms A/B)	Below 509	509–554	555–593	Above 593
Spelling	Below 461	461–509	510–546	Above 546
Environment	Below 537	537–578	579–612	Above 612
Listening	Below 530	530–569	570–605	Above 605
Language (Forms D/E)	Below 524	524–577	578–617	Above 617

Table P–2. Performance Levels Corresponding to Scaled Scores at Primary 2

Subtest	Level 1	Level 2	Level 3	Level 4
Total Reading	Below 555	555–605	606–654	Above 654
Word Study Skills	Below 561	561–610	611–660	Above 660
Reading Vocabulary	Below 534	534–583	584–629	Above 629
Reading Comprehension	Below 565	565–615	616–658	Above 658
Total Mathematics	Below 537	537–581	582–627	Above 627
Mathematics Problem Solving	Below 537	537–579	580–628	Above 628
Mathematics Procedures	Below 534	534–581	582–617	Above 617
Language (Forms A/B)	Below 561	561–597	598–635	Above 635
Spelling	Below 529	529–578	579–620	Above 620
Environment	Below 571	571–605	606–637	Above 637
Listening	Below 565	565–604	605–647	Above 647
Language (Forms D/E)	Below 559	559–608	609–654	Above 654

Table P–3. Performance Levels Corresponding to Scaled Scores at Primary 3

Subtest	Level 1	Level 2	Level 3	Level 4
Total Reading	Below 583	583–624	625–666	Above 666
Word Study Skills	Below 592	592–631	632–670	Above 670
Reading Vocabulary	Below 564	564–601	602–641	Above 641
Reading Comprehension	Below 589	589–633	634–674	Above 674
Total Mathematics	Below 566	566–613	614–656	Above 656
Mathematics Problem Solving	Below 567	567–608	609–655	Above 655
Mathematics Procedures	Below 561	561–614	615–651	Above 651
Language (Forms A/B)	Below 588	588–624	625–661	Above 661
Spelling	Below 566	566–608	609–650	Above 650
Science	Below 576	576–617	618–658	Above 658
Social Science	Below 583	583–619	620–653	Above 653
Listening	Below 584	584–622	623–656	Above 656
Language (Forms D/E)	Below 586	586–619	620–653	Above 653
Prewriting (Forms D/E)	Below 570	570–610	611–639	Above 639
Composing (Forms D/E)	Below 595	595–624	625–648	Above 648
Editing (Forms D/E)	Below 586	586–617	618–656	Above 656

Table P–4. Performance Levels Corresponding to Scaled Scores at Intermediate 1

Subtest	Level 1	Level 2	Level 3	Level 4
Total Reading	Below 600	600–639	640–677	Above 677
Word Study Skills	Below 594	594–633	634–675	Above 675
Reading Vocabulary	Below 584	584–627	628–665	Above 665
Reading Comprehension	Below 604	604–640	641–679	Above 679
Total Mathematics	Below 593	593–632	633–674	Above 674
Mathematics Problem Solving	Below 588	588–626	627–670	Above 670
Mathematics Procedures	Below 600	600–639	640–675	Above 675
Language (Forms A/B)	Below 601	601–634	635–672	Above 672
Mechanics (Forms A/B)	Below 596	596–629	630–667	Above 667
Expression (Forms A/B)	Below 604	604–637	638–673	Above 673
Spelling	Below 588	588–627	628–666	Above 666
Science	Below 595	595–634	635–669	Above 669
Social Science	Below 605	605–637	638–672	Above 672
Listening	Below 594	594–625	626–671	Above 671
Language (Forms D/E)	Below 595	595–629	630–660	Above 660
Prewriting (Forms D/E)	Below 599	599–631	632–661	Above 661
Composing (Forms D/E)	Below 591	591–621	622–645	Above 645
Editing (Forms D/E)	Below 586	586–623	624–660	Above 660

Appendix P

Table P–5. Performance Levels Corresponding to Scaled Scores at Intermediate 2

Subtest	Level 1	Level 2	Level 3	Level 4
Total Reading	Below 611	611–660	661–705	Above 705
Reading Vocabulary	Below 604	604–653	654–693	Above 693
Reading Comprehension	Below 614	614–662	663–708	Above 708
Total Mathematics	Below 623	623–658	659–698	Above 698
Mathematics Problem Solving	Below 617	617–651	652–697	Above 697
Mathematics Procedures	Below 629	629–666	667–694	Above 694
Language (Forms A/B)	Below 618	618–649	650–687	Above 687
Mechanics (Forms A/B)	Below 615	615–646	647–698	Above 698
Expression (Forms A/B)	Below 617	617–648	649–700	Above 700
Spelling	Below 601	601–646	647–690	Above 690
Science	Below 614	614–646	647–684	Above 684
Social Science	Below 616	616–650	651–690	Above 690
Listening	Below 601	601–643	644–683	Above 683
Language (Forms D/E)	Below 603	603–642	643–676	Above 676
Prewriting (Forms D/E)	Below 602	602–634	635–663	Above 663
Composing (Forms D/E)	Below 601	601–631	632–655	Above 655
Editing (Forms D/E)	Below 599	599–646	647–690	Above 690

Table P–6. Performance Levels Corresponding to Scaled Scores at Intermediate 3

Subtest	Level 1	Level 2	Level 3	Level 4
Total Reading	Below 627	627–674	675–720	Above 720
Reading Vocabulary	Below 621	621–669	670–715	Above 715
Reading Comprehension	Below 629	629–674	675–720	Above 720
Total Mathematics	Below 639	639–676	677–717	Above 717
Mathematics Problem Solving	Below 635	635–670	671–710	Above 710
Mathematics Procedures	Below 646	646–683	684–725	Above 725
Language (Forms A/B)	Below 631	631–672	673–710	Above 710
Mechanics (Forms A/B)	Below 625	625–666	667–710	Above 710
Expression (Forms A/B)	Below 633	633–674	675–703	Above 703
Spelling	Below 618	618–662	663–696	Above 696
Science	Below 620	620–649	650–683	Above 683
Social Science	Below 630	630–666	667–706	Above 706
Listening	Below 615	615–653	654–694	Above 694
Language (Forms D/E)	Below 617	617–655	656–693	Above 693
Prewriting (Forms D/E)	Below 621	621–642	643–674	Above 674
Composing (Forms D/E)	Below 615	615–647	648–676	Above 676
Editing (Forms D/E)	Below 611	611–658	659–703	Above 703

Table P–7. Performance Levels Corresponding to Scaled Scores at Advanced 1

Subtest	Level 1	Level 2	Level 3	Level 4
Total Reading	Below 636	636–681	682–729	Above 729
Reading Vocabulary	Below 630	630–681	682–720	Above 720
Reading Comprehension	Below 637	637–680	681–730	Above 730
Total Mathematics	Below 662	662–698	699–744	Above 744
Mathematics Problem Solving	Below 654	654–692	693–733	Above 733
Mathematics Procedures	Below 674	674–704	705–759	Above 759
Language (Forms A/B)	Below 638	638–677	678–715	Above 715
Mechanics (Forms A/B)	Below 635	635–676	677–720	Above 720
Expression (Forms A/B)	Below 638	638–672	673–703	Above 703
Spelling	Below 629	629–671	672–717	Above 717
Science	Below 634	634–666	667–695	Above 695
Social Science	Below 649	649–680	681–720	Above 720
Listening	Below 621	621–659	660–705	Above 705
Language (Forms D/E)	Below 627	627–664	665–700	Above 700
Prewriting (Forms D/E)	Below 631	631–662	663–691	Above 691
Composing (Forms D/E)	Below 623	623–656	657–686	Above 686
Editing (Forms D/E)	Below 619	619–659	660–703	Above 703

Table P–8. Performance Levels Corresponding to Scaled Scores at Advanced 2

Subtest	Level 1	Level 2	Level 3	Level 4
Total Reading	Below 645	645–689	690–737	Above 737
Reading Vocabulary	Below 642	642–689	690–735	Above 735
Reading Comprehension	Below 646	646–687	688–735	Above 735
Total Mathematics	Below 669	669–706	707–757	Above 757
Mathematics Problem Solving	Below 666	666–701	702–745	Above 745
Mathematics Procedures	Below 674	674–710	711–772	Above 772
Language (Forms A/B)	Below 649	649–688	689–721	Above 721
Mechanics (Forms A/B)	Below 644	644–686	687–722	Above 722
Expression (Forms A/B)	Below 650	650–685	686–715	Above 715
Spelling	Below 649	649–688	689–724	Above 724
Science	Below 652	652–685	686–717	Above 717
Social Science	Below 646	646–684	685–723	Above 723
Listening	Below 631	631–673	674–711	Above 711
Language (Forms D/E)	Below 633	633–673	674–715	Above 715
Prewriting (Forms D/E)	Below 635	635–670	671–710	Above 710
Composing (Forms D/E)	Below 623	623–656	657–688	Above 68-8
Editing (Forms D/E)	Below 630	630–673	674–719	Above 719

Appendix P

Table P–9. Performance Levels Corresponding to Scaled Scores at TASK 1

Subtest	Level 1	Level 2	Level 3	Level 4
Total Reading	Below 668	668–709	710–751	Above 751
Reading Vocabulary	Below 684	684–722	723–757	Above 757
Reading Comprehension	Below 658	658–699	700–745	Above 745
Mathematics	Below 693	693–732	733–782	Above 782
Language (Forms A/B)	Below 669	669–706	707–752	Above 752
Mechanics (Forms A/B)	Below 664	664–703	704–746	Above 746
Expression (Forms A/B)	Below 671	671–706	707–750	Above 750
Spelling	Below 666	666–708	709–752	Above 752
Science	Below 666	666–698	699–728	Above 728
Social Science	Below 655	655–699	700–735	Above 735
Language (Forms D/E)	Below 647	647–689	690–726	Above 726
Prewriting (Forms D/E)	Below 659	659–693	694–733	Above 733
Composing (Forms D/E)	Below 643	643–678	679–710	Above 710
Editing (Forms D/E)	Below 634	634–683	684–719	Above 719

Table P–10. Performance Levels Corresponding to Scaled Scores at TASK 2

Subtest	Level 1	Level 2	Level 3	Level 4
Total Reading	Below 681	681–720	721–761	Above 761
Reading Vocabulary	Below 688	688–725	726–759	Above 759
Reading Comprehension	Below 676	676–716	717–761	Above 761
Mathematics	Below 710	710–746	747–801	Above 801
Language (Forms A/B)	Below 672	672–708	709–749	Above 749
Mechanics (Forms A/B)	Below 678	678–711	712–755	Above 755
Expression (Forms A/B)	Below 663	663–700	701–734	Above 734
Spelling	Below 679	679–714	715–749	Above 749
Science	Below 681	681–717	718–749	Above 749
Social Science	Below 660	660–703	704–741	Above 741
Language (Forms D/E)	Below 660	660–703	704–735	Above 735
Prewriting (Forms D/E)	Below 684	684–723	724–745	Above 745
Composing (Forms D/E)	Below 640	640–675	676–715	Above 715
Editing (Forms D/E)	Below 652	652–700	701–727	Above 727

Table P–11. Performance Levels Corresponding to Scaled Scores at TASK 3

Subtest	Level 1	Level 2	Level 3	Level 4
Total Reading	Below 698	698–735	736–774	Above 774
Reading Vocabulary	Below 696	696–734	735–772	Above 772
Reading Comprehension	Below 697	697–732	733–770	Above 770
Mathematics	Below 728	728–759	760–809	Above 809
Language (Forms A/B)	Below 684	684–714	715–751	Above 751
Mechanics (Forms A/B)	Below 683	683–709	710–752	Above 752
Expression (Forms A/B)	Below 681	681–715	716–744	Above 744
Spelling	Below 682	682–710	711–748	Above 748
Science	Below 697	697–733	734–765	Above 765
Social Science	Below 672	672–713	714–753	Above 753
Language (Forms D/E)	Below 674	674–708	709–746	Above 746
Prewriting (Forms D/E)	Below 681	681–720	721–748	Above 748
Composing (Forms D/E)	Below 661	661–690	691–713	Above 713
Editing (Forms D/E)	Below 670	670–700	701–736	Above 736

Appendix Q:

Percentages of Students in Each Performance Level for the Stanford 10 Standardization Samples

Appendix Q

Table Q–1. Percentages of Students in Each Performance Level for Primary 1 – Spring, Grade 1

Subtest	Level 1	Level 2	Level 3	Level 4
Total Reading	10	30	40	20
Word Study Skills	5	43	33	19
Word Reading	17	24	29	30
Sentence Reading	14	30	31	26
Reading Comprehension	8	32	33	27
Total Mathematics	8	49	36	7
Mathematics Problem Solving	9	38	41	12
Mathematics Procedures	11	56	22	11
Language (Forms A/B)	12	46	28	14
Spelling	14	39	21	27
Environment	7	44	35	15
Listening	13	37	33	16
Language (Forms D/E)	12	54	26	8

Table Q–2. Percentages of Students in Each Performance Level for Primary 2 – Spring, Grade 2

Subtest	Level 1	Level 2	Level 3	Level 4
Total Reading	18	39	36	7
Word Study Skills	15	42	26	17
Reading Vocabulary	18	29	32	21
Reading Comprehension	26	40	26	9
Total Mathematics	16	41	33	10
Mathematics Problem Solving	13	37	37	14
Mathematics Procedures	21	41	22	16
Language (Forms A/B)	25	32	29	14
Spelling	15	33	29	24
Environment	9	43	33	15
Listening	16	39	34	10
Language (Forms D/E)	21	53	21	5

Table Q–3. Percentages of Students in Each Performance Level for Primary 3 – Spring, Grade 3

Subtest	Level 1	Level 2	Level 3	Level 4
Total Reading	17	37	33	12
Word Study Skills	20	35	30	15
Reading Vocabulary	14	27	32	28
Reading Comprehension	23	38	29	10
Total Mathematics	15	41	33	11
Mathematics Problem Solving	14	33	35	18
Mathematics Procedures	18	42	24	16
Language (Forms A/B)	24	34	31	12
Spelling	14	42	32	12
Science	11	37	38	14
Social Science	26	32	28	15
Listening	18	44	30	8
Language (Forms D/E)	31	39	24	6
Prewriting (Forms D/E)	25	41	24	10
Composing (Forms D/E)	47	27	14	12
Editing (Forms D/E)	32	30	28	11

Table Q–4. Percentages of Students in Each Performance Level for Intermediate 1 – Spring, Grade 4

Subtest	Level 1	Level 2	Level 3	Level 4
Total Reading	23	36	30	12
Word Study Skills	25	28	30	16
Reading Vocabulary	18	28	28	26
Reading Comprehension	27	30	32	12
Total Mathematics	29	34	29	8
Mathematics Problem Solving	23	32	33	11
Mathematics Procedures	35	32	20	13
Language (Forms A/B)	33	32	25	9
Mechanics (Forms A/B)	27	29	30	14
Expression (Forms A/B)	38	27	24	11
Spelling	19	38	29	14
Science	22	38	31	9
Social Science	39	28	24	9
Listening	22	37	37	45
Language (Forms D/E)	26	36	27	12
Prewriting (Forms D/E)	33	22	27	19
Composing (Forms D/E)	22	32	28	18
Editing (Forms D/E)	22	33	29	16

Appendix Q

Table Q–5. Percentages of Students in Each Performance Level for Intermediate 2 – Spring, Grade 5

Subtest	Level 1	Level 2	Level 3	Level 4
Total Reading	20	46	29	4
Reading Vocabulary	15	41	31	13
Reading Comprehension	24	43	28	5
Total Mathematics	32	39	23	6
Mathematics Problem Solving	29	36	27	7
Mathematics Procedures	36	33	19	12
Language (Forms A/B)	30	33	28	8
Mechanics (Forms A/B)	28	28	34	10
Expression (Forms A/B)	32	30	29	9
Spelling	15	44	31	10
Science	25	37	30	8
Social Science	35	34	28	4
Listening	14	51	30	5
Language (Forms D/E)	34	35	22	8
Prewriting (Forms D/E)	30	29	18	24
Composing (Forms D/E)	37	23	23	17
Editing (Forms D/E)	31	42	22	6

Table Q–6. Percentages of Students in Each Performance Level for Intermediate 3 – Spring, Grade 6

Subtest	Level 1	Level 2	Level 3	Level 4
Total Reading	19	45	32	5
Reading Vocabulary	11	42	32	15
Reading Comprehension	23	42	31	4
Total Mathematics	33	37	24	6
Mathematics Problem Solving	29	39	24	8
Mathematics Procedures	38	33	21	8
Language (Forms A/B)	30	41	24	5
Mechanics (Forms A/B)	24	37	35	5
Expression (Forms A/B)	34	34	17	15
Spelling	18	42	26	14
Science	16	33	37	14
Social Science	33	37	25	4
Listening	15	43	34	8
Language (Forms D/E)	40	35	22	4
Prewriting (Forms D/E)	39	19	28	14
Composing (Forms D/E)	40	27	21	12
Editing (Forms D/E)	36	41	19	5

Table Q–7. Percentages of Students in Each Performance Level for Advanced 1 – Spring, Grade 7

Subtest	Level 1	Level 2	Level 3	Level 4
Total Reading	22	40	32	6
Reading Vocabulary	14	41	27	18
Reading Comprehension	26	39	30	4
Total Mathematics	48	30	18	3
Mathematics Problem Solving	42	35	18	6
Mathematics Procedures	56	18	20	5
Language (Forms A/B)	35	31	24	9
Mechanics (Forms A/B)	29	34	28	9
Expression (Forms A/B)	39	25	19	18
Spelling	17	45	31	7
Science	26	33	26	15
Social Science	42	35	21	2
Listening	17	40	37	6
Language (Forms D/E)	41	35	20	4
Prewriting (Forms D/E)	36	34	18	11
Composing (Forms D/E)	37	34	18	12
Editing (Forms D/E)	41	33	22	5

Table Q–8. Percentages of Students in Each Performance Level for Advanced 2 – Spring, Grade 8

Subtest	Level 1	Level 2	Level 3	Level 4
Total Reading	20	40	35	4
Reading Vocabulary	13	34	41	12
Reading Comprehension	26	38	32	4
Total Mathematics	44	32	21	3
Mathematics Problem Solving	42	32	22	5
Mathematics Procedures	50	27	21	3
Language (Forms A/B)	34	35	24	7
Mechanics (Forms A/B)	24	39	26	11
Expression (Forms A/B)	39	28	22	11
Spelling	26	43	22	9
Science	30	42	23	5
Social Science	30	44	21	5
Listening	19	50	27	4
Language (Forms D/E)	31	41	24	3
Prewriting (Forms D/E)	31	26	34	9
Composing (Forms D/E)	17	28	33	22
Editing (Forms D/E)	36	33	24	7

Appendix Q

Table Q–9. Percentages of Students in Each Performance Level for TASK 1 – Spring, Grade 9

Subtest	Level 1	Level 2	Level 3	Level 4
Total Reading	28	42	25	4
Reading Vocabulary	32	39	16	13
Reading Comprehension	26	39	30	5
Mathematics	51	38	10	1
Language (Forms A/B)	45	29	20	7
Mechanics (Forms A/B)	32	36	21	11
Expression (Forms A/B)	51	22	15	12
Spelling	35	40	20	4
Science	35	38	21	6
Social Science	29	52	17	2
Language (Forms D/E)	43	43	13	1
Prewriting (Forms D/E)	52	30	12	5
Composing (Forms D/E)	40	32	19	10
Editing (Forms D/E)	35	47	15	3

Table Q–10. Percentages of Students in Each Performance Level for TASK 2 – Spring, Grade 10

Subtest	Level 1	Level 2	Level 3	Level 4
Total Reading	29	43	25	4
Reading Vocabulary	24	34	29	13
Reading Comprehension	31	42	25	2
Mathematics	62	30	8	0
Language (Forms A/B)	34	38	23	6
Mechanics (Forms A/B)	36	38	18	8
Expression (Forms A/B)	29	33	23	15
Spelling	35	35	20	10
Science	42	40	14	3
Social Science	20	61	17	2
Language (Forms D/E)	47	39	12	2
Prewriting (Forms D/E)	45	36	14	5
Composing (Forms D/E)	31	36	25	9
Editing (Forms D/E)	49	33	11	7

Table Q–11. Percentages of Students in Each Performance Level for TASK 3 – Spring, Grade 11

Subtest	Level 1	Level 2	Level 3	Level 4
Total Reading	40	34	21	5
Reading Vocabulary	24	35	26	15
Reading Comprehension	47	29	19	5
Mathematics	80	16	4	1
Language (Forms A/B)	48	26	20	7
Mechanics (Forms A/B)	44	26	22	8
Expression (Forms A/B)	48	22	18	12
Spelling	39	24	27	10
Science	60	31	8	2
Social Science	31	55	12	2
Language (Forms D/E)	56	32	11	1
Prewriting (Forms D/E)	47	36	12	5
Composing (Forms D/E)	51	24	10	14
Editing (Forms D/E)	54	26	14	5

Table Q–12. Percentages of Students in Each Performance Level for TASK 3 – Spring, Grade 12

Subtest	Level 1	Level 2	Level 3	Level 4
Total Reading	44	31	19	5
Reading Vocabulary	27	33	23	17
Reading Comprehension	53	26	17	5
Mathematics	83	12	4	1
Language (Forms A/B)	49	24	20	7
Mechanics (Forms A/B)	46	25	21	8
Expression (Forms A/B)	50	21	16	13
Spelling	42	20	25	14
Science	64	27	7	1
Social Science	36	50	12	2
Language (Forms D/E)	62	27	11	0
Prewriting (Forms D/E)	52	29	14	5
Composing (Forms D/E)	59	20	9	12
Editing (Forms D/E)	62	19	13	5

Appendix Q

Table Q–13. Percentages of Students in Each Performance Level for Primary 1 – Fall, Grade 2

Subtest	Level 1	Level 2	Level 3	Level 4
Total Reading	5	19	40	36
Word Study Skills	3	31	36	31
Word Reading	8	16	29	46
Sentence Reading	8	21	32	39
Reading Comprehension	5	21	30	45
Total Mathematics	5	37	43	16
Mathematics Problem Solving	5	25	47	23
Mathematics Procedures	7	50	26	18
Language (Forms A/B)	4	34	35	27
Spelling	7	26	24	43
Environment	3	33	43	21
Listening	6	26	36	32
Language (Forms D/E)	5	42	38	15

Table Q–14. Percentages of Students in Each Performance Level for Primary 2 – Fall, Grade 3

Subtest	Level 1	Level 2	Level 3	Level 4
Total Reading	10	31	44	15
Word Study Skills	9	31	31	29
Reading Vocabulary	10	22	35	33
Reading Comprehension	15	36	35	15
Total Mathematics	10	31	39	20
Mathematics Problem Solving	8	26	41	25
Mathematics Procedures	15	34	25	27
Language (Forms A/B)	14	27	33	26
Spelling	9	25	29	37
Environment	4	30	41	25
Listening	8	32	45	16
Language (Forms D/E)	11	48	30	10

Table Q–15. Percentages of Students in Each Performance Level for Primary 3 – Fall, Grade 4

Subtest	Level 1	Level 2	Level 3	Level 4
Total Reading	16	30	36	18
Word Study Skills	20	31	30	19
Reading Vocabulary	13	20	30	37
Reading Comprehension	22	30	34	14
Total Mathematics	17	35	35	13
Mathematics Problem Solving	16	26	36	21
Mathematics Procedures	17	40	25	18
Language (Forms A/B)	24	27	33	16
Spelling	14	35	35	16
Science	12	28	40	19
Social Science	23	26	31	20
Listening	19	38	32	11
Language (Forms D/E)	29	33	28	9
Prewriting (Forms D/E)	21	39	24	16
Composing (Forms D/E)	44	29	14	13
Editing (Forms D/E)	31	25	32	12

Table Q–16. Percentages of Students in Each Performance Level for Intermediate 1– Fall, Grade 5

Subtest	Level 1	Level 2	Level 3	Level 4
Total Reading	18	30	32	19
Word Study Skills	22	27	31	21
Reading Vocabulary	13	24	27	36
Reading Comprehension	21	25	36	18
Total Mathematics	22	31	35	12
Mathematics Problem Solving	18	27	39	15
Mathematics Procedures	29	31	23	17
Language (Forms A/B)	25	27	33	15
Mechanics (Forms A/B)	21	29	32	19
Expression (Forms A/B)	27	25	28	20
Spelling	16	33	32	19
Science	16	33	37	14
Social Science	31	28	27	14
Listening	15	34	42	9
Language (Forms D/E)	26	32	27	16
Prewriting (Forms D/E)	33	21	27	20
Composing (Forms D/E)	21	28	25	25
Editing (Forms D/E)	24	30	28	18

Appendix Q

Table Q–17. Percentages of Students in Each Performance Level for Intermediate 2 – Fall, Grade 6

Subtest	Level 1	Level 2	Level 3	Level 4
Total Reading	15	41	36	8
Reading Vocabulary	12	36	34	19
Reading Comprehension	19	39	35	8
Total Mathematics	29	36	26	9
Mathematics Problem Solving	24	31	34	11
Mathematics Procedures	36	33	18	13
Language (Forms A/B)	28	29	29	14
Mechanics (Forms A/B)	26	26	34	14
Expression (Forms A/B)	29	28	30	13
Spelling	15	39	31	15
Science	17	34	36	14
Social Science	26	30	34	9
Listening	12	45	37	6
Language (Forms D/E)	23	33	33	12
Prewriting (Forms D/E)	22	23	24	31
Composing (Forms D/E)	24	22	24	30
Editing (Forms D/E)	22	38	29	11

Table Q–18. Percentages of Students in Each Performance Level for Intermediate 3 – Fall, Grade 7

Subtest	Level 1	Level 2	Level 3	Level 4
Total Reading	21	40	33	6
Reading Vocabulary	16	35	32	17
Reading Comprehension	26	37	31	5
Total Mathematics	34	36	23	7
Mathematics Problem Solving	32	33	25	10
Mathematics Procedures	41	34	18	8
Language (Forms A/B)	36	36	22	5
Mechanics (Forms A/B)	32	33	30	5
Expression (Forms A/B)	40	29	16	15
Spelling	21	39	26	15
Science	16	32	32	20
Social Science	34	35	27	4
Listening	18	38	34	9
Language (Forms D/E)	29	33	32	6
Prewriting (Forms D/E)	28	17	29	26
Composing (Forms D/E)	28	29	25	18
Editing (Forms D/E)	29	38	28	5

Table Q–19. Percentages of Students in Each Performance Level for Advanced 1 – Fall, Grade 8

Subtest	Level 1	Level 2	Level 3	Level 4
Total Reading	19	35	36	10
Reading Vocabulary	12	36	27	25
Reading Comprehension	23	34	36	7
Total Mathematics	43	30	21	5
Mathematics Problem Solving	35	34	22	9
Mathematics Procedures	53	20	22	5
Language (Forms A/B)	34	29	27	10
Mechanics (Forms A/B)	29	33	29	10
Expression (Forms A/B)	37	21	21	20
Spelling	16	41	35	9
Science	24	30	26	20
Social Science	36	34	26	4
Listening	15	38	39	8
Language (Forms D/E)	26	31	35	9
Prewriting (Forms D/E)	22	32	20	26
Composing (Forms D/E)	26	25	26	23
Editing (Forms D/E)	24	33	31	12

Table Q–20. Percentages of Students in Each Performance Level for Advanced 2 – Fall, Grade 9

Subtest	Level 1	Level 2	Level 3	Level 4
Total Reading	17	42	35	7
Reading Vocabulary	13	31	40	16
Reading Comprehension	24	37	34	5
Total Mathematics	38	33	27	3
Mathematics Problem Solving	36	32	25	7
Mathematics Procedures	45	29	24	2
Language (Forms A/B)	37	37	20	6
Mechanics (Forms A/B)	25	44	24	7
Expression (Forms A/B)	44	27	16	13
Spelling	22	42	25	11
Science	28	42	23	7
Social Science	30	40	22	8
Listening	16	48	30	6
Language (Forms D/E)	20	37	37	6
Prewriting (Forms D/E)	24	23	40	14
Composing (Forms D/E)	15	23	30	32
Editing (Forms D/E)	28	31	31	11

Appendix Q

Table Q–21. Percentages of Students in Each Performance Level for TASK 1 – Fall, Grade 9

Subtest	Level 1	Level 2	Level 3	Level 4
Total Reading	29	45	22	4
Reading Vocabulary	34	40	15	11
Reading Comprehension	26	42	27	5
Mathematics	58	34	7	1
Language (Forms A/B)	52	30	15	4
Mechanics (Forms A/B)	36	39	19	6
Expression (Forms A/B)	59	22	12	7
Spelling	35	44	17	3
Science	34	41	18	6
Social Science	32	50	16	2
Language (Forms D/E)	39	45	16	1
Prewriting (Forms D/E)	49	33	14	5
Composing (Forms D/E)	34	33	21	12
Editing (Forms D/E)	29	49	19	3

Table Q–22. Percentages of Students in Each Performance Level for TASK 2 – Fall, Grade 10

Subtest	Level 1	Level 2	Level 3	Level 4
Total Reading	48	35	15	2
Reading Vocabulary	45	32	15	8
Reading Comprehension	49	32	17	2
Mathematics	78	18	4	0
Language (Forms A/B)	48	32	17	3
Mechanics (Forms A/B)	49	35	13	3
Expression (Forms A/B)	44	30	16	10
Spelling	52	30	12	6
Science	53	34	10	2
Social Science	33	56	10	1
Language (Forms D/E)	44	49	6	2
Prewriting (Forms D/E)	50	37	10	3
Composing (Forms D/E)	32	35	27	5
Editing (Forms D/E)	48	39	9	4

Table Q–23. Percentages of Students in Each Performance Level for TASK 3 – Fall, Grade 11

Subtest	Level 1	Level 2	Level 3	Level 4
Total Reading	54	28	16	2
Reading Vocabulary	38	36	20	6
Reading Comprehension	60	23	15	3
Mathematics	88	10	2	0
Language (Forms A/B)	51	29	17	3
Mechanics (Forms A/B)	51	27	18	5
Expression (Forms A/B)	50	26	15	9
Spelling	53	21	20	6
Science	67	27	5	0
Social Science	40	52	7	1
Language (Forms D/E)	79	17	4	0
Prewriting (Forms D/E)	71	21	6	1
Composing (Forms D/E)	75	15	4	6
Editing (Forms D/E)	78	13	7	2

Table Q–24. Percentages of Students in Each Performance Level for TASK 3 – Fall, Grade 12

Subtest	Level 1	Level 2	Level 3	Level 4
Total Reading	45	35	16	4
Reading Vocabulary	24	42	23	12
Reading Comprehension	55	26	16	3
Mathematics	84	12	4	0
Language (Forms A/B)	47	27	20	6
Mechanics (Forms A/B)	45	27	21	7
Expression (Forms A/B)	47	24	18	11
Spelling	41	24	27	8
Science	62	28	9	1
Social Science	36	51	11	1
Language (Forms D/E)	66	23	10	1
Prewriting (Forms D/E)	59	27	10	3
Composing (Forms D/E)	61	21	5	13
Editing (Forms D/E)	67	16	11	6

Appendix R:

Classification Probabilities: Accuracy and Consistency

Appendix R

Table R–1. Classification Probabilities for Primary 1, Grade 1

Subtest	Performance Level Boundary					
	1 : 2		2 : 3		3 : 4	
	Accuracy	Consistency	Accuracy	Consistency	Accuracy	Consistency
Total Reading	.98	.97	.95	.93	.96	.94
Word Study Skills	.97	.95	.88	.83	.93	.90
Word Reading	.92	.89	.93	.90	.94	.91
Sentence Reading	.94	.91	.90	.87	.92	.88
Reading Comprehension	.94	.91	.90	.87	.92	.88
Total Mathematics	.95	.93	.92	.88	.97	.95
Mathematics Problem Solving	.93	.91	.89	.85	.95	.93
Mathematics Procedures	.89	.86	.90	.86	.95	.93
Language Form A	.90	.87	.89	.85	.94	.92
Spelling	.87	.83	.92	.89	.95	.92
Environment	.95	.92	.85	.79	.91	.87
Listening	.92	.89	.89	.85	.94	.91
Language Form D	.89	.85	.91	.87	.97	.96

Table R–2. Classification Probabilities for Primary 2, Grade 2

Subtest	Performance Level Boundary					
	1 : 2		2 : 3		3 : 4	
	Accuracy	Consistency	Accuracy	Consistency	Accuracy	Consistency
Total Reading	.95	.94	.93	.91	.96	.95
Word Study Skills	.86	.82	.88	.83	.94	.91
Reading Vocabulary	.93	.90	.91	.87	.93	.90
Reading Comprehension	.92	.89	.91	.88	.96	.94
Total Mathematics	.94	.91	.92	.89	.96	.95
Mathematics Problem Solving	.92	.89	.89	.85	.95	.93
Mathematics Procedures	.89	.85	.90	.86	.94	.91
Language Form A	.91	.87	.91	.87	.95	.93
Spelling	.91	.88	.92	.88	.93	.91
Environment	.91	.88	.83	.77	.91	.87
Listening	.90	.86	.89	.84	.95	.93
Language Form D	.85	.80	.91	.88	.97	.96

Appendix R

Table R–3. Classification Probabilities for Primary 3, Grade 3

Subtest	Performance Level Boundary					
	1 : 2		2 : 3		3 : 4	
	Accuracy	Consistency	Accuracy	Consistency	Accuracy	Consistency
Total Reading	.96	.94	.94	.91	.96	.95
Word Study Skills	.86	.82	.89	.85	.93	.90
Reading Vocabulary	.92	.89	.89	.85	.92	.88
Reading Comprehension	.92	.89	.89	.85	.92	.88
Total Mathematics	.95	.92	.93	.90	.96	.95
Mathematics Problem Solving	.93	.91	.91	.87	.94	.92
Mathematics Procedures	.91	.87	.92	.88	.95	.93
Language Form A	.92	.89	.91	.87	.95	.93
Spelling	.89	.85	.91	.87	.95	.93
Science	.94	.91	.88	.84	.94	.91
Social Science	.92	.89	.90	.85	.94	.91
Listening	.88	.97	.42	.97	.98	.97
Language Form D	.89	.85	.91	.88	.97	.96
Prewriting	.86	.80	.83	.77	.90	.85
Composing	.83	.77	.89	.84	.93	.91
Editing	.82	.76	.84	.79	.93	.90

Table R–4. Classification Probabilities for Intermediate 1, Grade 4

Subtest	Performance Level Boundary					
	1 : 2		2 : 3		3 : 4	
	Accuracy	Consistency	Accuracy	Consistency	Accuracy	Consistency
Total Reading	.95	.93	.94	.92	.97	.96
Word Study Skills	.88	.83	.91	.88	.94	.92
Reading Vocabulary	.90	.87	.91	.87	.93	.90
Reading Comprehension	.90	.87	.91	.87	.93	.90
Total Mathematics	.93	.91	.93	.90	.97	.95
Mathematics Problem Solving	.92	.88	.90	.86	.95	.93
Mathematics Procedures	.90	.86	.91	.88	.95	.92
Language Form A	.91	.87	.92	.88	.96	.94
Language Mechanics	.87	.83	.88	.84	.93	.90
Language Expression	.88	.84	.90	.86	.94	.91
Spelling	.91	.87	.90	.86	.94	.92
Science	.91	.87	.89	.85	.95	.93
Social Science	.90	.86	.92	.88	.96	.94
Listening	.90	.85	.87	.82	.96	.94
Language Form D	.89	.85	.91	.88	.96	.94
Prewriting	.80	.72	.83	.76	.91	.87
Composing	.85	.79	.87	.81	.91	.88
Editing	.83	.77	.86	.81	.94	.91

Appendix R

Table R–5. Classification Probabilities for Intermediate 2, Grade 5

Subtest	Performance Level Boundary					
	1 : 2		2 : 3		3 : 4	
	Accuracy	Consistency	Accuracy	Consistency	Accuracy	Consistency
Total Reading	.95	.93	.93	.90	.97	.96
Reading Vocabulary	.91	.88	.89	.85	.93	.90
Reading Comprehension	.93	.90	.92	.88	.97	.96
Total Mathematics	.92	.89	.94	.91	.98	.97
Mathematics Problem Solving	.90	.86	.92	.88	.97	.96
Mathematics Procedures	.89	.85	.91	.88	.95	.92
Language Form A	.91	.88	.91	.87	.95	.93
Language Mechanics	.89	.84	.86	.81	.94	.92
Language Expression	.88	.83	.89	.85	.95	.94
Spelling	.91	.87	.91	.87	.95	.93
Science	.89	.85	.88	.83	.95	.93
Social Science	.90	.86	.90	.86	.97	.95
Listening	.92	.89	.87	.82	.96	.94
Language Form D	.90	.86	.91	.88	.97	.95
Prewriting	.82	.76	.81	.74	.87	.81
Composing	.85	.79	.87	.82	.91	.88
Editing	.83	.76	.90	.86	.97	.96

Table R–6. Classification Probabilities for Intermediate 3, Grade 6

| Subtest | Performance Level Boundary | | | | | |
| | 1 : 2 | | 2 : 3 | | 3 : 4 | |
	Accuracy	Consistency	Accuracy	Consistency	Accuracy	Consistency
Total Reading	.95	.93	.93	.90	.98	.97
Reading Vocabulary	.91	.87	.88	.83	.93	.90
Reading Comprehension	.93	.90	.92	.89	.98	.96
Total Mathematics	.92	.89	.94	.92	.98	.97
Mathematics Problem Solving	.90	.86	.93	.89	.97	.96
Mathematics Procedures	.89	.85	.92	.89	.96	.95
Language Form A	.91	.88	.91	.88	.97	.95
Language Mechanics	.87	.82	.87	.82	.95	.93
Language Expression	.89	.84	.91	.87	.94	.92
Spelling	.90	.86	.90	.87	.94	.92
Science	.92	.88	.86	.81	.93	.89
Social Science	.89	.85	.90	.86	.97	.96
Listening	.91	.88	.87	.83	.95	.93
Language Form D	.89	.85	.93	.90	.98	.97
Prewriting	.81	.74	.83	.76	.90	.85
Composing	.84	.79	.87	.82	.94	.91
Editing	.85	.79	.92	.88	.97	.96

Appendix R

Table R–7. Classification Probabilities for Advanced 1, Grade 7

Subtest	Performance Level Boundary					
	1 : 2		2 : 3		3 : 4	
	Accuracy	Consistency	Accuracy	Consistency	Accuracy	Consistency
Total Reading	.95	.93	.93	.91	.98	.97
Reading Vocabulary	.92	.89	.91	.87	.93	.91
Reading Comprehension	.93	.90	.92	.88	.98	.97
Total Mathematics	.93	.90	.95	.93	.99	.98
Mathematics Problem Solving	.90	.86	.93	.91	.98	.97
Mathematics Procedures	.91	.88	.94	.91	.98	.96
Language Form A	.92	.89	.92	.89	.96	.95
Language Mechanics	.88	.83	.88	.84	.95	.93
Language Expression	.90	.85	.91	.87	.94	.91
Spelling	.91	.87	.90	.86	.96	.95
Science	.90	.86	.89	.84	.94	.91
Social Science	.88	.83	.91	.87	.98	.97
Listening	.92	.88	.88	.83	.97	.95
Language Form D	.90	.87	.93	.90	.98	.98
Prewriting	.83	.76	.85	.79	.91	.87
Composing	.84	.78	.88	.83	.94	.91
Editing	.85	.79	.89	.85	.96	.94

Table R–8. Classification Probabilities for Advanced 2, Grade 8

| Subtest | Performance Level Boundary | | | | | |
| | 1 : 2 | | 2 : 3 | | 3 : 4 | |
	Accuracy	Consistency	Accuracy	Consistency	Accuracy	Consistency
Total Reading	.95	.93	.93	.90	.97	.95
Reading Vocabulary	.91	.88	.89	.85	.93	.90
Reading Comprehension	.94	.91	.91	.88	.97	.96
Total Mathematics	.92	.89	.95	.92	.98	.98
Mathematics Problem Solving	.89	.84	.93	.90	.97	.96
Mathematics Procedures	.90	.85	.93	.90	.98	.97
Language Form A	.92	.89	.92	.89	.95	.93
Language Mechanics	.88	.84	.88	.83	.93	.90
Language Expression	.90	.86	.90	.87	.93	.90
Spelling	.91	.87	.89	.84	.94	.92
Science	.88	.83	.89	.84	.95	.93
Social Science	.90	.86	.90	.86	.97	.95
Listening	.91	.87	.88	.84	.96	.94
Language Form D	.90	.87	.91	.88	.97	.96
Prewriting	.83	.77	.85	.79	.92	.88
Composing	.89	.84	.84	.78	.92	.89
Editing	.84	.78	.86	.81	.96	.94

Appendix R

Table R–9. Classification Probabilities for TASK 1, Grade 9

Subtest	Performance Level Boundary					
	1 : 2		2 : 3		3 : 4	
	Accuracy	Consistency	Accuracy	Consistency	Accuracy	Consistency
Total Reading	.93	.90	.93	.90	.98	.97
Reading Vocabulary	.89	.84	.90	.86	.95	.93
Reading Comprehension	.91	.88	.91	.88	.97	.96
Mathematics	.89	.84	.95	.92	1.00	.99
Language Form A	.92	.89	.94	.92	.98	.97
Language Mechanics	.87	.82	.89	.85	.95	.93
Language Expression	.91	.87	.93	.90	.96	.94
Spelling	.89	.85	.92	.89	.98	.97
Science	.87	.81	.90	.86	.96	.94
Social Science	.88	.83	.91	.87	.98	.97
Language Form D	.91	.87	.92	.88	.98	.97
Prewriting	.83	.76	.87	.82	.95	.93
Composing	.87	.81	.88	.83	.92	.89
Editing	.85	.79	.87	.81	.96	.93

Table R–10. Classification Probabilities for TASK 2, Grade 10

Subtest	Performance Level Boundary					
	1 : 2		2 : 3		3 : 4	
	Accuracy	Consistency	Accuracy	Consistency	Accuracy	Consistency
Total Reading	.93	.90	.93	.90	.98	.97
Reading Vocabulary	.88	.83	.89	.84	.93	.90
Reading Comprehension	.91	.88	.92	.89	.98	.98
Mathematics	.90	.86	.96	.95	1.00	1.00
Language Form A	.90	.86	.92	.89	.97	.96
Language Mechanics	.86	.80	.89	.85	.97	.96
Language Expression	.89	.85	.89	.85	.93	.91
Spelling	.90	.86	.91	.88	.96	.94
Science	.88	.83	.93	.90	.98	.97
Social Science	.87	.82	.91	.87	.99	.98
Language Form D	.91	.87	.92	.89	.97	.95
Prewriting	.82	.76	.83	.77	.86	.82
Composing	.89	.84	.84	.77	.88	.84
Editing	.86	.80	.91	.88	.94	.92

Appendix R

Table R–11. Classification Probabilities for TASK 3, Grades 11 and 12

Subtest	Performance Level Boundary					
	1 : 2		2 : 3		3 : 4	
	Accuracy	Consistency	Accuracy	Consistency	Accuracy	Consistency
Total Reading	.94	.92	.95	.93	.98	.96
Reading Vocabulary	.92	.88	.91	.88	.93	.91
Reading Comprehension	.92	.88	.94	.91	.97	.96
Mathematics	.93	.90	.98	.97	1.00	1.00
Language Form A	.92	.89	.93	.90	.97	.95
Language Mechanics	.88	.83	.89	.85	.96	.94
Language Expression	.91	.87	.92	.89	.95	.92
Spelling	.91	.87	.92	.88	.95	.93
Science	.88	.83	.95	.93	.99	.99
Social Science	.85	.79	.93	.90	.99	.98
Language Form D	.92	.89	.94	.92	.99	.98
Prewriting	.85	.80	.88	.84	.93	.90
Composing	.87	.82	.88	.83	.92	.88
Editing	.87	.82	.90	.86	.96	.94